SCIENCE AS POWER

S·C·I·E·N·C·E

A·S

P·O·W·E·R

Discourse and Ideology
in Modern Society

Stanley Aronowitz

University of Minnesota Press • Minneapolis

Published by the University of Minnesota Press
2037 University Avenue Southeast, Minneapolis MN 55414.
Published simultaneously in Canada
by Fitzhenry & Whiteside Limited, Markham.
Printed in the United States of America.

Library of Congress Cataloging-in-Publication Data

Aronowitz, Stanley.
    Science as power.
    Bibliography: p.
    Includes index.
    1. Sociology—Philosophy. 2. Ideology. 3. Science—
Philosophy. 4. Science—Social aspects. 5. Discourse
analysis. I. Title.
HM24.A73 1988   301'.01   88-4782
ISBN 0-8166-1658-2
ISBN 0-8166-1659-0 (pbk.)

CONTENTS

Preface   vii

Part I

1. Science and Technology as Hegemony   3
2. Marx 1: Science as Social Relations   35
3. Marx 2: The Scientific Theory of Society   60

Part II

4. Engels and the Return to Epistemology   91
5. The Frankfurt School: Science and Technology as Ideology   121
6. Habermas: The Retreat from the Critique   146
7. Marxism as a Positive Science   169
8. Soviet Science: The Scientific and Technological Revolution   201

Part III

9. The Breakup of Certainty: History and Philosophy of Modern Physics   239
10. The Science of Sociology and the Sociology of Science   272
11. Scientism or Critical Science: The Debates in Biology   301
12. Toward a New Social Theory of Science   317

Notes   355

Index   377

PREFACE

The power of science consists, in the first place, in its conflation of knowledge and truth. Devising a method of proving the validity of propositions about objects taken as external to the knower has become identical with what we mean by truth. Walter Benjamin called attention to this departure in the prologue to his study of the origin of German tragic drama. Plato's idea that truth is self-representing, adapted by Hegel to signify the process by which consciousness takes itself as its own object, is discarded by modern scientific thought. Disciplines outside the boundaries of science, consigned since the eighteenth century to the margins of intellectual life, are permitted to explore truth in the traditional, prescientific sense of the term. In philosophy this inquiry is today limited to ethics and aesthetics; art is conceived as self-representing, its beauty understood entirely in formal terms. But even in these humanistic disciplines, strong intellectual movements have emerged in the past century which, from time to time, challenge the notion of truth's autonomy, in either its ethical or its artistic forms. Some literary critics insist that the meaning of artistic texts can be construed only in relation to their social and historical contexts; ethical inquiry, after analytic philosophy's internalist, logical influence, has rediscovered the categories of moral agency and social interest, categories which social scientists have presupposed since Marx and Weber.

Since the "truth" claims of science are tied to the methodological imperative, it insists that science must be held immune from the influences of social and historical situations. Science, therefore, is truth and can, for this reason, represent itself by means of its procedures, by which the objects of investigation are apprehended. Hence, the self-criticism of science is conducted within the boundaries of its own normative structures. Further, science insists that only those inducted, by means of training and credentials, into its community are qualified to undertake whatever renovations the scientific project requires.

In this regime, all inquiry is obliged to direct itself to science, or, by inference,to distinguish itself from science. To claim for nonempirical inquiry the status of science is to violate the prescription that truth and methodological rigor are synonymous. In this book I claim that these norms are by no means self-evident but can be traced to specific historical and discursive influences, that science, no less than art or any other discourse, legitimates its power by claiming self-referentiality.

I do not undertake an inquiry into the nature of truth, nor do I claim that there is another discourse capable of self-representation. For us, the problem of truth has been relegated to issues entailed by knowledge acquisition and science, especially natural science, which occupies a privileged space in the pantheon of knowledges. In this book, I want to show that science is not free of historical and discursive presuppositions and that it has constituted itself as an autonomous power precisely through its convincing demonstration that it is free of such preconditions. Further, I argue that this power is constituted not only by the specific knowledges generated by scientific inquiry, but by its truth claims. For our culture requires belief in discourse that it takes outside the universe of social determination, whether this be religion, magic, or science.

The quest for the absolute has been shown by students of culture to possess universal, transhistorical characteristics. The genius of science, unlike religion or philosophy before Kant, has been its ability to identify the absolute with knowledge of nature, taken as a quantitatively apprehended series of appearances whose essential object is a particle that defies observation, that can be known only by mathematical inference. This inference tells us that we could see the particle, if we possessed the mechanical devise that would permit it. Therefore, while contemporary physics relies primarily on its method of rational calculation, it persists in searching for the machine that can yield the raw material for penetrating the secrets of nature. Clearly, the idea that we should be able to observe with the senses that which we know is a cultural posit, one that violates the practice of many sciences. These sciences rely, in their everyday activity, on quite different processes for determining what counts as reliable knowledge, particularly tests.

Of course, natural sciences paid no attention to Hegel's critique, which forms the core of his lengthy preface to the *Phenomenology of the Spirit*. Hegel does not dispute the validity of scientific knowledge as partial truth, since it relentlessly purges from the object any references to the subject, that is, the agent of knowledge production. Since Hegel posits the subject-object relation as the key to possible truth, his conception of science is identical with a historical, dialectical philosophy whose crucial presupposition is the totality of relations, not only the relation of the knower to the external object, but that of self to self where the "other" is grasped as both the internalized nature and the (temporarily) alienated self. For Hegel, the problem of consciousness is not the province of psychology, but the crux of the scientific project itself.

It would be excessive to claim that the development of quantum mechanics, especially the discovery that knowledge of the physical object entails bringing the observer into the observational field, represented a direct acknowledgment of the power of Hegel's attack. Yet, although physics has recuperated this admission within realist epistemology, some of the more philosophically minded theoretical physicists still have nagging doubts that the "correction" of the principle of indeterminacy is insufficient, that physics and truth are nonidentical.

The second major argument in the book is that power is not exercised sui generis, either as coercion alone or through institutional domination. While these forms are manifest in economic and political spheres, I argue that claims of authority in our contemporary world rest increasingly on the possession of legitimate knowledge, of which scientific discourses are supreme. So, although scientific communities may be described as emergent power centers which participate in the crucial decisions of the state, the power of science goes far beyond specific institutional sites. Science and its slightly degraded partner, technology, intrude into what we mean by economics, politics, culture. It is difficult to conceive of a single significant mode of representation that consciously or not fails to try to emulate scientific norms. Thus, even the critical intention of the present work is undermined by the ubiquity of scientific discourse. One is reminded of Wittgenstein's claim for language as the mode of life: even the process of attempting critical distance in order to discern its characteristic features constitutes a logical contradiction, for one can perform this inquiry only by means of language. Similarly, Jacques Derrida argues that "modern" discourse is caught in the centralizing power of logos, the language and thought modes according to which laws of identity, noncontradiction, and exclusion form the foundation of what we mean by knowledge. Using these points of logic, science introduces elements that are dispersed throughout discourses: First, the qualitative is excluded, or, more precisely, quality is occluded from the objective world; quantitative relations, expressed in the language of mathematics, have become the *lingua franca* of all discourse that claims knowledge as its content. Second is the imperative to empirical inquiry, which excludes speculation except at the outset. Third, it is claimed that exact knowledge is free of value orientations, of interest. Fourth, method is given primacy in the confirmation of scientific knowledge. Taken together, scientific power becomes coercive in the same way that the belief in the deity was received truth (and in some places still is) of many nations. Just as God is taken as axiom, so the four elements of scientific discourse are generally beyond dispute in our world.

This raises the problem of alternatives. Although this volume is not the place to outline a new science, it does argue that only a social theory of science that combines critical distance with historical analysis can provide the precondition to such a development. As long as science is not situated within the panoply of discursive practices, but retains its axiomatic character, its power will remain beyond challenge. In this connection, the fairly recent discipline of the history of science has, with notable exceptions, done little more than act as pegagogy (in Thomas Kuhn's most generous phrase) or outright publicity for science. The story of science as the conquest by "man" of nature, as the achievement of human dominion of the universe, and so forth, is precisely the romantic veil that has been draped over the history of science as social relations. And, as I show in chapter 9, philosophy has become the servant of the sciences, especially in its analytic incarnation. Within the sociology and political study of the scientific community, writers have taken different positions on the relation of scientific texts to contexts. As we shall see in chapter 10, the social study of science has tended to elide the relation of the constitution of scientific knowledge to the conditions under which it is produced. Again, there are some important exceptions, especially two groups of British investigators (the Radical science group and the Edinburgh group). Yet, what has been defined as the "mainstream" in the social and historical study of science remains deferential to its pristine status.

The power of science has survived and dominated the social study of science, not only in its ordinary academic precincts, but in the oppositional intellectual and political practices that group under the heading of Marxism. If I have felt obliged to deal with Marxism in considerable detail, it is largely due to my belief that the real evidence for the degree to which science has become dispersed as an a priori cast of mind in contemporary life is the fealty of Marxism to its canonical precepts. Scientificity not only dominates the rich traditions of Marxist-inspired history, political economy, and sociology, but characterizes Marxism's own self-reflection. At the same time, I am constrained to treat the Marxist theories of science and technology because they are, at least in tendency, concerned with the relation of scientific text and the social conditions under which it is produced. No other tendency in modern social thought achieves even this intention, much less provides a thorough social theory of science.

This book is the result of more than thirteen years of writing and reflection. When first conceived, it was to encompass a critique of science and technology, a study of the influence of science on culture, and an exploration of the relation of work as social activity to the development of technology. But in the course of writing the book, it developed into a study of the discourses about science. The introduction, following this preface, may be understood in the context of the broad themes envisioned in my earlier plan; it points out what is at stake in the study of science. In future works, I shall deal with these themes in more detail.

This book originated in a Ph.D. project conducted in 1974–75 at Union Graduate School. It was preceded by my book *False Promises* (1973), which is a study of the American working class. The thesis was an outgrowth of that work insofar as I became aware that the questions associated with technology, science, and social theory bore crucially on the fate of the working class in contemporary life. Major portions of this project have been considerably altered and condensed for the present book and appear as the substance of chapters 2, 3, and part of chapter 4. Len Rodberg, then of the Institute for Policy Studies, was my adviser; his warmth and positive responses to this earlier work encouraged me to look more deeply into the problems I raised in that thesis. The thesis was read and criticized by the Labor Process study group at the institute: Jeremy Brecher, Keith Dix, Robb Burlage, Joan Greenbaum, and others who helped me clarify my views.

In 1978–79 I circulated my article "Science and Ideology" to friends. Among them, Doug White, my colleague in the School of Social Sciences at Irvine, was extremely helpful and discussed the implications of its central arguments for the present book. Largely as a result of his comments, I was able to conceive a larger study. This article, substantially revamped, appears as the core ideas of chapters 5, 6, and 7.

Chapters 8 through 12 as well as the introduction were written in 1986–87. Major portions of the entire book have been read and criticized by Ellen Willis, Bill DiFazio, and the Science study group at The Graduate Center City University of New York. Among these, David Chorlian, Jeff Schmidt, and Lou Amdur commented extensively on the chapters on natural science and sociology of science and made major contributions by correcting errors of the first drafts, improving my arguments, and providing examples to support my more controversial contentions. Margaret Yard read the sections on the medical industry with special care and made helpful suggestions. As a member of the Science group, she and David Sinclair contributed helpful comments for improving the manuscript. These people gave unstintingly of their time and energy. Michael Brown read several sections of the manuscript, and his approbation was deeply important to me. Ivan Szelenyi read and criticized chapter 8, and shared his wide knowledge of eastern Europe. I want to especially thank Terry Cochran and Victoria Haire of the University of Minnesota Press for their meticulous work in making the manuscript into a book. Lynn Chancer performed the word processing with accuracy and dispatch and also offered detailed suggestions for improving the manuscript. Her participation in both respects is deeply appreciated. Of course, none of these people bears responsibility for errors and weaknesses of the book. These fall exclusively on my shoulders.

PART I

CHAPTER 1
SCIENCE AND TECHNOLOGY AS HEGEMONY

When my daughter Nona was two years old, she frequently exclaimed, after a fall, "the chair did it," or, as she bumped into the wall, "the wall did it." On September 11, 1986, the New York stock market plunged eighty-six points. The next day, after a drop of thirty four points, a *New York Times* story read, "wide use of computers contributed to slide." According to the writer, trading on the stock market is often detonated by signals supplied by a computer program. It's a "split second" automatic process; in appearance, at least, large institutional traders such as pension plans and major banks respond to the slightest movement of interest rates without significant reflection. On this particular day, when stock traders came to their offices, "they were met by a big jump in interest rates . . . (which) immediately dragged down the price of futures contracts for stocks."[1] This triggered sales of current stocks by the money managers and widespread purchases of futures contracts. As the large investors unloaded stocks, others followed suit and stock prices plummeted. The investors respond to the computer as my daughter might explain bumping into the wall. Of course, the *Times* and the analysts they consulted allowed that national and international economic troubles might be an underlying influence on the serious fall of stock prices that week, but from the point of view of investors, it was their computer that seemed to make the decision to sell.

That we live in a computer age no one seems to doubt. Yet, along with the paeans of praise heaped upon this electronic device, there is

3

also a growing chorus of criticism and a pervasive mood of doubt about its redemptive features. Actually, the computer is now widely viewed as more than a tool, after three decades during which many of its proponents attempted to assure us that, despite its considerable power, it is really an extension of the human head and hand and has no autonomy. However, in recent years, this view has been sharply attacked by experimental computer research, on the one side, and by the practice of those who interact with it, on the other. There are those like Herbert Simon who impute to it almost mystical powers, the pinnacle of which is the ability of computers to think (of course, in order to prove this proposition, Simon was forced to define humans as "information processors"—a more complex conception would certainly defeat his theory).[2] Artificial intelligence research seeks to find ways to solve the age-old dilemma of uncertainty in human action which has plagued the labor process, politics, and vast areas of social life. Machines that think are more reliably subject to prediction and control.

But the computer seems to have a mind of its own, especially if the controllers are guided by its information. Many, including some computer scientists, have already begun to compare computers to the Golem of the medieval ghetto or the monster created by Dr. Frankenstein. Far from remaining a stunning but subordinate tool, the computer frequently jumps the track, subverting human purposes that set it in motion. Like the machines that characterized the Industrial Revolution, computers are just the latest occasion for the displacement of fears that "things" are out of control, that their human origin has been lost, and that it is too late for salvation.

My stock market example is by no means the heart of the matter. For those who would not speak of Chernobyl or Three Mile Island should also keep silent about the wonders of technology. Naturally, those who continue to defend the use of nuclear technology for supplying power attribute the problems at these sites to "human error," just as Union Carbide company officials blamed workers and inadequate supervision for the disaster at Bhopal, India, in December 1984 which killed several thousand people and injured another 200,000.[3] The phrase "human error" usually refers to those who operate equipment; when an airplane crashes, the pilot or maintenance mechanic is nearly always blamed. Almost never is "human error" blamed on the design of the aircraft or the basic judgment that nuclear energy is a safe bet. When technology is involved, managers and the media rarely ask whether the premises of the machine in question are valid. For example, government officials are prepared to ask whether an adequate evacuation plan has been devised by a company operating a nuclear reactor; the technique of generating nuclear energy or producing nuclear weapons is beyond the competence of regulatory agencies because promoting nuclear energy is national policy.

There are many reasons why nuclear power and air travel became privileged technologies in fuel production and transportation. I do not wish to dispute the economic and political arguments employed by corporations to persuade the many governments, including that of the United States, that these technologies were more efficient than existing means of energy production. For our purposes, the criterion of efficiency is closely linked to concepts such as cost savings, whose major component is saving time—time in the extraction of raw materials from the earth, a labor-intensive activity, or, concomitantly, the time saved by traveling 650 miles per hour rather than 100 or 200 miles per hour, the current maximum of rail transportation. Yet, rails have suffered at the hands of trucks which are, by any conventional standard, more energy-wasteful than trains. The contradictory arguments made on behalf of various methods of transport belie pure efficiency criteria: it may be that choices of technology are made entirely independent of "rational" production decisions but obey a different rationality, the power imperative.

Some students of the introduction of nuclear energy technology in advanced industrial societies have raised an entirely different set of arguments why this dangerous, even disastrous, way of obtaining energy has received such powerful support. The argument is that the decisions to "go nuclear" transcend any possible rational criteria, measured by economic or technical considerations. Instead, these critics claim that the basic impetus for the introduction of nuclear-powered energy is rooted in the hierarchical structure of society. Those at the pinnacle are able to impose a *logic of domination* on the rest of us by simply repeating their falsehoods through every avenue of public debate and discourse. This explanation removes the discussion of causality from the domain of instrumental reason. It is not this or that interest that has triumphed, tragically, in the corridors of power. Power itself carries its own demand, which exceeds, with impunity, the boundaries of reason. In effect, power sets new rational codes without reference to ideology.

If this is true, none of the by now conventional rational criticisms are adequate to the situation. One convention is to show that some interests can expect to reap huge profits from the introduction of certain technologies. Another appeals to the cult of efficiency as the supreme instrumental arbiter of social choice. Each of these positions generates a concomitant ideology or justification of action. The notion of power as the arbiter requires no independent ideology to accompany it. By virtue of its command over knowledge, it can manufacture justification to the extent that the need for justification remains necessary in mass democracies. For example, human survival is threatened by another major scientific/technological development: genetic engineering, the technological concomitant of discoveries generated by molecular biology. Gene splicing is a eugenic technoscience; its justification is almost invariably built on the proposal to improve the survival traits of the species by designing the gene rather than passively accepting genetically related traits as unavoidable. Genetic engineering is an intervention into evolution, an attempt to gain control over life processes.

These examples are not, strictly speaking, commensurable but are invoked to illustrate the range of issues that are emerging to question technology and its twin, modern science. Now the statement that science and technology have become inseparable is certainly controversial, especially among those who would insist that science is autonomous from the concerns of power and ideology. The division between science and technology is meant to protect science from its implication in the matrix of economic and political considerations, which are generally recognized to influence—if not determine—the course of technological development and its dispersion. It is still true, however, that most students of science, while acknowledging the influence of what is often labeled "cultural factors" on the process of knowledge acquisition, insist that economic, political, and ideological questions must be strictly demarcated from considerations bearing on the content of scientific knowledge. Instead, the term "scientific community" has become identical with "social context." Indeed, recent developments in the social study of science have narrowed this context to the laboratory, leaving other "external" influences aside. Despite developments over the past thirty years which try to establish a relation between scientific discourses and the historical and other social conditions within which they function, support for the proposition that science and the scientific milieu is relatively autonomous is still powerful. In Part III, I shall explore developments in the philosophy, history, and social study of science that have challenged the idea that what counts as knowledge of the external world is attained by means of scientific procedures alone. As we shall see, most investigators of science remain tied to the concept of science as a distinct knowledge sphere and have barely touched its relation to technology. The great exception, of course, is Critical Theory, the foundation of which is to establish this link and to assert the dominance of technology over science as well as its domination over contemporary social relations.

The assertion that Western culture is thoroughly technological derives from the German "romantic" critique of the Enlightenment, a critique which members of the Frankfurt School share with strains of late nineteenth- and early twentieth-century neo-Kantian thought, especially Edmund Husserl and Martin Heidegger. After arguing that the "rationalism of the Age of Enlightenment is now out of the question," Husserl nevertheless is quick to add, "but their [the Enlightenment philosophies] intention, in its most general sense, must never die out in us"; the intention is a "humanity based on pure reason."[4] Max Horkheimer and Theodor Adorno follow a similar line of argument. They lament the Enlightenment, which created a new science and technology that could dominate nature in order to promote the flowering of reason, but that led instead to its eclipse. The end of reason was rooted in the belief, current even today, that nature could be reduced to pure object, possessing mere quantitative extension. The mathematicization and mechanization of the world picture was undertaken by seventeenth-century science in the service of prediction and control. In the process, according to Husserl, we lost contact with the "life world" in a series of dualisms: mind and body, quantity and quality, mathematic relations and human relations. For the Frankfurt School, the logical result was positivism—the alienation of human reason from itself.[5]

Whereas Husserl's critique remains at the level of philosophic discourse, Horkheimer, Adorno, and Herbert Marcuse ascribed these dichotomies to the rending of society by social domination. The domination of nature fulfills a human project, the domination of people by people. Science and technology are practices that mirror the social world.

Still, the Frankfurt School was criticizing the enlightenment *from within*. Although Marcuse was ambiguous concerning science, claiming at one point that a new science free of technological domination needed to be created, he proclaims in the final chapter of his *One Dimensional Man* that science and technology "are great vehicles of liberation" if only they can be subordinated to new ends to replace those of domination. These ends "would operate in the project and construction of machinery and not only in its utilization." Technology can be instrumental "in the reduction of toil—it remains the very base of all forms of human freedom."[6] Thus, the sharpest critics of science and technology hesitate at the door of irrationalism and will not cross the threshhold, however harsh their evaluation of the gloomy record of domination that lies in the wake of modern science and technology. On the other side, of course, lies reason's nemesis: religion, mysticism, myth.

The warfare between what has been called "science" and other discourses that purport to explain the natural and social worlds is a story many times told. As Gaston Bachelard has argued, science is constituted by its break from competing discourses that claim to explain the same objects.[7] Physics, for example, breaks with forms of "irrationalism" but also with metaphysics. As Bachelard's contemporary, Karl Popper, has commented, metaphysics offers meanings that may be helpful to other aspects of human affairs, but not science.[8] Science, for Popper and Bachelard, is established by statements that can be subjected to refutation (in Bachelard's terms, "empirical validation"). It is the spirit and practice of relentless self-criticism that marks science off from other discourses, including traditional philosophy. For recent philosophers of science, even these procedures do not guarantee that the results are identical with "truth," only that they are demarcated thereby from other discourses. Within true science, there may be serious and often profound disputes of interpretation. But what all scientists share is a *community* rooted in method. The primacy of shared methods guarantees the reliability of what counts as science. In Parts II and III we shall explore these assertions in more detail. For now, it is enough to suggest that what those dedicated to Western scientific ideology mean by the term "science" reduces to two procedures: mathematical calculation and experimental validation/falsification of results. Mathematics ensures the rigor of investigation, establishing measurable relations and, in Bachelard's words, "gives body to pure thought." Mathematics is "the realization of the rational." However, lest science fall victim to Descartes's unwarranted "refusal to base thought on experience," experimentalism restores to observation its role as final arbiter of knowledge.[9]

Presumably, neither philosophy nor religion fulfills either of these criteria. Modern science demarcates itself, not by reconstituting the object, but by defining rationality in a specific way. At the same time, given the power of all metaphysical discourses in everyday life, it is obliged to make room for the extrascientific so long as these spheres are clearly subordinated to scientific rationality. In the knowledge hierarchies of postfeudal societies, modern scientific rationality is the privileged discourse, and all others are relegated to the margins. As a result, institutions of the state as well as the economy—education systems, government bureaus, the law and criminal justice systems—emulate scientific procedures within the constraints imposed by their own traditions and exigencies. Art and religion hold their places at the margins of human endeavor and become extracurricular, or, to use Freud's term, "deflections," for the frustrations produced by the inhibition of desire by the reality principle.[10]

The rise of Protestantism in leading industrializing countries in the eighteenth and nineteenth centuries seemed to provide moral sanction for the preeminent position of modern science as knowledge. Religious institutions now viewed themselves as supplicants in an increasingly secular age and understood their role not so much as deflectors but as moral guardians for individuals damaged by the blandishments of money and power. The type of knowledge offered by religion was confined to the ethical sphere; it concerned itself with matters of family life, personal grief, and, when it felt obliged to take social action, it was remedial rather than transformative. The great denominations of Protestantism relinquished that which Catholicism had struggled to retain: a claim on epistemological as well as ontological truth. However, in the nineteenth and early twentieth centuries the triumph of world capitalism over the remnants of the old feudal aristocracy in eastern and southern Europe forced even the recalcitrant Catholic and other orthodox churches to accommodate to the new order.

By the late nineteenth century, industrial production depended on scientifically based technologies; the craft traditions, of which early manufacturing was merely a form of rationalization, were themselves subordinated to the new technology; the motive force of production, energy, was no longer mechanical—really an extension of hand or water power—but became electrical, the principles of which derived from "pure research"; engineering replaced artisanal knowledge in designing the mode of transformation of raw materials into end products; in turn, the intellectual foundation of engineering became physics and chemistry, which themselves were institutionalized into large laboratories sponsored by and controlled by the state and large corporations. Thus, science itself no longer is only a hegemonic ideology of the new social order of capitalism and its industrial stage, but becomes integrated into the practices and discourses of production. The interchangeability of science and technology is, of course, either denied or ignored by most philosophers and scientists, but their growing convergence extends beyond the workplace.[11] As scientific discourse permeates state and civil society, scientific culture spills over beyond the laboratory. Business dares make no decisions that are not grounded on mathematical calculation that provides projections; legislators enact laws based on "data" generated by scientifically trained experts. Raymond Callahan has noted, referring to education, that technological criterion of efficiency becomes the new cult of public and private schools.[12] In schools, the idea of the liberal arts slowly gives way to occupational education. "Functional" literacy becomes the criterion of success for no-frills state school systems that are stripped of their music and drawing curricula and which reduce English and history to service departments for the technical-oriented programs. Therefore, in several dramatic moments since World War II, beefing up science and math in schools has become a matter of high government priority since these disciplines are understood to be vital for a country's economic and military position in an increasingly competitive and dangerous world.

Science education is accorded, in the current anxiety over the loss of America's erstwhile preeminence in industrial production, a primary role in the long road to restoration. In the 1980s, as the defense budget of the United States has increased dramatically, engineers and scientists are in demand in military-related industries. Consequently, despite enormous pressure from government and industry on schools, teacher recruitment in science and math has lagged, primarily because salaries are far below those offered by private industry and research institutions. Reversing its steadfast policy of reducing federal expenditures and advocacy of scaled-down public services at all levels, the Reagan administration saw the virtue of making an exception for the teaching of science and math.

These illustrations do not exhaust the extent to which technological culture penetrates the social world. Right after World War II, as America assumed a position of economic, political, and military dominance in the world, secularism—always closely related to the growth of modern science and technology—seemed to have triumphed, irrevocably, in the cultural sphere as well. Church attendance dropped sharply, and religion in this most religious of modernized societies seemed finally on the wane, a century after its final marginalization in most industrialized European countries.[13]

Technological culture may not provide salve for the spirit, and it may do nothing to fill the void left by the marginalization of religious morality, but, as even its most severe critics are forced to admit, it gives rein to the pleasure principle. As Marcuse observed, in advanced capitalist societies, especially the United States, "the defense structure makes life easier for a greater number of people and extends man's mastery over nature."[14] Although his description of technological society as a "comfortable, smooth, democratic unfreedom" was adopted by some as a reason for opposing the prevailing setup, millions who had been condemned to deprivation during the decades of industrialization welcomed technology as a savior. Culture was massified, overtaking high art and marginalizing the intellectuals for whom it was sustenance, but providing a respite from the drudgery of even the most automated labor. Marcuse may have spoken for more than the intellectuals. His remorse for the contradictions of the Enlightenment finally paled against the achievements of science and technology. As a child of reason, he was unable to draw the logical conclusion from his critique that science and technology had become obstacles to freedom. Writing in the early 1960s, he could only allow that a transformed science and technology might serve the emancipatory interest better than the version that had emerged from the Dark Ages.

Goethe once quipped, "He who possesses science and art also has religion; but he who possesses neither of these two, let him have religion."[15] The vision is a thoroughly pluralist one. Science and art are to exist alongside religion, which as Freud noted, echoing Goethe, was the possession of everyman, defined as someone lacking the means to sublimate the irrepressible pleasure principle. For the nineteenth-century poet or scientist, the idea that religion would one more time raise its claim to yield reliable knowledge of the external natural world was unthinkable. Freud is quite clear: the belief in the supernatural is an "untenable" doctrine but better for the common folk than alcohol or drugs. Just as long as it is not taken seriously as more than an opiate.[16]

But this is precisely what has transpired in the past decade. Despite the unambiguous triumph of the scientific worldview and the totalizing effects of technology, science and technology are experiencing unprecedented attack. For the past fifteen years, counterenlightenment religious movements have taken on worldwide and revolutionary proportions. Far from disdaining politics as beneath their ethical missions, some Jews, Moslems, and Christians alike have laid siege to the state, demanding that public life conform to religious precepts and that the separation of church and state be ended. The theocratic state, once consigned by scholars, politicians, and jurists to the past, is loudly proclaimed as the present and future by the new movements, and has already captured important outposts in the Middle East—within the Moslem, Christian and Jewish worlds—and is rapidly becoming the *sine qua non* of political victory for parties in the United States.[17] Here, it is not only that lawmakers adopt biblical criteria for establishing the rules of punishment; this is an old story in America. The doctrine of retribution opposes itself to the idea that the criminal may be subject to rehabilitation. In retrospect, liberal assumptions about crime seem a brief interlude in an unbroken history of deep-seated, religiously based conservatism in American law and political life. Fundamentalist theology is reentering the public schools as well. And among its adversaries, none is more significant than modern science. As biologist Douglas Futuyama has warned, science is "on trial" and reason is "under fire."[18] A new science of creationism has challenged the exclusive teaching of evolution in public schools and demanded that its "science" be given equal treatment. This is no longer merely a case of the ministry getting involved in politics, supporting candidates who favor state aid to parochial schools. Such requests, made by the Catholic Church for decades, challenge the separation of church and state but do not confront the ideological hegemony of modern scientific doctrine. Indeed, many parochial schools of Catholic and Protestant origins teach modern science and mathematics as necessary elements of a relevant curriculum, relevant, that is, to the job and career aspirations of their own constituencies. The fundamental distinction between the old political intervention of religious institutions, either on behalf of social justice or their own parochial interests, and the new fundamentalism is that the latter condemns enlightenment science as ideology, as one viewpoint in a plurality of discourses possessing no privileged knowledge of the external world. Thus, its demand for an equal place in the curriculum. In essence, fundamentalism has stepped forward where enlightenment critics of science have stepped back. For the purposes of this discussion, it is not relevant to point out the absurdities of the creationist account of the origins of the universe, life, and the evolution of species. At issue is the claim of enlightenment science to certainty and its refusal to acknowledge its own discourse as a form of ideology.

Take a case in point. In the course of his attack against creationist accounts of development, Futuyama makes two arguments: "anyone who believes in Genesis as a literal description of history must hold a world view that is entirely incompatible with the idea of evolution, not to speak of science itself . . . where science insists on material, mechanistic causes that can be understood by physics and chemistry, the literal believer in Genesis invokes unknowable supernatural forces."[19]

The second argument is, for Futuyama, more important: "if the world and its creatures developed purely by material, physical forces, it could not have been designed and has no purpose or goal. The fundamentalist, in contrast, believes that everything in the world, every species and every characteristic of every species, was designed by an intelligent, purposeful artificer, and that it was made for a purpose."[20] Futuyama defends evolution and the whole of science as holding that "mechanisms," not teleology, govern events in nature. These mechanisms free science from reliance on Aristotelian final causes and rely instead on the sufficiency of efficient causes. Changes within and among species can be ascribed not to ultimate purposes of God, but to the struggle for existence, to adaptation by organisms to a changing environment. Futuyama's underlying view is that of "material, mechanistic" explanations for natural events: he identifies this view as identical with that of "science."

Surely, the propositions of evolutionary theory are incompatible with creationism. For modern biological science, there is no question of plan or purpose in nature (although as Alexandre Koyre shows, Newton was convinced that physical laws corresponded to God's plan even if teleological explanation had no place in physics).[21] However, many evolutionists and philosophers of biology have come to insist that Aristotle's idea of final cause, so vehemently rejected by the early evolutionists, had a place in modern scientific theory. Stephen J. Gould remarks:

> It is still unfashionable in biological circles to use such words as 'design,' 'purpose,' or 'teleology.' Since final cause is so indispensable a concept in the elucidation of adaptation, and since natural selection can produce a well-designed structure without any conscious intervention of God's superhuman wisdom or the sub-human intelligence of the animal in question, one would think that these terms would again be admitted into orthodoxy. Evidently, however, we are still fighting the battle with theologians that we won in deeds almost a century ago.[22]

Futuyama may be one of those still fighting the battle, but in his zeal to distance evolution from theology, he succeeds only in making his ideas into the mirror image of his opponent's. Francois Jacob puts the issue colorfully. Commenting on the role of sexual reproduction as the "aim" for each organism and for the history of organisms, he concludes, "For a long time, the biologist treated teleology as he would a woman he could not do without, but did not care to be seen with in public. The concept of the programme has made an honest woman out of teleology."[23]

Today, the metaphors of program and system to describe the processes of heredity and organic evolution have opened a new debate concerning the question of causality in biological science. The earlier belief that science could dispense with causal explanation in favor of description was intended to remove from natural science what Laplace termed the "hypothesis" of God's design. What has been termed "essentialism" is not so easily removed. The religious counterattack against modern science may be misplaced, but it resonates with a widespread feeling among ordinary people that science has become a kind of priesthood and that scientists, like the old priests, are not always forthcoming about their own doubts, at least in public. The new attack on modern science, however, is not mounted simply on the authority of the Bible as the true word of God: it claims the mantle of science. Controversies within science that reveal deep fissures within that community on questions such as causality and truth have, until recently, been conducted behind closed doors. In biology, the mechanical world picture is challenged; in physics, the image of nature as pure objective extension, subject to prediction and control, is questioned by those whose work functions at the level of theory, at the same time practitioners insist that the old news is still best.

Perhaps the most broad-based skepticism has arisen in the field of medicine. There are several issues here: at the most fundamental level is the question of the traditional medical model of diagnosis and treatment. According to this way of seeing, the human body is a machine (in the older version) or a computer program (in the up-to-date model). The parts are relatively autonomous, making possible a treatment regimen that ignores the relations of the parts to the whole, except in cases of side effects which can be counteracted. In the old mold of medical practice, the relation of the person to the environment and the emotional state to the physical state are irrelevant. In the first instance, unless severe hunger is observed, nutritional issues are not factored into diagnostic practice. The context of medical practice itself—treating individuals—does not permit the effects of such "external" phenomena as pollution, work environments, quality of life based on economic circumstances, and stress related to working or family life, to become objects of treatment.

Medicine, based on the findings of molecular biology, adopts a systems approach to both diagnosis and treatment. In the more ecological mode, the human body is an open-ended system that is *determined* by its genetic code, but may be *influenced* by the environment with which it has homeostatic relations. In the less interesting version, the body is a closed system consisting of information networks that are only peripherally subject to external influences. Francis Crick's version of the basic requirements for life includes the Darwinian idea, adapted from Thomas Malthus, that the organism must compete for water, food, and other life-sustaining resources. But, in the final accounting, its capacity for survival, he argues, is its internally generated information system which is built up from its molecular structures.[24] Second is its capacity to mutate in the wake of changing conditions, a capacity that is given internally. Recent medical science has taken its picture from Crick and his associate James Watson, who together played key roles in developing molecular biology. The organism itself remains the object of study, pristine in its singularity but seen not so much as a series of parts but as a system of "organized complexity" whose driving force is the molecular structure of the gene.[25] We treat individuals differently than before. Since "the problem of life is a problem of organic chemistry,"[26] measures taken to repair damaged organs should seek both diagnosis and treatment in dysfunctions in chemical combinations that affect the organ's information program. Thus, such diverse pathologies as psychosis and cancer become subject to chemical treatment. One seeks the reasons for various emotional illnesses not in the stresses produced by the life situation but in chemical malfunctions. Similarly, the most widely used treatment for some cancers is chemotherapy.

Of course, medical applications of molecular biology are only one of the major changes initiated by this new discipline. More controversial, and surely more problematic, have been the new technologies of genetic engineering which present themselves, in their visionary manifestations, as fulfilling the dream of both nineteenth-century eugenics and modern medicine to eliminate disease by altering those traits that make humans less competitive in a changing physical environment. According to Jeremy Rifkin, "we are virtually hurling ourselves into the age of biotechnology." Molecular biology has succeeded in changing the heredity of the gene, transferring genes from one organism to another and, perhaps, more powerfully, has synthesized cells through engineering. Its most recent applications in food technology promise to eliminate the need for chemical preservatives.

Molecular biology is a science with a purpose and design: it seeks to transform life by recombining DNA which it has postulated as the core of life. This aim surely violates a wide spectrum of religious principles but also raises profound ethical issues. For, even if altering or remaking life were to be judged a worthy social purpose, the question is to whom are scientists performing these tasks to be accountable? This issue is in the process of adjudication by law and the courts in the United States, and the results are, to some, horrifying. The industrial state of mind has dominated these discoveries, and, in the United States, this means private ownership and control of the means of production (splicing) and of the product (altered organisms). Indeed, MIT has recently forged an extensive series of arrangements with some major biotechnology firms. In return for huge grants for research, the university has agreed that the patents for discoveries will belong to the corporations. Courts have decided that these arrangements are legal because they have viewed bioengineering as simply one more industry. In a free society, so the argument goes, we may sanction a free market in gene splicing just as we recognize the free market in the production of steel or cars. Measures to prevent corporations from engaging in commerce violate their rights under law.

Two major issues are raised by these developments. The first is whether we can afford to free the new discoveries associated with biotechnologies from rigorous public control, assuming we could identify an appropriate regulatory agency. The second strikes to the heart of the conflict between modern science and religion, but has also been framed by nonfundamentalists who, nevertheless, hold to ethical beliefs that are counter to some types of scientific investigation. Simply stated, the issue is whether modern science has a right to alter our relationship with nature, *sui generis*. On the one side, the fundamentalists argue against any scientific discoveries that may violate the word of God as they have interpreted it. For those who would reject such objections on the grounds that they are "irrational," the alternative does not imply accepting the new biotechnologies. For if the aim of bioengineering is to transform our relation to nature, especially our own "nature," then the claim of science to be free of purpose, to be engaged in some anterior conception of "pure" science is most suspect. If modern science does not imply the intervention of supernatural powers, it nonetheless cannot escape interrogation of its underlying ethical assumptions, neither with respect to the content of its propositions, which it calls "laws" of nature, nor with respect to its responsibility for the results of its investigations.

What is at stake here is the question of *entailment*. If molecular biology chooses to link its discoveries in DNA and RNA research to industrial tasks, are we not justified to ask whether such discoveries do themselves entail ethically problematic purposes? This becomes especially relevant at a time when universities, the traditional seats of scientific research, are making agreements with large corporations and the government, which provide the bulk of the funds for their work. Under these circumstances, the claim of free inquiry becomes difficult to sustain.

Yet, it would be a mistake to declare that science and technology are entirely determined, in their content as well as their uses, by the hands that feed them. Science is a complex, multivaried activity. Its relation to what it "observes" is never unmediated: that is, the economic, political, and social environment in which people "do" science and technology intervenes between cognition and its object. Alexandre Koyre, who vehemently rejects explanations for scientific discovery that depend on the practical conditions that may spur it on, substitutes metaphysics and philosophy as mediations between theory, experiment, and nature. Similarly, Paul Davies ascribes the continuing popularity of astronomy to the fact that the study of the heavens is not far from our common and still passionate desire for transcendence, for the supernatural. At the same time, he emphasizes the rebelliousness of nature, its "inherent uncertainty," its "randomness," and the significance of "subjectivity" in the new physics. According to Davies, the "observer is beginning to play a rather essential role in the nature of the world," making absurd objectivist concepts of time and space, including those that impute to nature an inherent order.[27]

Indeed, difficulties experienced by partisans of a picture of nature as a unified field have generated major debates within physics. It is generally recognized by physicists themselves that few of its discoveries are exempt from the variability brought to them by interpretation. The facts do not speak for themselves and, through this door, marches religion and other metaphysical doctrines as well as philosophy. In Part III, I shall discuss the relation between philosophies and the social conditions that may give rise to them, challenging Koyre and others who wish to preserve the internalist account of science by limiting the context of discovery to purely intellectual influences. Suffice it to mention here that many who speak of nature as a unified field tread dangerously close to teleology, and even God, as the ultimate purpose. David Bohm, a theoretical physicist, asserts the formal unity of all natural phenomena, defining physical laws as tending toward the absolute through the study of partial, relative phenomena; most important, he defends causality in nature as an unlimited principle. That is, for Bohm, order, not randomness, is inherent in nature, the truth of which can be grasped by science. Although he has no recourse to the supernatural, Bohm insists that there are "hidden variables" in quantum mechanics that belie the picture of randomness and indeterminacy put forward by such thinkers as Niels Bohr and Werner Heisenberg.[28]

In short, the struggle between mechanistic science and religion reappears in different form within science itself. As I have shown, teleology is almost respectable in biology, and one of the great debates initiated by Einstein (unwittingly) and Heisenberg (with some hesitation) is whether God plays dice with the universe. Some of the key interpretative struggles within physics are whether we may posit *real* time and space, and whether the intervention of "subjectivity" removes these categories from the realm of objectivity. More broadly, the question remains: what do we know when we know something? If our knowledge is ineluctibly bound up with the processes of observation and experiment, and if these are permeated with a priori assumptions about the character of the observed or the uses to which knowledge is destined to be put, in what sense can we say that science as a type of knowledge is superior to metaphysics, including religion?

Social ecology and radical feminism, two social movements of considerable political and ideological weight in many advanced industrial societies have, from an ethical standpoint, challenged the content and results of contemporary science and technology. Both have their roots in the 1960s and early 1970s when a general skepticism about the autonomy of science was invoked because of the subordination of much of the scientific community to military requirements, and the propping of scientific research and technological innovation by industry. Yet, the content of their critique differs from the assertion that the problem of science and technology consists essentially in the uses to which discovery and innovation are put. Instead, social ecology and radical feminism have mounted a fundamental critique of the scientific worldview, especially the contention that science and technology are neutral instruments that can be separated from the context in which they are developed. Social ecology argues that science and technology, by virtue of their subordination to the interest of the domination of nature, share responsibility with the state and capitalist corporations for the "death of nature," a metaphor that describes the increasing danger to life wrought by scientifically based technology. Rather than blame the growing problem of environmental pollution merely on the lack of state regulation, ecologists argue that technologies of boundless domination are leading to ecological disaster. On the one hand, the idea of nature as pure "object" has been a normal presupposition of practical physics insofar as it has enabled nature to be defined by categories of pure extension in the service of prediction and control. On the other hand, the imperiousness of molecular biology consists in its eugenic ideology, its eagerness to play God by altering the distinctive characteristics of our species. Ecologists explain the rebelliousness of nature as a reaction to human interventions that disturb its internal relations and its homeostatic relations with various species of life. The consequences of the industrial logic that marks contemporary biology can be dangerous to life. But the explanation given by social ecology for this state of affairs transcends the purview of social policy. What enables science to engage in life endangering activities is a belief that humans may insure their future by subjecting the gene to engineering, just as earlier industrialists exploited nonrenewable raw materials to advance various ends, including profits and the promise of liberation from material hardship and other socially approved values. Here we are engaged in a drama of both intention and unintentional consequences. The ideology of nature as "other" or "matter" undifferentiated by quality and infinitely fungible is a necessary condition for a science that has integrated itself into technology, in which the difference between them becomes analytic rather than ontological. Theoretical physics and biology may try to hold themselves aloof from the uses to which their discoveries may be put; Einstein certainly did not mean to lend his discoveries to the invention of weapons of nuclear destruction, and seventeenth-century astronomers may not have performed their work exclusively for the sake of improving instruments of navigation. Nor is it necessary to make direct connections between developments in wave mechanics and the rise of communications technology. The arguments made by those who try to discover, both by empirical research and by inference, a commercial or industrial telos in processes of discovery are certainly misdirected. The point is, rather, to examine the tendency of scientific ideology, the stuff of which discovery is made, that is, to situate science as a discourse within a larger system of social relations in which economic and political influences do not necessarily appear directly in the laboratory.

Ecology takes two directions on this issue. One is to argue from what might be called a historical standpoint: this is no longer the age of the individual thinker puttering in the laboratory under conditions similar to those of the artist. While scientific procedures may be considered quite distinct from those of the artist (except perhaps the composer of music), the creative side of science was, at the time of Galileo and even Lavoisier, commensurable with that of art. Scientific and artistic work requires months, perhaps years, of drudgery. But the process of scientific discovery in the seventeenth and eighteenth centuries, and even well into the nineteenth, was marked by individual effort. The scientist, like the artist, worked alone or with very small groups. Then comes Pasteur, in the later years of the century, a professor who, as Bruno Latour reports, attaches himself to forces of industry and the state and makes science a social force.[29] Latour's narrative of Pasteur's claim that he had made a serum that cured an anthrax epidemic in France was not validated through reproducible experiments made by scientists but by three institutions possessing the power to certify such a claim: the scientific community, of course; milk processors and producers requiring some relief from a costly and nearly devastating outbreak of the disease among cattle; and the state. Latour's conclusion is not that "society" influences the course of science, but that the laboratory becomes a model for social power. Drawing on Foucault's thesis that the knowledge/power relation becomes characteristic of modern society, not only does Latour posit this conjuncture as the condition of scientific truth, but he conversely argues that traditional institutional power draws its strength from the laboratory which in the twentieth century becomes a producer of commodities and of social power. Latour wishes to demonstrate that science is not subordinate to capital, as Marx and Marxists argue, but has become a crucial form of capital, with the laboratory a site of its production. Science retains some of the features of art, but the process of scientific discovery has been permanently removed from anything resembling an autonomous sphere.

Science lends itself to integration because it shares, with the rest of society, the teleology of domination over nature. Its incorporation into industry and the state may be necessary for the survival of nations, considering the world economic and political order. But, as these ecologists contend, such a close relationship between science and the forces of order means that industry has been able to multiply its productive powers by geometric proportions over industrial production in the (relatively) prescientific age. Specifically, the transformation of physical, chemical, and biological knowledge into instruments of economic, political, and military power is the foundation of our ecological crisis.

It does not matter that the scientific community ritualistically denies its alliance with economic/industrial and military power. The evidence is overwhelming that such is the case. Thus, every major industrial power has a national science policy; the United States military appropriates billions each year for "basic" as well as "applied" research. The National Institutes of Health provide vast sums for research that will be transformed into industrial technologies by private corporations which, in turn, make products for industry and for mass consumption that often spell disaster for the environment. What distinguishes this indictment from the criticisms of environmentalists is the demand by ecologists for new social arrangements that embody a radically new principle in our collective relationship to nature. Their demand for an ecologically pacific natural environment threatens the organized complexity of the marriage of science and industry in every country. The collective nature of scientific research is made possible only by the investments of the state and large corporations. Better to halt this form of science and return to science as art. Under such circumstances, a way of life would come to an end, for scientific and technological development underlies rising standards of living in the most industrially developed countries. Beyond the standard of living is a whole concept of culture based upon consumption, our notions of the relation-

ship between work and what we call "leisure," in short, our relationship to experience. That the "technological sensorium" dominates
everyday life seems hardly disputable.[30] The question is whether this is
what we want. Ecology normatively challenges this sensorium rather
than focusing only on its effects. It has formed and inspired movements, political parties, and massive protests against those who
employ scientifically produced technologies that, in its view, inhibit
freedom. What is at stake in its alternative is the possibility of constructing a social ethic that proposes different assumptions about what
constitutes the good life. For the philosophy of ecology, a world without domination, domination of nations, of women, of nature, of cultures, requires a radically different conception of science and technology since what it entails is a radically new idea of the everyday.

Ecology's critique of science and technology, while not ostensibly
religious, suggests that ethical neutrality with respect to the results of
scientific enterprise is not justified. The ecological position, in its most
sophisticated formulation, challenges the idea that any form of human
knowledge can be separated from its consequences. If one finds that
science has made an alliance with prevailing forces of destruction to
further the interest of domination of humans, its position as an unassailable, almost sacrosanct discourse must be overturned. Murray
Bookchin, perhaps America's most trenchant and global ecological theorist, puts the case:

> In our discussions of modern ecological and social crises, we
> tend to ignore the more underlying mentality of domination of
> each other and by extension of nature. I refer to an image of
> the natural world that sees nature as "blind," "mute," "cruel,"
> "competitive," and "stingy," a seemingly demonic "realm of
> necessity" that opposes "man's" striving for freedom and
> self-realization. . . . This all-encompassing image of an
> intractable nature that must be tamed by a rational humanity
> has given us a domineering form of reason, science and
> technology.[31]

To this image, Bookchin opposes "an ecological standpoint: nature
as a constellation of communities . . . freed of all anthropocentric
moral trappings, a participatory realm of interacting life forms whose
most outstanding atttributes are fecundity, creativity and directedness
marked by complementarity."[32] Bookchin clearly identifies traditional
notions of reason, science, and technology with the identification of
nature with the capitalist marketplace (itself linked to conventional
evolution theory, which, as I have noted, is adapted by Darwin from

Malthusian conceptions of natural history). In effect, Bookchin counterposes nature in metaphors derived from art to those of enlightenment science, which has employed the metaphors of commodity relations. In both cases, we have recourse to ethical discourses in which telos is present as a regulative principle. Enlightenment science, which believes itself free of references to purpose, and aim in its conception of nature, has, as Bookchin shows, simply portrayed nature in the image of capitalist social relations. In contrast, reflecting the views of a wide spectrum of ecological thought, Bookchin frankly admits that ecology pictures nature as an active, purposive agency, clinging to standard notions of objectivity when he tries to distance ecological thought from "anthropocentric moral trappings." This anachronism in Bookchin's discourse reveals not so much the weakness of ecological ideas as the strength of science as a hegemonic ideology. For the meaning of "hegemony" consists precisely in its presence within the discourse of opponents of the dominant ideology. At the same time, it can be argued that the ecological and feminist critiques of modern science are based on exposing science as ideology by means of showing its subjectivity, its character as ambiguous moral agent.

The explosion of the modern feminist movement since the late 1960s has generated what is perhaps the richest body of social theory in the last twenty years. This is not to ignore the extraordinary work of Simone de Beauvoir or earlier theorists of feminism such as Charlotte Perkins Gilman,[33] or the many feminist novelists whose critiques of marriage and the family were framed within conventional literary forms. Feminist theory has sought to explore the position of women in the West as well as the Third World in a great number of intellectual fields. Feminist historians have sought to debunk the myth that there is no "her" story in economic, political, and cultural terms. Since, as Walter Benjamin remarked, history is written by the victors, the hidden story of women's struggles for equality and escape from the yoke of male domination has to be told by other women who will explore the underside of dominant narratives. This history joins the general movement of social history that has succeeded in uncovering the lives of working people, black slaves, and other "invisible" groups denied a public identity by those who are the appointed guardians of official history. Sociologists and anthropologists have explored the lives of women engaged in homemaking and have sought to restore to that activity the name "labor."[34] Others have theorized the equal social weight of the labor of biological and social reproduction—especially child rearing—with that of material production, arguing that women participate as workers in both spheres but have been systematically expunged from considerations of questions of social relations and power. In short, the lives of women, their forms of self-production and self-organization, their successes and failures, have been the loud silence in historical and social scientific studies. A whole army of feminist scholars have made enormous efforts to recover women's voices, to uncover the specificity of women's discourse.

The attempt to find women's voice in science, especially in scientific discovery, proceeds according to at least two major roads of inquiry. First is to assert that women have participated in science and technology, if not as equals, at least in considerable depth. The figures of geneticist Barbara McClintock, organizational and technological theorist Lillian Gilbreth, physical therapist Sister Elizabeth Kenny, and physicist Marie Curie are invoked only to illustrate the range of contributions that women have made to scientific and technological discovery.[35] Citing Erik Erickson's question, "What will happen to science if and when women are truly represented in it—not by a few glorious exceptions, but in the ranks of the scientific elite?" Evelyn Fox Keller is quick to remind us that despite McClintock's contribution to understanding the mechanism of inheritance, she is still the exception, and that science is dominated by men.[36] Extensive study of McClintock's difficult travail as a genuine theorist rather than practitioner of normal science was meant to illustrate that, as a visionary, she faced resistance from a scientific elite ensconced in its own assumptions and prejudices against women, and that the moral neutrality of scientists could not be assumed despite the possibility that McClintock offered new knowledge. Keller's interest is to show that sex and gender relations bear on the exclusion of women from scientific communities and that these relations are deeply rooted in the ideology of Western culture, inscribed in its most eminent philosophical works—that woman is other, is identified with nature, and is an object of domination. Francis Bacon, one of the formative figures in the development of modern scientific ideology, was also a key proponent of the identity of science with male power. Nor was he exceptional. Keller's reading of Plato's *Symposium* and *Phaedrus* reveals the close relationship of male sexuality and the production of knowledge. Knowledge, for Plato, is erotically based, but closed to female participation.[37]

While affirming that gender has become a crucial element in the exclusion of women from science, Keller rejects the idea advanced by some feminist writers that Western science is male science. In feminist theory, this concept seeks to remove from Western science the mantle of objectivity by proving that its antiecological practices—and, even more significant, the mathematicization of nature and the scientific ethic of prediction and control—are steps toward the exclusion of women and undermine modern science's universalist claims. Whereas Keller confines her critique to the exclusion of women from the scientific elite, Carolyn Merchant contrasts the modern scientific worldview which is formed by the dual concepts of "mechanism and the domination of nature" with an earlier "organic-oriented mentality" that held to a nurturing image of the earth, an image of which the womb, woman/mother are the central referents.[38] According to Merchant, the replacement of this imagery, which survived as late as the Renaissance, by the mechanical world picture is linked to commercialization and industrialization, which remain the animus of our contemporary social world. These images had been a crucial feature of Western civilization, functioning as an "ethical constraint" on mining and other activities that destroyed the earth. Thus, Merchant historicizes modern science and, far from conceiving it as "progressive," argues that it represents a regression from an earlier science that was prevalent in ancient civilization. "Violence" against the earth is consonant with violence against women, since earth relates to humans as mother to child. Man's conquest of nature, then, is analogous to dominating women. For Merchant, it is impossible to abstract the development of modern science and technology from this process.

This is an example of one vision of what has become known as the "feminist standpoint." In the materialist account, the sexual division of labor in which women have been made subordinate to men in all aspects of their existence—material production, social reproduction, especially in their enforced role of chief, if not sole, child rearer—has consequences for the character of knowledge. Knowledge is gender differentiated, according to Nancy Hartsock, because the sexual division of labor forces women to occupy a different position in the social structure.[39] This position is not induced by the biological differences between men and women but is, instead, socially constructed. She acknowledges these differences but argues that we can never know the degree to which they might influence, if not determine, the forms of knowledge until sexual divisions are abolished.

This is the "objectivist" account of the feminist standpoint insofar as it does not rely for its argument on the actual consciousness of women. Rather, following Marx and Georg Lukács, the structure of Hartsock's argument reiterates the view that social being determines social consciousness but substitutes women for the proletariat. Otherwise, one would have to adopt the viewpoint that knowledge is free of its material preconditions, that there is an unmediated relationship between the knower and known. Hartsock's central metaphors are those of political economy, although she draws liberally on the psychoanalytic literature of object relations as well. Yet, the ideas of commodification, appropriation, exploitation, and oppression are the crucial analytic categories in her attempt to establish a specifically feminist discourse or standpoint.

Hartsock does not spell out the implications of such a view for the relation of gender and science, but her account bears on the production of knowledge. Even though Hartsock distances herself from culturalist versions of feminism such as those offered by Merchant, Mary Daly, and Susan Griffin—all of whom represent the fall of civilization in terms of the transformation from maternal, or female, images and myths to those of masculinity—she adopts their theory of knowledge while framing it within Marxist concepts of historical materialism. Her criteria for the concept of feminist standpoint is totally derived from the presuppositions of Marxist theory in its Lukácsian form, including the idea that the standpoint of the exploited or the oppressed makes possible the unmasking of prevalent social relations that produce oppressive conditions of life. That there is a specifically woman's experience of the world is beyond question. What needs to be interrogated, however, is whether this experience produces a feminist epistemology. Like Merchant, Hartsock reads masculine epistemology into the history of Western philosophy. Hegel's master-slave dialectic, in which the self is formed through a "life and death struggle," is ascribed by Hartsock to class and masculine relations, thus denying its universality. Males are afflicted with dualistic consciousness; they live in, but are not of, the family. Their formative experiences are of an abstract character "and a denial of the relevance of the material world to the attainment of what is of fundamental importance: love of knowledge, or philosophy (masculinity). The duality of nature and culture takes the form of a devaluation of work, or necessity, and then primacy instead of a purely social interaction for the attainment of undying fame."[40] Beginning with a materialist account of women's position in social reproduction, Hartsock finds common ground with cultural feminism's attack against science as the realm of the abstract and the identification of women with nature. But she goes further to link masculinity to the pursuit of knowledge as an abstract ideal. In this paradigm, women are concerned with the affective realm, men with the cognitive; women are consigned to the concrete, men to the abstract. Women are nature, males culture. Of course, these antinomies are the elements of masculine ideology. But Hartsock does not stop there. For her critique of the antinomies of sexist thought has a point: it is to construct an image of women that is represented as the real in contrast to the male imaginary:

> Women's construction of self in relation to others leads in an opposite direction—toward opposition to dualism of any sort; valuation of concrete, everyday life; a sense of a variety of connectedness and continuities both with other persons and with the natural world. If material life structures consciousness, women's relationally defined existence, bodily experience of boundary challenges, and activity of transforming both physical objects and human beings must be expected to result in a world view to which dichotomies are foreign.[41]

In other words, there is a feminist standpoint which radically transforms the nature of knowledge. This is so because science derives from its philosophical presuppositions, which, if we are to follow Hartsock's logic, are rooted in the conditions of material existence. Hartsock omits discussion of the concrete knowledge that can be derived from this standpoint and of the degree to which it would differ from that proposed by the male worldview; she does not provide a method for arriving at that knowledge. For the present, we must confine ourselves to the structure of her argument. Hartsock's critique, unlike those of ecofeminism and cultural feminism, stands firmly on the twin sciences of psychoanalysis and Marxist theory of historical change, historical materialism. These are, of course, derivative of the abstract, universal premises of enlightenment philosophy of science. For both, the self is constructed out of the manifold relations into which it enters; it is not a thing, *sui generis*. Freud and Marx share the idea that the self is a reified construction masking deeper relations that are to be uncovered by scientific investigation. In each case, we are constrained to employ abstract, a priori categories such as the psychic structure with its three parts (Freud) or the mode of production of material life with its two elements being forces and relations of production (Marx). In each case, the domination of nature becomes the precondition for the formation of self; it is the structuring relation within which class or psychic relations fight it out. So, if we are to follow Hartsock, her own theoretical sources are intimately entwined with masculine forms of knowledge, and the price of appropriating these modes of knowing must be calculated. Her own description of the categories of the feminist epistemology are strikingly similar to those produced by the cultural feminists. Women are concrete; their relation to nature is continuous rather than dualistic; "dichotomies are foreign" to their way of knowing. These conclusions also appear in the discourses of Merchant, Daly, and Griffin, although they make no claim to follow the precepts of male-oriented science. Instead, they offer an alternative myth to that of science: the ancient times were a golden age when feminist ways of relating to nature were dominant. Then, "commercialism and industrialism" signaled a fall from grace or, in Daly's account, patriarchy appeared as a violent counterrevolution to maternal culture.[42] Merchant's apocryphal, utopian history becomes a weapon in the feminist struggle against male science. Premodernity, regarded by the Enlightenment as a static and reactionary worldview, becomes, for this version of the feminist standpoint, a science that views the "cosmos as a living unit" in which humans were linked both to the heavens and to other animals. Aristotle, whose cosmology became the object for refutation by seventeenth-century science, is here regarded as a flawed but essentially ecological thinker. Merchant does not render the history of the transformation of science as a struggle of ideas. As I have noted, she points to material, economic reasons for the changing assumptions of science. But since she adopts a frankly utopian and political standpoint, her account of the "death of nature" does not rely on material categories of causality. Instead, she weaves myth and science together in a mosaic that takes ideas seriously as historical agents. Hartsock, on the other hand, manifests the contradiction of a logocentric method with feminist ideology of organicism.

In a future work, I explore the substance of feminist and ecological claims about science. For the present, it is sufficient to open the discussion. Religious, ecological, and cultural feminist objections to modern science differ in many ways. What unites them is their condemnation of the failure of science to come to terms with its own social and political commitment. The virtue of feminist critiques of science is that they try to go beyond the tendency, prevalent among ecologists, environmentalists, and others, to confine their objections to the uses without challenging the philosophical presuppositions of science, the content of contemporary scientific theories, or the methods of science. Merchant has provided a critique, parallel to that of the Frankfurt School, which focuses on the metaphysical foundations of modern science and suggests, following the lead of Adrienne Rich, that the pejorative metaphor of women as "other" to male reason, "divested of the trappings of patriarchy, gives rise to a distinctively . . . female bond with nature."[43] If the feminist critique of science as *a* but not *the* form of reason is right, it may occupy no privileged position with respect to knowledge of nature. It is a way of knowing, burdened with presuppositions permeated with the interest of (male) domination of women/nature. These presuppositions, derived from the social world, configure its results. Thus, we may subject the scientific method to scrutiny as well as the overwhelming tendency of some strains of physics and biology toward reductionism, especially their denial of the concepts of levels according to which biology, psychology, and social science are discourses whose object of knowledge is irreducible to physical and chemical properties. For feminism, the problems of social relations are not subsumed by basic biological drives, which, in turn, are derived from "organic chemical reactions." At the same time, the feminist standpoint invokes metaphors of women's physiology to describe nature.

In sum, the mechanical and reductionist worldview of natural science is under siege from three major movements that have questioned the gulf separating scientific knowledge from moral and ethical considerations. Among the weapons of criticism employed by these movements, none is more powerful than the readings of the canon of scientific philosophy since the post-Socratics. The evidence of what Max Black calls "models and metaphors" in the laws of science reveal the extent to which the categories of myth still pervade science and, particularly, its inability to eliminate human purpose from the multiple issues surrounding scientific discovery.

To be sure, science itself has provided the impetus for much of this work. For, as historians and philosophers of science have tried to demonstrate, recent profound changes in the scientific picture of the world have reinserted the subject into scientific discovery, have revived once hated theories, and have introduced uncertainty about the question whether science "needs" a teleological hypothesis to explain its discoveries. The reinstatement of teleology and neo-Lamarckian ideas in biology, uncertainty in physics, parapsychology and the religious connection to the big bang theory in cosmology points to scientific communities that are philosophically and even metaphysically rent. While the public face of science remains resolutely rationalistic, doubt is creeping in; some scientists have launched their own movement against corporate and military control over research, demanding once again a return to the autonomy of scientific work. And, as was true of physics and biology in the past, the merger of theoretical science with philosophy has reappeared, even as some philosophers of science insist that they have nothing to offer the acquisition of positive knowledge—their role is merely to clear up misunderstandings.

Since the relativity and quantum mechanics debates of the first third of the twentieth century, in which theoretical physicists openly clashed on questions of interpretation of the results of discovery, historians, sociologists, and even a few philosophers of science have been given "permission" to interrogate science and to suggest, however meekly, that there are ideological elements of scientific theory. I shall discuss these developments in some detail in Part III. For now, it is enough to remark that there exists widespread skepticism concerning the autonomy of science within the scientific community itself as well as social movements and a section of the populations of industrialized countries. Some scientists, philosophers, and social scientists are exploring the historical and social constitution of science, especially the "facts" that constitute scientific knowledge. They have concluded, in various ways, that what might be called the "social relations" of science, including the worldviews that are constitutive of scientific knowledge, are inseparable from the results of discovery. The philosophy, history, and sociology of science explores the claim of modern science (a) to be independent of the social/historical context within which it works and (b) to discover "facts" that, even if theory-dependent, correspond to the external world. The first debate concerns the autonomy of science from the conditions of the production of knowledge. Since Thomas Kuhn's suggestion (1962) that paradigm shifts in science were dependent not only on "shifts in perceptions," a concept borrowed from N. Hanson (1958),[44] but also upon changes in the context within which scientific discovery took place, historians and sociologists of science have tried to get a handle on these presumably "external" influences upon scientific development. Those who wish to preserve the fundamental autonomy of science, such as Koyre, Hanson, and Kuhn himself, confine these extraideational influences to the scientific community, defined variously in broader terms, or, as in some recent sociological histories and studies, to the laboratory.[45] The exceptions (see Paul Forman, 1971)[46] speak of an epistēmē, or cultural environment, that may influence the content of scientific theory. However, only Marxism has provided a social/historical interpretation that encompasses both science and technology, explaining their development in relation to large economic, political, and ideological transformations.

Perhaps it was the mechanical correspondence theory of scientific truth employed by Marxism after Marx that resulted in the deafening silence with which mainstream philosophers, historians, and social scientists greeted its accounts. When Marx and the Marxists are discussed, it is only to dismiss them as in Koyre's summary refutation cited above. The mainstream tradition in the sociology of science, represented by Robert Merton and the Columbia school, confines its explorations of the social context of science to studies of the scientific community, particularly "institutional and ethical factors" such as the character of scientific education, and networks of personal relations based on schools attended and scientific organizations.[47] Even the "new social studies in science,"—grouped around the Edinburgh School (Barry Barnes, David Bloor), Michael Mulkey, Karin Knorr-Cetina, and, in a somewhat different vein, Steve Woolgar and Bruno Latour—have, in the name of abolishing the dichotomy between internalist and externalist explanations, fashioned a microsociology of knowledge in which the actual interactions of working scientists, their discourse concerning observations, and their negotiations as to what they actually "saw" constitute scientific facts.[48] These accounts differ from Mertonian sociology insofar as the social studies school argues in various ways that scientific facts are constructed socially, but the scientific community is often made nearly identical with the social. In this connection, one of the dominant strains of the social study of science has become ethnomethodology, a derivative of phenomenology; recent sociology of science seeks to actually locate the point of production of scientific knowledge as the outcome of intersubjectivity. In recent times, only the Edinburgh School and Bruno Latour attempt to link scientific knowledge with social interest, owing much in this respect to the Marxist and neo-Marxist conceptions of ideology.

Although social studies of science, including the older work of Merton, do not ignore Marxism, it is also true that the historical perspective on science and technology that has been closely identified with the Marxist tradition is marginal to this work. As I shall show in Part III, historiographic works on science, following Koyre and Kuhn, remain embedded in the problem of intertextuality, that is, the relation of scientific ideas to philosophy, or, more broadly, the prevailing mentality of a historical period. Missing in these accounts are concrete studies in the social relations of science that go beyond the laboratory or professional contexts within which scientists act.

In Part I, I discuss the theory of scientific discovery emanating from Marx and Engels. As we shall see, controversies concerning this question within Marxism reveal differences of interpretation as wide as those in the philosophy of science and technology. Moreover, as I shall show, Marxism, following Marx's own ambivalence, mirrors the debates in every discipline concerning the general question of what constitutes science; the relevance of social relations to the form and substance of scientific knowledge, that is, to what counts as knowledge; and, more broadly, the relation of science to what is called "society."

I follow with an examination of the Marxist tradition, in its "orthodox" version emanating from the theorists of the Second and

Third Internationals as well as from those who may be designated as "neo-Marxist." My main point in Part II is that the transformations within Marxism tend to follow changes in mentalities that accompany both social and cultural developments and the ideologies embedded within them. Like natural science itself, Marxism is not immune from the controversies of the time(s) within which it theorizes social and natural phenomena.

At the same time, Marxism is constantly confronted with a disruption not experienced by other perspectives. This consists in the problem of reconciling its own claim to be a "science," in the sense of nineteenth-century physics or chemistry, with its equally powerful axiomatic proposition that nature and history are constituted by the social relations of production, and that the production of the material means of social existence is at the same time the production of humans themselves (including their mental life). Thus, if social production is not merely the instrument through which humans survive the vicissitudes of their external environment but constitutes the multiplicity of their social relations, Marxism provides the clue to a radically different conception of scientific knowledge than is contained in its own aspirations. For, under this axiom, not only science, but Marxism itself, must be comprehended within the framework of social relations. Both its axiomatic and its theoretical structures must be understood as aspects of prevailing relations of production. In this regard, the referent of scientific knowledge is not only the object of investigation or, as in recently discovered quantum mechanics, the observer, but also the social matrix within which modes of thought are constituted. That most of the leading theorists of science and technology in the Marxist perspective have been unable to situate their own "paradigm" reflexively, that is, to understand the extent to which Marxist science contains ideological elements or, at the very least, is dependent upon the character of those relations that constitute it, attests to the power of the enlightenment faith that nature (and human nature) may be comprehended directly. As I shall show in Part II, on neo-Marxist science and Soviet science, even when they acknowledge the social constitution of scientific facts, the historical relativity of scientific knowledge is attenuated by a strict adherence to a realist epistemology, according to which the correspondence of scientific propositions to the material world may be established through experimental or mathematical proofs. There are, of course, exceptions, but these have had little lasting influence on Marxism as an intellectual movement.

One may notice striking commensurabilities with this development in the non-Marxist historiography, sociology, and philosophy of science as well as among scientists themselves, especially those who are also philosophers. To acknowledge that the production of knowledge is a social process has not been commonplace in discourses on science; however, since Kuhn, the idea is becoming more respectable. Yet, of the major movements of thought originating in the nineteenth century, only Marxism can account theoretically for this development (the exceptions, poststructuralist philosophy and historiography and the sociology of knowledge, are in fundamental respects derivative of Marxism, even if by negation). At the same time, the countervailing tendency within Marxism toward totality bars a wholehearted historical relativism.

Nevertheless, whereas Kuhn, for example, *alludes* to social and cultural influences on paradigm shifts in science, Marxism specifies these in terms of economic, political, and ideological determinants. And, it is precisely because Marx himself generates categories that make a social analysis of science and technology possible, that Marxism, despite its ambivalence, has produced the only coherent social theory of science, a theory which has scandalized its opponents (for example, Karl Popper). Only recently, however, have some working in the Marxist tradition been willing to extend theory to the production of scientific knowledge. Others, notably the structuralist and analytic schools, have directed their energies to affirming Marxism's scientificity by declaring that the social aspects in its development are of little or no interest, except insofar as they illuminate the degree to which Marxist propositions conform to canon of scientific evidence and proof.

In chapters 2 and 3 I examine, in detail, Marx's theory of science and technology. Except for comments in letters and his notes for *Capital* published as the *Grundrisse*, Marx offers no separate theory of science and technology. Rather, his theory is imbedded, almost coded, in the rich description of the labor process developed in the first volume of *Capital* and in the crucial sections on accumulation. As we shall see, Marx understands science in terms of the domination of capital over labor. Science is subsumed under capital in the period of the transformation of the labor process from manufacturing to modern industrial production. Chapter 4 traces the subtle shift in Marxist theory of science from its role in production to an epistemological inquiry closely connected to the status of Marxism itself as a science, and also considers the role of the productive forces in the transition from capitalism to socialism, a preponderant concern of leading theorists of early twentieth-century socialism, who, taking Marx literally, foresaw that the role of science in the new social order would be even more central than it had hitherto been in advanced capitalist societies.

In chapter 5, I take up the work of the Frankfurt School, which, having taken seriously Lukács's insistence that nature was a social category and science ineluctably entwined with social relations, returns to the historical focus of Marx himself. But now, the shift away from considerations of the labor process is definitive. Rather—like Lukács, Horkheimer, and Adorno—Marcuse and Sohn-Rethel, whose theory of science is deeply influenced by the Frankfurt critique of science, undertake an examination of the ideological aspects of scientific assumptions and discoveries. However, whereas Sohn-Rethel retains Lukács's focus on the commodity form, the Frankfurt School presupposes Marx's categories and adopts a Weberian emphasis on forms of rationality. Chapter 6 shows how later theorists, particularly Habermas, want to find pathways back to reason but no longer find them through the route of science. In effect, for Habermas, science and technology have become part of the taken-for-granted world of instrumental reason. His is a postindustrial theory of communication that seeks to address the problems of social relations through language. In Chapter 7 I discuss the full-throated emergence of a Marxist epistemology introduced by Althusser through the influences of Jacques Lacan and particularly Gaston Bachelard. I also show the affinity of this tendency with the Italian school emanating from Galvano Della Volpe. Finally, perhaps the apogee of this new Marxist scientism is the most recent development of analytic Marxism, which wants to articulate Marxism with normal, positivistic scientific method. Chapter 8 explores the history of Soviet ideology of science, an ideology that has become official state doctrine. I show that Soviet views of science, while ensconced in orthodox Marxist presuppositions, parallel tendencies in non-Marxist theories of science as well.

Part III traces parallel developments in non-Marxist discourses on science. Here, we shall see the battle joined principally at the epistemological level, since those perspectives that I loosely group under the tentative rubric of liberalism have no specific *social theory* of economic, political, or scientific discourse. In fact, I argue that a distinction between Marxism and liberalism at the ideological level lies precisely in the absence of a distinctive liberal social theory of science. Since enlightenment ideology, especially its scientific and technological modes, proceeds from the presuppositions of individually driven market relations, on the one hand, and claims concerning the universality of reason, on the other, social theory is occluded from this antinomy. At most, liberalism appropriates conservative communitarianism as a social site. But indigenous liberalism implies that society is constituted by individuals and that individual choice is the foundation of collective association. Thus, the sociological conception of the "scientific community" as the site of scientific discovery and the court of scientific truth. In turn, the scientific community is composed of associated individuals united by their training and knowledge, a unity which makes possible the determination of whether any assertion is scientifically valid.

Clearly, this conception of the social derives from possessive individualism. There are really no "structures" of social life, no relations that transcend individual determination. More recently, Michel Foucault has advanced the idea of discursive formation which links social groups to discourses arranged spatially. Foucault's insistence on the inextricable link between knowledge and power suggests that various discursive communities are also political/economic formations and, more generally, that what counts as knowledge is entwined with domination. Although Foucault is usually catalogued within post-Marxism, there is no question that the referent is still historical materialism, even if the primacy of the economic is denied.

However, there can be no returning to the letter of historical materialist theories as they were articulated by Marx and Engels. For as we shall see, both are imbued with the enlightenment ideal of science as somehow resistant to the infusion of the ideologies produced in the course of the production and reproduction of capitalist social relations. This view is attributable not only to the context in which studies of language and discourse as sources of ideology were still in their infancy, but also to the problems posed by evolutionist ideology. According to this ideology, humans stand at the pinnacle of the natural order; their unique stature is owed to their capacity to produce their means of subsistence, and thereby to produce their life. In the self-production of all aspects of existence, science is the master discourse and Marxism underscores itself as the authentic representation of that mastery in the social field, and, as metascience, in the so-called natural field as well.

At issue in this book are both propositions and evolutionary ideology: I argue that science is a labor process like others; that its practices constitute an intervention of a specific kind, whose contrast with other types of social and natural interventions cannot be arranged hierarchically on a scale of truth or adequacy; and finally that science is a discourse that narrates the world in a special way.

CHAPTER 2
MARX 1:
Science as Social Relations

Orthodox Marxism has often portrayed all ideas within the bourgeois epoch as nothing more than reflexes of the dominant material relationships, that is, all ideas except those embodied in science and technology. To a large extent, Marxism has shared the capitalists' worship of scientific understanding and industrial technique as reified, eternal truths. Whereas ruling ideas are clearly grasped as reflexes of material relationships and have been not been accorded independent existence by Marxism after Marx, the "forces of production" have been almost universally regarded as relatively autonomous from the social relations of production. Moreover, much of the Marxist tradition has come to view the productive forces as the motive power of historical change. For much of the tradition, the social relations of production, that is, the class structure, has itself been a response to the imperatives of the productive forces.

Indeed, in his preface to the *Contribution to the Critique of Political Economy*, Marx lends some credence to this interpretation. The core of the revolutionary process, according to Marx in this text, is the conflict between the developing productive forces and the regressive social relations of production which have, at a certain juncture, become a fetter on the forward march of social development. For Marx, the productive forces consist in the stage of development of human knowledge and skills of the producers, on the one hand, and the corpus of the means of production, on the other.

35

It is clear in the Preface, at least, that Marx regarded the productive forces as determinant, in the final instance, of the relations of production. "The material powers of production," the specific, progressive expression of the degree of human domination of nature, constitute the central category of history. To be sure, social changes do not occur unless "men fight it out" to advance or retard the development of the productive forces. But,

> no social order ever disappears before all the productive forces for which there is room in it have been developed; and new higher relations of production never appear before the material conditions of their existence have matured in the womb of the old society. Therefore, mankind always takes up only such problems as it can solve; since, looking at the matter more closely, we will always find that the problem itself only arises when the material conditions necessary for its solution already exist or are at least in the process of formation.[1]

The inevitability of socialism, given new material productive forces, does not follow even though some have given such a reading to the text. What is plain, however, is that Marx accords primacy to the struggle between the new productive forces which arise out of the womb of the old society and the old relations of production. Yet, Marx leaves the matter as a problem rather than giving the emergence of the "new higher relations of production" the status of inevitability. All Marx does here is to outline the objective possibility of historical change. That "mankind always takes up only such problems as it can solve" does not imply more than that solutions are at hand owing to the development of new human productive powers. The historic error within the Marxist tradition is to equate Marx's formula, that the existence of the problem presupposes the existence of the *necessary* conditions for its solution, with the existence of *sufficient* conditions. Marx recognized that the sufficient conditions for revolutionary change were not present in the contradiction between the forces and the relations of production. The sufficient conditions for historical transformation, the development of a revolutionary class capable of "fighting it out," could not be predicted with the accuracy of "scientific law." In truth, Marx's historical works, especially *The 18th Brumaire* and the works on the Paris Commune, constitute an eloquent testament to the fallacies of formulaic Marxism. In these works, Marx was constrained to treat concrete historical events as problematic, for they depended on the actions of persons whose interests could not be adequately described by the paradigm of social classes to be found in *The Communist Manifesto*. Marx was forced to follow the actual course of social struggle rather than imposing his own a priori scheme upon it. In so doing, he acknowledged the empirical inexactitude of the two-class model and confuted the scribes of socialist inevitability.

Among the lessons of Marx's historical writings, as distinguished from his philosophical and political-economic work after the defeat of the 1848 revolution, is the difficulty of reducing Marxism to an exact science of history. The complexity with which Marx treated economic and political developments of his own time is belied by his own theoretical summations that have constituted the formation of Marxism after Marx as an ideology.

"Marxism"[2] has betrayed a tendency to exempt scientific and technological knowledge from the general prescription offered by Marx himself that, in general, the means of mental production are controlled by those who possess the means of "material" production. Marxism has often made the distinction between the social conditions that permit scientific discoveries of new techniques of production that become social forces, that is, to be incorporated into the production process, on the one hand, and the integrity of the discovery or production technique itself as a reflection of generalized human knowledge, on the other. For many of Marx's epigones of the late nineteenth and twentieth centuries, science and technology have been regarded as forms of knowledge that are neutral with respect to social relations. Although the appearance of new scientific theories or technologies depend upon social conditions, the content of these discoveries are in no way dependent upon social relations. Marxism has been able to recognize the ideological character of bourgeois philosophies, including the philosophy of science. But, following Engels, empirical science, that is, knowledge of the external natural world derived from observation and experiment, has been endowed with the status of truth relative only to the stage of human mastery over nature, expressed in the development of the productive forces.[3] The dialectical relationship between social production and scientific knowledge is often asserted by Marx and Engels and most of their orthodox followers, as we shall see in chapter 3. Reduced to its simplest expression, Marxism has regarded science as much as an outgrowth of spontaneous developments in the sphere of production as the other way around. But what is important here is the care with which Marxism has treated empirical scientific theory itself and the development of technology as an outgrowth of those sciences that arise from observation and experiment rather than from ideologically mediated institutions.

Conventional Marxist theory recognized that science and technology are characteristically pressed into service by the bourgeoisie for the purposes of domination. Such examples as the use of modern physics for the development of new techniques of warfare are frequently cited by Marxist critics as evidence that the social uses of science are not exempt from political or economic determination. Politics influences the transformation of science into certain types of technologies. But with few exceptions, which I shall describe, Marxist theory has not generally acknowledged the way in which bourgeois social relations influence the relations of humans to nature, determine or condition the nature of scientific discovery, and provide a basis for a critique of the methodology of the sciences. On the contrary, Marxism has, on the whole, been content to describe the social circumstances that advance or retard the development of scientific theory and its transformation into the techniques of social production. Even as Marxism is prepared to show that certain techniques of production have been suppressed by the capitalist relations of production, it has not been willing to undertake a critique of the technologies themselves. Both explicitly and implicitly, Marxism has tended to regard technology as a function of science in the modern era. Invention may have influenced or even determined scientific theory in the period of early competitive capitalism, as many historians of science and technology who work in the Marxist tradition have demonstrated; but in the period of late or monopoly capitalism when both science and technology have been centralized by capital in accordance with its own centralization, the dialectical relationship seems to have been exhausted. The history of twentieth-century capitalism illustrates the degree to which technology is dependent on science and both have been subordinated to the state and the large corporations.

Thus, the critical issue remains whether and to what extent that which counts as scientific knowledge is subordinate to ideology/power not only in relation to the social conditions necessary for its emergence and development or its political and social uses, but also with respect to the constitution of scientific facts, laws, and methodology. For, if scientific knowledge is the central force of production in modern capitalism and in those "really existing" socialist countries that appropriate technologies and sciences inherited from the capitalist tradition, the question is whether both science and technology in the nineteenth and twentieth centuries correspond in their substance and methods to the imperatives of the bourgeois worldview rather than constituting an epistomologically neutral element in the mode of production. It follows that a socialist society cannot uncritically incorporate the inheritance of science and technology developed in the capitalist epoch any more than seventeenth-century science could have followed medieval versions of science. Further, the conception of the traditional relationship between the forces and relations of production, in which the latter are reflexive of the former, may require substantial reexamination. The bourgeois worldview has, as an essential element, the notion of instrumental rationality. All objects of experience are regarded as manipulable for human ends. The transformation of objects by human practice may be the gateway to knowledge, as Marx argued in his *Theses on Feuerbach*, but the tendency of the bourgeoisie to construe objects in its own image demonstrates the problematic character of the notion of the neutrality of scientific knowledge. More specifically, it is my contention that both science and technologies that arise from it correspond to the domination of capital over all aspects of human existence within society. The bourgeois worldview is an essential element of scientific discoveries and technology, illustrating that ideology is constitutive of social reality and is not merely a reflection of existing social relations.

The dominant role of the forces of production in determining historical and social configurations and the reflexive and subordinate role of ideologies of the material relations of production are cornerstones of traditional Marxist theory. It is these formulas I propose to examine. Paraphrasing Marx and Georg Simmel, Georg Lukács pointed out that the commodity form penetrates all corners of the social world; to this maxim, I add Marx's comment that at a certain stage in the development of capitalism, the imperative of expansion requires not only imperialist conquest, as Rosa Luxemburg, V. I. Lenin, and other Marxists have asserted, but also the development of science and technology. For Marx, the accumulation of capital, whose qualitative side is the vast expansion of production, entails the appearance of new needs made possible by this production: "a qualitatively different branch of production must be created, which satisfies and brings forth a new need. . . . Hence the exploration of all of nature in order to discover new useful qualities of things; universal exchange of the products of alien climates and land; new artificial preparation of natural objects, by which they are given use values . . . the development hence of natural sciences to their highest point."[4]

The development of science, therefore, arises out of the accumulation of capital, mediated by burgeoning needs for new use values. Having been brought forth by capital, science actually corresponds to the emergence of human needs that require it, but it is not the nature of scientific knowledge that is determined by these needs. Marx is concerned to account for the appearance of certain sciences at a particular time. "Capital creates the bourgeois society, and the universal appropriation of nature as well as of the social bond itself by members of society. . . . For the first time, nature becomes purely an object for humankind, purely a matter of utility; ceases to be recognized as a power for itself; and the theoretical discovery of its autonomous laws appears merely as a ruse to subjugate it under human needs."[5] Thus, even though science discovers the "autonomous laws" of nature, this activity is not for the sake of knowledge but for the sake of utility; nature becomes nature "for us." Human purposes are implicit in the development of scientific activity and these are historically produced.

Following Hegel, Marx gives a thoroughly historical account of the development of knowledge, and it is this that marks the distinctive characteristic of Marxist theories of science and technology. However, although science cannot escape capital and has been subsumed under the dialectic of the production of needs and capital (use value and exchange value), the content of its discoveries is not implicated in this relation. For, even if it can be shown that the relation of scientific institutions and capital has become inextricable, a proposition I shall discuss in chapters 10 and 12, there is no question for Marx of rendering scientific knowledge problematic. Capital accumulation generates new theories verified by practice. These practices range from experiment to technique which early in the capitalist development were the basis for theory but later derived from it in proportion as production became increasingly scientifically driven. Marx's historical perspective does not suggest that his is an "external" theory of scientific discovery, as is claimed by Kuhn and others. Marx is concerned to root the emergence of scientific concepts in "the historical forms of their practice." These are conceived as "objectively existing laws of nature."

Indeed, as Alfred Schmidt points out, practice mediates knowledge of objects, freeing Marx from the charge of naive realism. As we shall see below, this has not always been true of Marxism. The central text buttressing this claim is the second thesis on Feuerbach:

> The question whether objective truth can be attributed to thinking is not a question of theory but is a *practical* question. Man must prove the truth, *i.e.*, the reality and power of his thinking in practice. The dispute over the reality or non-reality of thinking that is isolated from practice is a purely scholastic question.[6]

The correspondence of ideas with the external world is historically mediated by human activity. "Wrong" ideas—ideology—are rooted in speculation linked to social interests that are antithetical to the project of historical transformation. Thus, even the bourgeoisie, against its will, is obliged to foster the advance of science in order to accumulate capital, create new needs, etc., even as its philosophy and theologies lapse into mysticism and irrationality. Implied in this perspective is the notion, most forcefully advanced by Emile Durkheim, that ideas that fulfill human needs cannot be false.

Therefore, Marx purports to have no need for epistemology as such. The theory of knowledge proposed by British empiricism and given shape by Kant turns out to be nothing more than a scholastic endeavor grounded in ahistorical premises. For this reason, there is no intention in Marx of founding a philosophy of science and technology that is not historical. In the capitalist epoch, human practice appears entirely sub-sumed under capital and its qualitative side, industrial production. Although not explicitly stated, Marx now adds experiment to techno-logical change as the proof in practice of the truth of propositions about the material world, just as revolutionary political activity serves as the demonstration of the veracity of propositions that constitute social science. Yet, as we shall see, despite intentions to the contrary, Marxism was unable to dispense with the problem of knowledge. In their later writings, Marx and Engels retreated to epistemology, while retaining their view that the origins of science are not to be found in the exigencies of pure knowledge, but in technique.[7] Scientific discov-ery depends increasingly on the sophistication of the machines of experimental science. For, to the extent that science believes it relies on observation as much as mathematical calculation, its collective experience is mediated by the accuracy of the data collected by means of mechanical interventions into the nature it constructs in the laboratory.

I shall argue in chapter 9 that the historical dimension in the account of scientific discovery remains a necessary component of an account of the social relations of science, but is insufficient. Marx's profound reli-ance on an evolutionary concept of the growth of knowledge demar-cates his perspective from that of mainstream philosophy of science and was consonant with the epistēmē of his own time. The idea of the continuous accumulation of scientific knowledge paralleling capital-ism's own increasing dominion over the social world was no more implicated by the idea of progress than the twentieth-century proclivity in scientific philosophy toward ideas of indeterminacy, discontinuity, and relativism. Each, of course, expresses a partial, but indispensable, "truth."

It is not my intention here to examine Marx's views on the question of the relationship between the productive forces and the productive relations in order to buttress Marx against the Marxists. The tendency of Marxists to use Marx as an authority in order to conduct current ideological disputation is a conservative use of his ideas. Yet, Marx bears some responsibility for the scientistic tendency of many of his followers.

Marx gives no clear-cut, unambiguous guidance to those who would assert one line of argument over another. In this regard, I believe it can be shown that many of his statements are made on the question in the heat of polemic and necessarily emphasize one aspect of the problem over another, depending on the circumstances. Further, whereas Marx deals explicitly with the relationship between the two aspects of production in his earlier writings, many of the clues to his position during the time he wrote *Capital* (1857 on) must be inferred from the concrete examination he made of the historical development of the capitalist mode of production. In *Capital* and other later writings, he was less interested in the purely philosophical mode of explanations. Marx expressed himself immanently: that is, the chief Marxian categories may be approached in *Capital* in terms of a critique of capitalist production rather than a critique of ideology.

To be sure, all of Marx's major early writings, such as *The Economic and Philosophic Manuscripts of 1844*, *The German Ideology*, and the *Poverty of Philosophy*, introduce elements of the concrete investigation. Nevertheless, it is plain that in the 1840s, Marx was preoccupied with clearing away much of the idealist Hegelian legacy as well as treating the limitations of Feuerbach's materialism. The purpose of my investigation is to discover what can be learned from Marx on the problem of the productive forces and the development of capitalist labor process and to define some issues for further investigation. Against the orthodox claim that everything Marx said, however, problematic, can be regarded as holy scripture, I begin with a cardinal axiom of Marx's method—relentless criticism of everything. In the course of pursuing this way of working, some of Marx's formulations may have to be scrapped and others affirmed. As I have already argued, Marx's own theory of science and technology is coded in his account of the labor process rather than serving as a concrete history and philosophy of the sciences. The reason is fundamentally tied to his conception of history, according to which the dialectic of labor (the double relation of humans to nature and to each other) is the key to understanding the forms of knowledge. Therefore, rather than examining the later epistemological theory of science first, I want to begin by rendering Marx's historically and theoretically situated description of the labor process since it is here that the clue to Marx's theory of science is to be found.

With the exception of his preface to *A Contribution to the Critique* *of Political Economy* (1859), Marx made no schematic statements about his conception of history, especially the process of social transformation. For the most part, he discussed his theory of history in relation to specific controversies or critiques of the works of other writers. However, there is little doubt that *The German Ideology* constituted his most important statement on the historical process until the publication of the *Critique* almost fifteen years later.

Written in 1845, *The German Ideology* was intended as an attack on the "young" Hegelians, many of whom were erstwhile associates of both Marx and Engels. The central theme of the work is the assertion of the primacy of social being, the "real" relations between human beings, and between human beings and nature over their "ideas, conceptions, etc.," which at times Marx seems to assign the status of "reflexes" of the actual life processes, i.e., production. For Marx, as humans produce their actual means of life, so they produce themselves. Much of the polemic is directed against such various notions current among the young Hegelians as the idea of God, the possibility of "pure criticism" on the ideological plane, and the notion that ideas themselves change things.

Marx notes that the real object of historical investigation is not the ideas humans have of themselves, but the concrete existence of real individuals engaged in producing their own means of subsistence. He argues that their ideas and theories arise from the "definite relations" they enter into with each other in the process of production. This is the origin of the base-superstructure model of society that is stated with dogmatic clarity in the later *Critique*. The social base of human society, the sphere of production, expresses the life of individuals. "As individuals express their life, so they are,"[8] says Marx. Here, Marx is directing himself against the possibility that the truer self can be separated from "life-activity," namely, production.

At first glance, Marx's conception of the relationship between the productive forces and the relations of production appears to assert a correspondence theory between the two, with the productive forces constituting the motor force for social change. Indeed, orthodox Marxism has relied on a few key passages from *The German Ideology*, the

Poverty of Philosophy, and especially the *Critique* to support the claim that Marx held to technological determinism as much as economic determinism. For example, in *The German Ideology*, we encounter a formulation that appears to support this view:

> Men are the producers of their conceptions, ideas, etc., real active men as they are conditioned by a definite development of their productive forces and of the intercourse (social relations) corresponding to these up to its furthest forms.[9]

In his letter to P. V. Annenkov, in the following year, Marx was even more emphatic on the point:

> Assume a particular state of development in the productive faculties of man, and you will get a particular form of commerce and consumption.[10]

The letter to Annenkov provides a sketch of Marx's long refutation of Proudhon's *Philosophy of Poverty* published the following year as the *Poverty of Philosophy* (1847). Again, Marx accords primacy to the productive forces as a determinant of changes in social relations:

> In acquiring new productive forces, men change their mode of production and in changing their mode of production, in changing their way of earning a living, they change all social relations. The hand mill gives you society with the feudal lord, the steam mill society with the industrial capitalist.[11]

The hierarchy of determination here seems clear enough. Changes in the productive forces generated changes in the mode of production, which in turn compel changes in other social relations including class forces, ideologies, and so on.

But what are the productive forces? If they are understood merely as instruments of production, i.e., mechanical inventions of varying degrees of complexity, then the development of science and technology as reflections in human knowledge of natural laws would appear to hold the ultimate trump card against the resistance of ruling classes to new social forces. Marxist orthodoxy assigns to the proletariat the status of "bearers" of the new productive forces destined to burst the bounds of the old social relations. Similarly, the bourgeoisie becomes the social class that "bore" the new productive forces within the "womb" of feudal society and "represented" them in the social struggle. The historic tendency to reduce class struggle to epiphenomenal representations of the struggle between the forces and relations of production considered as objective things is embedded in the literature of orthodox Marxism of both the Second and the Third Internationals.[12]

This objectivist interpretation of the concept of the forces of production was lent authoritative credence by Engels after the death of Marx. On the one hand, in a letter to Karl Kautsky written in 1884, just one year after Marx's death, he admonished Kautsky that agriculture and technology could not be considered apart from the political economy in any historical period. "You must not separate agriculture and also technology from political economy," he said. "Crop rotation, artificial fertilizer, steam-engine and power loom are inseparable from capitalist production just as the tools of the savage are inseparable from his production." On the other hand, Engels argued that "production" determines social institutions. "As soon as you speak of means of production, you speak of society, specifically society determined by these means of production."[13] Engels continued the line of Marx's technological determinist position in asserting the determining character of the means of production over social relations.

While it is my contention that his view reifies the means of production by dissociating them from human activity except by external relations, there can be no doubt that Engels drew his conception from such statements as are found in the Critique, the Poverty of Philosophy, and The German Ideology.

However, even if the orthodox school has extrapolated from certain tendencies in Marx himself, it would be inaccurate to assess Marx's theory of historical development simply in terms of the kind of technological determinism indicated by Engels's post-Marx writings and the epigones of the Second and Third Internationals. Marx holds a dialectical view of the motor forces of social development, both in his earlier writings and, by implication, in many of his later writings as well. As early as The German Ideology, Marx advances the idea that the productive forces themselves are not an autonomous determinant of history nor are they separable from human activity.

According to Marx, the production of life, both of one's own in labor and of fresh life in procreation,

> now appears as a double relationship, on one hand, as a
> natural, on the other hand, as a social relationship. By social,
> we understand the cooperation of several individuals no matter
> under what conditions, in what manner, and to what end. It
> follows from this that a certain mode of production or
> industrial stage is always combined with a certain mode of

cooperation, or social stage, and this mode of cooperation is a 'productive force.'[14]

The mode of cooperation has its other side, the division of labor. Marx says, "How far the productive forces are developed is shown most manifestly by the degrees to which the division of labor has been carried."[15] Whether the divisions about which Marx speaks refer to the natural divisions between the sexes or the social divisions (such as those between intellectual and manual labor and between town and country which give rise to social classes), these are, in his conception, part of the productive forces. The division of labor expresses the degree of human mastery over nature and the forms of exchange among different groups of individuals.

The productive forces are not to be understood merely as the sum of the tools or instruments of production and the level of science and technique, as the orthodox Marxists have claimed. Embedded in the productive forces are the forms of social life which are themselves productive forces. However, "the social power, i.e., the multiplied productive force, which arises through the co-operation of different individuals as it is determined by the division of labor" appears to them as an "alien power."[16]

Marx was aware that the relation between the forces of production, which arise out of the concrete forms of social labor, and the social relations of production take on a reified, that is, a "natural," form. Owing to the mystifying character of class relations, in which class antagonisms are suppressed and reappear only as the "common interest," the object world thereby becomes divorced from social practice so that it appears independent of human consciousness. The productive forces then also appear not as the "united power" of individuals, according to Marx, but

> as an alien force existing outside them, of the origin and goal
> of which they are ignorant, which thus they cannot control,
> which on the contrary passes through a peculiar series of
> phases and stages independent of the will and action of men,
> nay, can even be the prime governor of these.[17]

Here, Marx's intent was clearly to dispel the apparent autonomy of the productive forces and to show that they are nothing but the objectification of human activity mediated by social relations. In concert with Feuerbach's attempt to demystify religion by showing that all religious images are human productions, Marx was performing the same critique on the theory of society. The forces of production take on the aspect of a natural alien power over humans increasingly as the latter "forget" that the labor process has created all these forces of production. This reification is especially evident between generations. According to Marx, capital (raw materials and machinery) is nothing but stored-up labor, but each new generation "acquires" the forces of production generated by previous modes of cooperation as if they were external to human activity. Similarly, individuals view money and commodities as independent forces rather than as symbols of stored-up labor, on the one hand, or as the product of the rationalized labor process, on the other. The forces of production seem to determine the new productive relations in the sense that they proscribe the history-making capacity of humans within class-divided society. "It is superfluous to add that men are not free to choose their productive forces — which are the basis of all their history — for every productive force is an acquired productive force, the product of former activity."[18] In this passage, Marx establishes plainly his dialectical view of the question. Echoes of this formulation are to be found later, in his historical writings, particularly in the famous passage from *The 18th Brumaire*, where he asserts that "men make their own history" but are limited by the historical accumulation of productive forces and productive relations.

Seen in this context of his effort to dispel the notion that the process of production is in any way natural, in the sense of independent of historically situated human activity, it is probable that Marx speaks at this point of the productive forces only as the motive power of history in terms of a "moment," rather than in any way suggesting mechanical causality. The productive forces, having been generated by the stage of development of the division of labor, appear to determine social relations. Marx has suggested that the actuality is something quite different. The objectification of productive powers as formed by social relations becomes part of the new social relations of science and religion which, in bourgeois society, appear to stand over and above human powers in order to show their social content.

Orthodox Marxism has taken literally what, for Marx, is a critique of reification of productive forces as an objective thing standing outside human control. Reification, according to Marx, arises from the division of labor, particularly the division between intellectual and manual labor which, it may be argued, is the foundation of the division of society into classes. In capitalist society, the achievements of science and technology are invested with the mantle of absolute truth, the products of experts, who become the final arbiters of the verisimilitude of scientific discoveries. The social relations which underlie scientific discoveries, the exigencies of capital which may determine whether a particular scientific breakthrough will be incorporated into technology and finally into modern industry, are omitted by those who impute to science and the forces of production of which it is a part the status of autonomous power. Scientific and technological power replaces social power, or, more precisely, becomes a social power. Thus, it is possible for orthodox Marxism to speak of the "scientific-technological revolutions" as autonomous developments, much as bourgeois historians spoke about the Industrial Revolution as the creation of machines and inventions rather than of human beings.

The Marx of *The German Ideology* sought to penetrate the veil of science and religion which, in bourgeois society, appears to stand over and above human powers in order to show their social content. What of the later Marx?

Albrecht Wellmer has tried to show that the later Marx fulfills the requirement of positive science by taking the objective world as given.[19] In Wellmer's view, the positivist element in *Capital* consists in Marx's incipient abandonment of the critical method of penetrating appearance in order to reveal the human essence of the social world in favor of a science that attempts to discover objective "laws" of social development. Wellmer argues that *Capital* and other later writings betray a tendency toward discovering quantifiable regularities portrayed as "laws of development" within capitalist economy and between the economic base and the ideological superstructure. Thus, for Wellmer, the late Marx increasingly ceased to be a dialectical theorist and tended to social science of the positivist variety.

The next section will attempt to assess the validity of Wellmer's claims. However, in order to examine the question under scrutiny, it will be necessary to deal directly with the relations of the work of Marx and Engels to science and technology.

Prior to the appearance of *Capital*, Marx examined the development of capitalism as specific mode of production in several writings. In *The Communist Manifesto*, history was seen as structured by class struggles. This formulation suggested a fundamental continuity with his earlier views, which assigned to the division of labor and the social classes arising from it a central place in the historical process. Marx showed that the struggle of classes was rooted in the struggle over the division of social production between those who owned the means of production and those who were forced to give over their labor power or, as in the case of feudalism, to share the product of their labor power with a master. The formation of classes was a direct outgrowth of the earlier division of society into intellectual and manual labor, male and female,

and town and country. Marx seems to have excluded the sexual division of labor from the class structure presumably because it is "natural," whereas the other divisions are social-historical developments.
That women were deprived of ownership and control over the means
of production suggests, in Marx's philosophical anthropology, that the
sexual division of labor is subsumed under broader social divisions
that arose historically. In any case, by assigning the sexual divisions to
"natural" as opposed to social causes, Marx sidestepped the significance of this issue by assigning woman the role of the procreation of
fresh life within a social structure determined by other relations.

If all history is the history of class struggles, that is, if the division of
labor is placed at the center of the social process, the motor force of
revolutionary change is not "man" in general, but humans at the point
where society has been divided not only into spheres of production,
say, the division between men and women in terms of the biological
reproduction of life itself or between those who hunt and those
engaged in agriculture, but also between those who do manual labor
and those who are engaged in intellectual labor. At this point, it is
important to note that all manual labor in precapitalist times involves a
considerable degree of intellectual labor. The division has strong ideological connotations since we are dealing here with the question of
what counts as knowledge.

Many of the most important inventions and scientific discoveries of
the preindustrial era are made by handicraft workers. The water mill, to
which Marx ascribes a crucial role in feudal production, is the invention of craftsmen in the period of slave society but becomes a characteristic source of power only in feudal society.[20]

What does Marx mean by mental, or intellectual, labor if manual
work involves the choice of materials, the fashioning of tools, and
most important, the conception of the shape of the thing to be produced? For Marx, intellectual labor describes only those who produce
ideas, religious conceptions, etc., that is, the ideologists who serve the
rulers by legitimating their domination over the under-classes and,
second, those who administer society either in the process of production and exchange or in the state. Intellectual labor is the expression
for the various functions of rulers and their professional retainers
throughout class-divided societies, in exercising social and ideological
domination. Marx shows that in capitalist society, in the process of production, intellectual labor also consists in the administration of persons once a large number of laborers are employed under a single capitalist in the manufacture of commodities.[21]

"All combined social labor on a large scale requires more or less a directing authority," says Marx, "in order to secure the harmonious working of the individual activities and to perform the general functions that have their origin in the action of the combined organism. . . . The work of directing, superintending, and adjusting, becomes one of the functions of capital, from the moment that labor, under the control of capital, becomes cooperative."[22] The role of capitalist and the role of manager do not become distinct from each other in terms of the process of production until the mid-nineteenth century. In any case, capital established, according to Marx, its directing function over production and over science even though these functions are no longer performed directly by capitalists but instead have become separate jobs of "managers" and "technicians." The directing motive itself is discussed by Marx as a "force" of production:

> In proportion to the increasing mass of the means of production now no longer the property of the laborer, but of the capitalist, the necessity increases for some effective control over the proper application of these means. Moreover, the cooperation of wage-labourers is entirely brought about by the capital that employs them. Their union into one single productive body and the establishment of a connection between their individual functions, are matters completely foreign and external to them, are not their own act but the act of the capital that brings and keeps them together.[23]

Under capitalism, capital becomes a directing force for the process of production, as indispensable as a general who commands on the field of battle. For Marx, the directing motive of production is "to extract the greatest possible amount of surplus value and consequently to exploit labor power to the greatest possible extent."[24] The administration of persons and things by the capitalist or by his appointed subordinates is, according to Marx, the work of control. On the one hand, he treats this work as a "natural" function arising out of the nature of cooperative labor; on the other hand, he shows that this work has a specifically "capitalist character" and "is rooted in the unavoidable antagonism between the exploiter and the living and labouring raw material he exploits."[25]

Marx seems anxious to endow upon the mode of cooperation itself the quality of a productive force with intrinsic characteristics that are relatively independent of the capitalist mode of production. In his view, the work "of control," e.g., superintendence and direction, is made necessary by the cooperative character of the labor process, of the division of labor. In these passages, he takes pains to distinguish between cooperation as such and its specifically class character. But the mode of cooperation described by Marx has embedded within it the logic of capitalist rationalization and is presupposed by it. Cooperation within capitalist production is not neutral, as Marx shows. It is essentially determined by the capitalist mode of production, whose purpose is the creation of surplus value. In my view, intrinsic to Marx's critique of the neutrality of the labor process within capitalist society is a critique of the neutrality of technology as well.

Here, we can observe the historical ambivalence in Marx's writings. For, despite his attempt to bestow upon "cooperation as such" unique qualities which are somehow distinct from the social structure within which production is carried on, his concrete examination of the mode of cooperation within the capitalist context always involves the most intimate relation with the class context in which production occurs.

For example, Marx shows that within capitalist social relations, the mode of cooperation is directed by the capitalist with the aim of achieving the greatest surplus value from the labor power employed. Further, this mode of cooperation is construed as the early stage of the capitalist mode of production, whose methods degrade the worker to the status of a detail laborer from a relatively dignified position of artisan. In the chapter "Division of Labor and Manufacture," Marx demonstrates that, among other features, the specific character of capitalist production entails the fractionation of the labor process into small units in order to meet canons of capitalist efficiency, i.e., the extraction of the maximum possible surplus value in the least possible period of time. In the labor process, including the mode of cooperation, from the very beginning degrades formerly skilled craftspersons to the level of detail worker.

Marx's description of the development of the capitalist production process reveals clearly that administration is an intrinsic element of capital and that it directs the course of the labor process, the possibility of incorporating scientific discoveries into social production, and, in the last instance, the configuration of scientific discoveries themselves. Capitalist rationality, in Marx's exhaustive descriptions of the economic, political, and legal aspects of the labor process, dominates the course of social production in the later bourgeois epoch.

Rather than establishing the autonomy of the productive forces, the burden of Marx's actual analysis is to demonstrate their dependence on social relations. Specifically, the hegemony of the capitalist purpose and ideology over the form as well as the content of the productive forces begins with the first truly capitalist form of production—manufacture.

The most important productive force available to capital as a legacy of the feudal period are the artisans, who possess the skills and the instruments of production. Marx shows that during the feudal period, merchant capital is separated from the process of production. The sphere of the merchant is distribution, and, in feudal society, it is in this realm that it reigns supreme. In contrast, the artisans are limited by feudal law in their ability to accumulate capital. Therefore, the critical merger of artisan skill with merchant and even some agricultural capital accumulated by the manorial lords constitutes the basis for new forms of production. In his chapter "Genesis of the Industrial Capitalist," Marx shows that even though some artisans become capitalists, the great majority of capital is accumulated by means of commercial supremacy by some Western European countries, particularly England, France, and Holland, over colonies, less developed European countries, but especially Asia and America.

The key steps toward the creation of a proletariat are (1) the transformation of artisans into wage laborers, and (2) the expropriation of agricultural laborers and their migration to the towns.

In the period of manufacture, which Marx describes as the "assemblage in one workshop, under the control of a single capitalist, of laborers belonging to various independent handicrafts, but through whose hands a given article must pass on the way to completion," the skills of the craftpeople are not yet degraded by reducing them to machine hands or to assemblers of a small part of the product. "The tailor, the locksmith and the other artificers, being now exclusively occupied in carriage making, each gradually loses, through want of practice, the ability to carry on, to its full extent, his old handicraft."[26]

The multivalent worker—that is, the person able to produce any kind of garment, the blacksmith able to make horseshoes as well as an axle, the locksmith capable of making a lock for all purposes—disappears from the center of the productive forces, and is compelled to occupy the economic margins of society. We see the same phenomenon today. The old building crafts such as carpenter, plumber, and electrician have been degraded in proportion as each craft is brought into the factory and applied only to automobile production and machinery, or to any other particular industry.

A second form of production is introduced by the manufacturers by

One capitalist employing simultaneously in one workshop a

number of artificers, who all do the same, or the same kind of work, such as making paper, type or needles . . . the commodity from being the undeveloped product of an independent artificer, becomes the social product of a union of artificers, each of whom performs one, and only one, of the constitutent operations.[27]

Manufacture achieves a high productivity of labor by means of the social power of cooperation, but, according to Marx, at this stage of capitalist production, the handicraft continues to be the basis of the production process. It is the technical division of labor itself, as opposed to the skill of the artisan, that engenders the efficiency of this method of production. The capitalist transforms the skilled worker, by successive stages, into a detailed laborer now capable of "converting" his whole body into the automatic specialized implement of a specific operation.

The capitalist becomes the inventor within the production process, because the main force of production at this stage is not machinery but the organization of hand labor. The differentiation of the instruments of labor to permit them to transform a single perfected function constitutes a second aspect of manufacture. The tool, like the craftsperson, ceases to be capable of all-round work. It, too, becomes specialized and therefore unsuited for a wide variety of functions. Marx says:

The manufacturing period simplifies, improves and multiplies the implements of labor, by adapting them to the exclusively special function of each detail laborer. It thus creates at the same time one of the material conditions for the existence of machinery, which consists of a combination of simple instruments.[28]

Since the "division of labor is the distinguishing principle of manufacture," according to Marx, it gradually achieves the degradation of skilled labor and therefore the power of capital over the production process. The multiplied power of labor is mobilized to fit the aim of the capitalist for the greatest possible quantity of surplus value, but labor's control over the production process is immeasurably weakened. Labor has become specialized, now unable to grasp intellectually the totality of the labor process. In Marx's words, "the division of labor within the workshop implies concentration of the means of production in the hands of the capitalists."[29] Thus, by implication, Marx suggests that, at least at this stage of capitalist production, the division of intellectual and manual labor takes the form of the domination by the capitalist of the process of production by (a) concentrating labor of different kinds under a single roof; (b) constantly reducing the scope of handicrafts to a production of a single commodity, on the one hand, and finally to a single operation, on the other; (c) assuming command, in the manner of a military leader, of all decisions in the battlefield of production. The commanding position of the capitalist is further assisted by the (d) "perfection of tools" to fit the requirements of a highly rationalized division of labor.

These, then, are the main productive forces in the manufacturing period, with labor reduced in skill but multiplied in social and productive power, and the capitalist exercising more and more control and coordination over all aspects of the labor process with the force of a general. The tools of production have become the objectified possession of the capitalist's rationality to the extent that they, too, lose their multivalence. At the manufacturing stage of the capitalist mode of production, the imperatives of capitalist rationality are the decisive influences in the configuration of the production process. The main "productive forces" are the objectified social relations in which labor is degraded to another instrument of production by means of specialization. In this stage, even the tools appear as objectifications of the specialization characteristic of the workshop. They lose their global application and are, instead, fitted to the particular functions of the detail laborer. Thus, the concept of efficiency is given a particular meaning: on the one hand, it has a quantitative aspect. This method of production can yield the greatest amount of surplus value, which, in turn, depends on the ratio of output to the quantity of labor expended in the production of the commodity, at least in the competitive phase of capitalism.

On the other hand, "efficiency" can be viewed in a qualitative context. It is measured by the degree to which capitalist domination of the labor process is enhanced and the social challenge of collective labor to that dominance is reduced.

The manufacturing period of United States capitalism overlaps with the period of machine production, or what economic historians have described as the Industrial Revolution. Even though machines were employed in the U.S. textile industry as early as the 1840s in the United States, and even earlier in Britain, a considerable number of industries carried on production on the basis of hand production, with only the minimum employment of machinery until well into the twentieth century. For example, the iron and steel industry depended on the ability of the capitalist to bring a large number of craftspersons under one roof until the invention and implementation of the Bessemer process made possible the degradation of a considerable portion of the work force in the 1880s. Similarly, car production was essentially a continuation of carriage manufacturing, in which the degree of work skill was considerable, even if degraded from handicraft production and only partly rationalized until the advent of the assembly line. The assembly line was not introduced widely in the automobile industry until the years just prior to World War I.

The assembly line is not marked by a high degree of machine production in the sense that Marx uses the term. For Marx, machine production is now characterized by the use of power tools such as the sewing machine and the electric welder. In his view, the motive power of production may be an animal, a person, or such devices as steam or electricity. Of these alternatives, there is only a quantitative difference among the options. One yields greater motive power than another, but the process of working up the material is not thereby transformed. In Marx's words:

> The industrial revolution begins as soon as machines are
> employed where from ancient times the final result has
> required human labor: hence, not where as with tools, the
> material actually to be worked up has never been dealt with by
> the human hand, but where, in the nature of things, man has
> not from the very first acted merely as power.[30]

In Marx's view, the stage of modern industry, i.e., machine production, begins with the widespread employment of machines to replace human labor in the working up of material. From ancient times, humans have employed animals as the motive power of production. As I have noted, the waterwheel, invented hundreds of years before the birth of Christ, was widely used in the feudal mode of production. In iron manufacturing, water and steam and fire were employed as power resources, but tools alone—no machines—were used to work up the ore, make it into different shapes, and so on.

The Bessemer process for producing steel and the widespread introduction of machinery to produce various forms of steel such as wire, sheets, tubing, and structural shapes destroyed the traditional manufacturing methods and, with them, a large number of crafts that had established, prior to the American "Industrial Revolution," a substantial measure of job control.

According to David Montgomery:

> In the last four decades of the l9th century, the industrial
> craftsmen exercised a significant degree of what today would
> be called workers control. The means of production were in
> the hands of the employed, and therein lay their power.

Control over the actual use of the implement rested with the skilled workers. There was the source of their power and of their ideology.[31]

Montgomery shows that in the United States, iron craftsmen worked in teams and produced on a contract basis with the employers. The employers provided the workshop some tools and raw material, that is, the means of production, and were responsible for the sale of the commodities produced. The workers controlled the process of production, dividing tasks among the various crafts. The workers received a set fee for each ton of iron they collectively produced. In Montgomery's words, "The pay-roll structure was thus devised without interference from the boss, in open meeting of all those involved (save the buggy-man's helper). Similarly, overtime and special work assignments were allocated by union decisions."[32] The broad division of labor into crafts within production was established by custom.

But within these general parameters, there was still considerable control by rollers, heaters, and other craftsmen in dividing the work within their crafts between themselves and helpers. In fact, helpers belonged to the union and were able to participate in enforcing rules about the allocation of tasks. Further, the workers regulated upgrading, hiring and firing, and all benefits such as sick pay and death benefits.

In *Capital*, Marx speaks of the contract system during the manufacturing period of British capitalism. As with cottage industry, another variant of manufacturing, Marx retains the belief that the essential degradation of labor consists, in the first place, of the limitation of crafts to the production of a single commodity determined by the capitalist. In this sense, the old blacksmith's craft, the ironmaker's craft, and others related to metal working were fractionated as well by the introduction of the manufacturing process into American industry. But Montgomery's investigation of the iron industry reveals a considerable degree of control over the production process by the workers themselves even when the employers owned the means of production and its distribution.

The introduction of machinery into the iron and steel industry was accompanied by sustained combat between the workers and employers over the question of control over the production process. The Homestead Steel Strike of 1892 was a symbol of many struggles undertaken by the craftsmen to preserve their prerogatives in the wake of mechanization. The insistence by the employers that the work rules codified by the collective bargaining agreement had to be scrapped as the concomitant of the new technological innovations was resisted by the workers. F. W. Taylor, who made systematic the rationalization of the productive process and facilitated the spread of rationalization as a force of production, undertook his major studies at the Midvale Steel works. There, he noted the extraordinary degree of job control in the hands of the operatives and developed "scientific management" as a means explicitly to assist management in wresting the "traditional knowledge" from the workers hands, and secure management's own power over production.

Taylor pointed out that as late as 1905, management's control over production was severely limited to exhortation rather than genuine coercion owing to the monopoly of knowledge of production methods lodged in the craftsmen. In fact, management's control was largely confined to selling the product and coordinating of the best endeavors of the workers with the requirements of the market, such as meeting delivery dates, producing according to specifications of a customer, and attempting to keep the cost of production within the price structure. "Scientific management" was a means to give management absolute control over the most important factor governing production costs—the labor process.

Here, it is important to make the distinction between the process as a whole and the system of wage determination which is only one of its many parts. Montgomery's study of iron manufacturing is significant because it presents evidence of workers' control over the entire labor process. Taylor himself provides contemporary testimony of the accuracy of this claim. In several industries, particularly the building trades, the notion of workers' control has been reified by the transference of many prerogatives once held by craftsmen to a bureaucratic union apparatus. Unions controlling the supply of labor have slowed down the pace of technological innovation and have established and enforced work rules. The job superintendent is still sharply limited in his command function, which is among the main reasons for the survival of the social relations characteristic of the manufacturing period, and production has seen the survival of these methods in the building industry and the resistance of the workers to management's efforts to introduce machine production to a greater degree by constructing prefabricated parts in factory settings. Labor struggles within the United States building industry have been largely determined by the workers' resistance to modern industry and their efforts and ability to retard its spread. But recently, the remnants of the artisanal method of building construction have suffered serious erosion. Prefabrication of housing by machine now dominates residential construction. Moreover, the rationalization of labor has accelerated even within the "artisanal" methods, so that, for example, the multivalenced carpenter is rapidly disappearing. The division of labor in the electrician's trade is becoming more rationalized, and plumbing and iron work are now confined almost entirely to the tasks associated with installation rather than fabrication.

Almost all recent battles in the printing industry have stemmed from employers' efforts to introduce technological improvements that result in the loosening of job control by printers and lithographers. In these crafts, such decisions as overtime assignments, work tasks within the trade, and workloads were traditionally decided by the workers themselves. Work rules were, in the main, subject to workers' determination. The enlargement of management's power has been almost entirely related to its ability to introduce machinery that renders the old skills redundant. Employers' determinations to "automate" production were the basis of nearly all the strikes and job actions that broke out in newspaper publishing in the 1970s.

The history of American manufacturing as contrasted with machine production suggests, therefore, that social consciousness within the class structure determined the labor process with regard to both the efforts of management to rationalize it as well as the efforts of workers to limit the extent of rationalization. Second, during the manufacturing period capitalist ownership of the means of production was radically separated from control. Instead, at least in the United States, control over the labor process rested largely with the workers because of their possession of traditional knowledge upon which their contractual right to determine the conditions of production was based. Whereas modern industry is marked by the reversal of control to the employers and their managers, with workers serving as the force that tries to limit the arbitrariness of managerial power, much of American manufacturing was characterized by a high degree of workers' autonomy in determining the conditions of producing within a relatively rationalized framework. Employers, even the early paid managers, tried to cajole, persuade, and coerce craftsmen who possessed a near monopoly of production skills, and thereby a large measure of control over the production process, but they were powerless to impose that authority arbitrarily as they later did under machine production.

The introduction of machinery on a wide scale, accompanied by the systematic development of management as a technology,[33] was the central strategy employed by capital to gain control over the production process in the same measure as it had already won ownership of the means of production and exchange. The main issue to be examined in the next chapter is the degree to which management as a force of production, and machinery as a technological innovation, were wrought out of the imperatives of the class struggle and embodied within their internal processes of a specifically bourgeois, rather than a neutral, "scientific" character.

CHAPTER 3
MARX 2:
The Scientific Theory of Society

The consensus of those who have studied the Industrial Revolution—the period of the transformation of the labor process from manufacture to machine production on a wide scale within the capitalist mode of production—has been to attribute this development to the imperatives of the capitalist marketplace, particularly competition. David Landes has paid particular attention to the ability of capital to successfully incorporate science and invention into the production process.[1] Landes shows that the concentration of scientific discoveries in Europe after the sixteenth century represents a historical reversal of the relationship between East and West. Previously, he argues, Islamic science had been far ahead of European science. But because Islamic culture was characterized by a quietistic philosophy and pervasive religious conformity; and because science represented deviance of thought in relation to the prevailing ideology, these discoveries were never assimilated into the economic sphere. In the West, the rise of science was the cause of the breakdown of the powerful religious institutions, and the simultaneous rise of the mercantile class with its ideology of secular autonomy and intellectual pluralism. It is not my intention here to revive the controversy opened up by Max Weber's famous thesis of the significance of Protestantism for the rise of capitalism.

Suffice it here to point out that the relationship between ideology, culture, and changes in the mode of production is not solely determined by one factor or another. Engels was so exorcized by the tendency of orthodox Marxism to place exclusive emphasis on economic "causes" for all social phenomena that he was constrained to spend a good deal of his effort after Marx's death in the task of correcting those writers who were content to construct partial totalizations from isolated quotes of Marx and Engels.[2] In terms of the central theme of this study, "economic" explanations for the spread of modern industry are clearly insufficient. Landes's useful discussion of the comparison between European development of industry and the failure of Eastern societies to successfully assimilate their substantial scientific and technical lore within the mode of production rests to a large degree on an insight of Marx, as much as Weber's celebrated work on the role of ideology in the transition from feudalism to capitalism. Referring to the domination by England of India, Marx described the latter as

> a country not only divided between Mohammedan and Hindu, but between tribe and tribe, between caste and caste; a society whose framework was based on a sort of equilibrium resulting from the general repulsion and constitutional exclusiveness between all its members. Such a country and such a society, were they not the predestined prey of conquest? . . . India . . . could not escape the fate of being conquered, and the whole of her past history, if it be anything, is the history of the successive conquests she has undergone.[3]

Marx's writing on British rule of India reveals his acute observation of the interplay between the impact of the colonial policies of the British government, which prevented the Indians from using machinery in the production of cotton, and the weight of tradition upon the character of Indian society. Both were largely responsible for the original British conquest of India.

The enormous military superiority of the British over those peoples whom it conquered in Asia was closely related to its industrial development. Marx's insight into the cultural sources of the reasons India was unable to effect the transition from a more or less primitive agricultural economy to an industrial one in the middle of the nineteenth century must give pause to those who would rest the process of historical change exclusively on unmediated economic development or seek the chief sources of the transition from manufacturing to machine industry in external causes such as economic competition, the availability of scientific invention, and the capacity of a country to aggressively pursue colonial conquest.

Such an explanation is offered by Eric Hobsbawm in his study of the Industrial Revolution in England. The forces igniting the revolution in

English manufacturing, according to Hobsbawm, were not the drive toward profit in general or exogeneous factors such as population, climate, geography, and so on. Hobsbawm explains the rapid growth of industry in England in the eighteenth century by the availability of domestic and foreign markets for its goods. He places considerable emphasis on the ability of the bourgeoisie to mobilize government assistance in the sphere of foreign trade, particularly the government's exploitation of raw materials and commodities markets in the less industrially developed regions of the world and its action to make available to industrial capital a large supply of cheap, unskilled labor for the textile, metallurgical, and other industries. The cheapening of the elements of capital by force as well as by edict placed Britain in a most favorable position in the world market. "Behind our Industrial Revolution," says Hobsbawm, "there lies this concentration on the colonial and underdeveloped markets overseas the successful battle to deny them to anyone else."[4] Thus, for Hobsbawm, the capacity of Britain to capture world sources of raw materials and world markets for its manufactured goods was sufficient to account for its rapid expansion into the leading industrial power in the world during the eighteenth and nineteenth centuries.

H. J. Habbakuk has attempted to account for the extraordinary growth of technological invention in the United States in the nineteenth century by the crucial factor of labor scarcity.[5] Contrary to some views, according to which abundant population, scientific and material resources, such as raw materials, are largely responsible for the rapid development of industry, Habbakuk argues that the labor scarcity in the United States during the first part of the nineteenth century was far more important as a factor stimulating technological change than even its considerable natural and scientific resources. On the other hand, England's progress in the nineteenth century is portrayed as slowing down, largely because it was an older industrial power with large amounts of invested capital in older methods and was economically and ideologically unprepared to scrap these technologies in the wake of new inventions that were available to it. Habbakuk shows that other countries such as the United States moved more rapidly than Britain precisely because of their relative underdevelopment.[6] Specifically, Britain's advanced machine production rendered much of its labor supply redundant and thereby reduced the incentive for further mechanization, wheras in the United States, labor scarcity was among the major arguments for the employment of machine methods. Laborsaving technologies are prompted by high labor costs. According to Habbakuk, labor costs were relatively cheap in Britain after the initial phase of industrialization, reducing the spur to further industrialization. Louis Hacker, in criticizing Habbakuk, also argues that no single factor can explain the entire course of American economic growth and development.[7] His critique extends to all "single"-factor explanations for the dynamic emergence of American industrial capital as a world power in the later nineteenth century. The virtue of Hacker's description is the weight he places on the internal development of the labor process to account substantially for the transformation of the manufacturing stage of capitalist production into modern industry. In this respect, he is among a tiny fraction of economic historians in the United States who have put sufficient emphasis on the social relations of production conceived in the sense of the labor process. Perhaps the most remarkable of these is Edward Kirkland, who places responsibility for the rise of the factory system in the United States "as much on the managerial, as well as the mechanical."[8]

Kirkland gives organizational, or management, ideologies and methods their due in explaining the complete victory of the factory system over United States industry. In American parlance, the factory system was defined by one theorist as "raw material that can be converted into finished goods by successive, harmonious processes carried along by 'central power.' " Kirkland wishes to define the central power as managerial rather than mechanical in nature; that is, he places considerable emphasis on the hierarchical organization of production by capitalists and their ability to establish dominance over the "pure functioning" of science in such a way as to make science useful to industry. Simultaneously, he argues for the importance of the transformation of the "arts" into science, that is, the codification of inventions by experimental methods within industrial laboratories.[9]

It is not my intention to minimize or deny the role of government measures to assist industry such as subsidies for transportation and communication facilities, the role of a banking system able to extend large sums to industrial capitalists, the role of natural resources, or of the world market. These elements can help us understand why some countries were able to seize upon the machine rather than remaining ensconced within the manufacturing system. But exogenous factors, even those that rely on economic rather than demographic considerations, are inadequate for explaining the sufficient conditions for the transformation itself, however uneven the pace.

For example, it is fairly clear that the early development of the textile industry within British capitalism and within the United States as compared to other industries relied heavily on its ability to utilize women and children as the chief source of labor. That is, there were few obstacles to the transformation of the labor process within an older method of production, such as heavy investments in equipment and strong unions or other associations of workers who maintained control over the production process. The rationalization of labor was accomplished with relative ease, and the mechanization of the production process was facilitated by the availability of a large number of earlier inventions that were modern, in the sense employed by Marx. They replaced human labor in the specific task of working up material, in machines that were ordered sequentially, that is, were rationalized in the manner of manufacturing but without the intervention of human labor, except for the material-handling function. Marx calls this function the "transferring mechanism," and he does not regard it as an element specific to the industrialization process.

Already in the sixteenth century the mining and metallurgical industries employed a large number of machines, particularly pumps for extracting ore. Lilley observes the existence during the same period of the ribbon loom: "The ribbon loom is an adaptation of the loom to weave several ribbons simultaneously, a single movement of the weaver causing each operation to be performed on all of them."[10] By the end of the seventeenth century, after state authorities had tried for 100 years to suppress the ribbon loom and other machines, it was in use in Holland, Germany, Switzerland, and France. According to Marx:

> The sporadic use of machinery in the 17th century was of the greatest importance, because it supplied the great mathematicians of the time with the practical basis and stimulant to the creation of the science of mechanics.[11]

As Engels pointed out in a letter to Starkenburg, "If, as you say, technique largely depends on the state of science, science depends far more still on the state and requirements of technique. If society has a technical mind, that helps science forward more than ten universities."[12]

But scientific discoveries await the social preconditions for their assimilation into the productive forces, even if they arise from prior social and technical circumstances. For this reason, science and invention, particularly those relating to the development of machinery, were in existence long before capitalism could use them. On the contrary, the evidence points overwhelmingly to the requirements of economic and social relations in the feudal and capitalist manufacturing period. The revolutionizing of the instruments of production depends, therefore, on the social relations of production, especially the evolution of the labor process.

Although the previous labor process is only "one of the material conditions for the development of the next stage," it is clear that Marx rejects the notion that the impulse for technological change arises from purely external factors. If taken together with the previous analysis in which I have shown that the forces of production of the manufacturing period are largely those of capitalist management and the social power of a progressively degraded labor force whose essential skills are derived from those of the handicrafts, it becomes clear that the notion of technological determinism is foreign to Marx's conception of historical change up to the transition from manufacture to machinery.

Machinery is no mysterious result of technological genius or scientific law. It is an extension of the social relations of production and, at least in the early period, results from the conjuncture of the division of labor in manufacturing with the inventions of craftsmen who are themselves the products of an earlier mode of production. As we have seen, Marx argues that modern science presupposes the use of machinery. Science itself is an abstraction from inventions that occur largely within production, and becomes an influence upon the labor process only when capital determines its utility to the accumulation process. These conditions include capitalist competition as well as the breaking down of workers' resistance to the attempt of capital to capture absolute power over production and consequently over their labor. The mutually determining relationship between technology, science, and social relations is evident in Marx's view of the rise of modern industry.

The logic of capitalist production accords with the worldview of the bourgeois, which reduces all objects to their quantitative extensions, regards the whole question of quality as an ethical rather than a theoretical issue, and subjects scientific inquiry to the exigencies of domination over nature and over other humans.

The character of science is illustrated by the medical model of the human organism. With the growing understanding by Americans of Chinese medicine, we are becoming aware that bourgeois medicine has developed a practice that reflects capitalist rationality, that is, corresponds to the fragmentation of the worker in the labor process (by the specialization of medicine), and has viewed the human being as an object isolated from its social context. The scientific object, the human body, is viewed as an autonomous organism whose functions can be separated from its real existence. The historical abstraction of the organism from its environment has yielded some useful insights for the conquest of disease but at the same time has prevented a fuller understanding of the sources of disease. Within the social relations of prevailing medicine, where disease is treated individually and not socially, such a methodology functions to reinforce the ideology of the body as a self-regulating machine.

The problems with Marx's formulation of the link between productive forces and productive relations, and consequently the social character of scientific theory and technology, appear in relation to his conception of the outcome of industrialization resulting from the widespread introduction of machine technologies.

Marx accepted the characterization of the introduction of machine manufacture as a kind of Industrial Revolution, at least a new stage of capitalist production. "The tool or working machine," he says, "is that part of the machinery with which the Industrial Revolution in the 18th century started,"[13] but its continuity with the manufacturing stage of production means that the revolution in its early period did not deny the essential characteristics of the previous stage. According to Marx, "The machine proper is . . . a mechanism that, after being set into motion, performs with its tools the same operations that were formerly done by the workmen with similar tools."[14]

At first, the machine merely makes self-activating the "same operations" and "similar tools" that were once performed by hand labor using specialized hand tools. "In those branches of industry in which the machinery system is first introduced, manufacture itself furnishes, in a general way, the natural basis for the division and consequent organization of the process of production."[15] And "here then we see in manufacture the immediate technical foundation of Modern Industry," based on both the division of labor within the workshop that was adapted to the factory system, and "the inventions of Vaucanson, Awkwright, Watt and others [which] were . . . practicable only because these inventors found, ready at hand, a considerable number of skilled mechanical workmen placed at their disposal by the manufacturing period."[16]

But, for Marx, the manufacturing period "was an inadequate foundation" for the factory system because it still rested on hand labor; that is, the machine owed its existence to personal strength and personal skill.

The adequate foundation upon which modern industry had to rely was an "organized system of machines to which motion is communicated by the transmitting mechanism from a central automaton." Marx writes that in the mid-nineteenth century, the power loom, the hydraulic press and the carding engine furnished the examples of the new instruments of production. These machines permit the production of machines by means of machines in which the 'laborer' becomes a mere appendage to an already existing material condition of production.[17]

According to Marx, whereas the manufacturing system was purely subjective because it comprised a "combination of detail laborers," modern industry is purely objective. Labor is no longer the subject of the labor process: the new subject appears to be the machine. At first reading, Marx's relentless description of the way in which machinery begins to dominate the workers and replaces them as the core of the productive process seems to rely on technological forces. But, on closer inspection, it is clear that Marx understands the introduction of machinery in terms of the degradation and suppression of labor by the capitalist. "It would be possible to write quite a history of inventions made since 1830 for the sole purpose of supplying capital with weapons against the revolt of the working class," he observed.[18]

Approvingly, he refers to Alexander Ure, who makes the same point. "Among the purposes of the introduction of the self-acting machine," according to Ure, "was to restore order among the industrious classes." Ure also makes the trenchant remark that "when capital enlists science in her service, the refractory hand of labor will always be taught docility."[19]

Referring to Ure, Marx shows that workers did not accept their new subordination under the factory system as appendages to the machine. The degradation of labor to the margins by reducing their skills, relegating them to material handlers and watchers of self-activating machines, was accompanied by "violent collisions and interruptions" which Ure describes as "erroneous views."[20] At this point, Marx argues that these violent upheavals, for example, Luddism (the breaking of machinery by enraged workers), are the result of the capitalist conditions of labor. For Marx, the "machine considered alone" can be a blessing to labor because it can shorten the hours of labor, lighten tasks, and potentially free up the laborer for self-development through the provision of leisure. Instead, under the capitalist relations of the production, the introduction of machinery results in the intensification of labor through heavier work loads and speedup, and the potential lengthening of the work day because machines relieve the worker of physically taxing work. The machine is turned against the immense majority of laborers in the service of capital. Labor is enslaved by the forces of nature, according to Marx, because by increasing wealth through mechanized production, capital reduces the workers to paupers.

Here, for the first time, Marx becomes a celebrant of the industrial period of capitalist production for its potentially liberating technology. His critique is confined, almost entirely, to the specific uses to which the capitalist class in control of production puts the new productive forces. The productive forces themselves seem to lose their internal connection to the social relations of production. They *are* objective in the sense that their productive powers no longer rely purely on the division of labor. Machinery, for Marx, represents human mastery over nature, and only the social relations of production prevent its liberatory potential from being realized. The new principle of production "carried out in the factory system, of enlarging the process of production into its constituent phases, and of solving the problems thus proposed by the application of mechanics, of chemistry and of the whole range of natural sciences, becomes the determining principle everywhere."[21] According to Marx, the new forces of production, science and technology, are inexorable and are truly revolutionary because the logic of modern industry is one of perpetual revolutionary transformation of the process of production, including the role of human labor within it. The new methods of self-activating production, especially the seeming breakdown of previous processes and their integration by means of science and technique, replace the dominant role of manual labor as the chief productive force. Machines are the objective form of the emergence of "general scientific labor"—the new productive force. In the *Grundrisse*, Marx is careful to reiterate the dependence of the new productive forces on a historical progression from the social and technical divisions and combinations characteristic of earlier periods. He also makes clear that capital "envelops machinery,"[22] insofar as it presses the exploitation of labor in order to extract the surplus value which has been reduced by the introduction of machinery, since "the degree to which large industry develops, the creation of real wealth comes to depend less on labor time and on the amount of labor employed than on the power of the agencies set in motion during labor time."[23] But it is plain that the period of modern industry marked by the significance of scientific and technical principles, and not by the labor that carries these principles into production, has become for Marx the hope for the emancipation of labor because these are productive forces antagonistic, in the last instance, to capitalism.

Marx does not separate the productive forces in general from the social individual or from social relations. He regards forces of production and social relations as "two different sides of the development of the social individual."[24] The so-called autonomy of the forces of production in the period of advanced technology, which for Marx consists in the development of large-scale industry through science, is in one sense only the reification of the actual social relations individuals enter into through the production process. The qualitative change from the early period of modern industry (when machine production still is dependent to a large degree upon the skills and physical power of the manual worker) to the new scientific-technical period of production marks the triumph of intellectual labor over manual labor in the process of production itself, just as the manufacturing period is characterized by the victory of the capitalist in organizing the labor process. "The development of fixed capital," says Marx, 'indicates to what degree general *social knowledge has become a direct force of production, and to what degree hence the conditions of the process of social life itself have come under the control of the general intellect and been transformed in accordance with it.*" (my italics).[25]

The capitalist's domination over the production process in the period when hand labor still prevails is accomplished by the degradation of labor. In the period of large-scale industry, when knowledge has become the decisive productive force, in Marx's view, capital subordinates labor through the mediation of scientific and technical knowledge, but also creates the conditions for its own transcendence.

> The savings of labor time [resulting from the creation of machinery] is equal to an increase of free time, i.e., time for the full development of the individual, which in turn reacts back on the productive power of labor as itself the greatest productive power. From the standpoint of the direct production process, it can be regarded as production of fixed capital, this fixed capital being man himself.[26]

On the one hand, capital reduces labor to the function of watcher of the production process. On the other hand, intellectual labor, as the form of the productive power of all labor, is immeasurably enlarged by advanced technology.

Management, as the form of intellectual labor within the framework of capitalist social relations, is now joined to science and technology, representing not the old capitalist social relations but a new emancipatory form of labor, a form which can free the laborer from the production of necessities in order to undertake what Marx calls "higher activity." Higher activity, according to the immanent meaning of his discussion in the *Grundrisse*, is the freeing of the individual from the production of the necessities of life as much as possible so as to undertake educational, recreational, and scientific work.

Marx calls "experimental science" a higher activity that has been subordinated to the will of capital under the prevailing mode of production. He regards science in the late period of capitalist development as representing the general powers of humans over nature, in his words, "the general product of social development."[27] Even though Marx acknowledges that the origins of science were embedded in capitalist social relations, particularly in the labor process generated by them, there is an unmistakable separation of scientific and technical knowledge from the manifestations of these relations in the process of production, as evidenced in the *Grundrisse* and in *Capital*. Marx never abandons the insights of *The German Ideology* that the forces of production are objectivated human labor, including both manual and intellectual labor. But there is a tendency in the late Marx to regard the new forces of production as antagonistic to, and thus autonomous from, the class structure, at least potentially. Experimental science is not a class science: it is social knowledge belonging to the entire society. Ideology does not seem to be embedded in, much less a determinant of, scientific knowledge and invention in the period of large-scale industry, even if the motive force of modern industry was the suppression of labor and the securing of capital's absolute power over production.

Although Marx ascribes to modern industry the objective possibility of providing free time to at least some sectors of humanity, he rejects Fourier's effort to make labor into play; Marx argues that instead of making labor itself a form of pleasurable idle time, the free time engendered by the introduction of scientific and technical innovations transforms the laborer from a dependent object into the subject under a more advanced mode of production. He emphatically asserts that "human being itself and its social relations" is the real subject of history and of social life. All else, including the object world in the form of fixed capital, the product itself, and every objectification of the social process of product, is but "a vanishing moment" in the movement of bourgeois society toward its negation. One of the most significant implications of Marx's view of science and technology in the modern era is his belief that the emancipation of labor consists in the capacity of the new science to achieve the emancipation *from* labor of the mass of humanity.

Even as he attempts to show that laborsaving technologies permit the "leisure" necessary for the generation of new knowledge, which transforms the productive forces so as to free the laborer even more from subordinate, often mindless tasks, Marx argues that the promise of a new mode of production is to reduce "necessary" labor to a minimum. In this way, the laborer is freed for higher activities that are not connected directly with the production of goods.

Labor, in the traditional sense of the term, is transcended by the full use of the new forces of production by the proletariat. In communist society, the worker becomes a scientist rather than an idler, using her/his time to complete the scientific and technical revolution in order to realize its emancipatory potential. Here, Marx's radical distinction between "idle time" and time for "higher activity" plays a crucial role. In Marx's view of the scientific-technical revolution, there is no implication that he endorses the idea that the laborer will spend the time saved by the large-scale introduction of laborsaving machinery in mere idleness. Instead, he understands the possibility created by science for the use of "free time" as self-activity of the laborer in pursuing "work" that may not necessarily be connected with the production of goods.

In the later writings, the new productive forces are not so much the subordinated laborers as the highly educated technical and scientific workers who have grasped the accumulated knowledge of society in their hands but have been subjected to the domination of capital. But there are serious problems in this formulation. First is the problem of how degraded labor can become a historical subject, if capital divests it of its centrality to the production process.[28] Second is the reversal of Marx's posing of the historical role the labor process plays in the self-formation of humans. According to Marx, the immense majority of humans have been relegated to details of the production process, both in the manufacturing phase (owing to the domination of the capitalist) and in the period of modern industry (owing to the rise of scientific knowledge). Having been divested of power over the labor process, the manual worker becomes an object of the capitalist mode of production even as labor time remains the measure of the exchange value of commodities. The process of degradation is intensified by the victory of science and technology, because only a small fraction of the working class possesses the knowledge to participate in knowledge production in the industrial era. For the rest, access to the production of material necessities is limited by the social relations of production, in which labor is alienated from the ownership of the means of production; labor is also deprived of a central role by virtue of the transformation of the production process through the creation of self-activating machines generated by science and technology as the main productive force under the general control of capital. In this instance, only a minority of workers can be active agents within the labor process in future society.

Unless the productive forces can themselves allow for the intervention of the mass of humanity, how is it possible to create a society without hierarchy, in which social ownership of the means of production is matched by social control of the production process, much less all other social institutions? Marx's foreclosure of the possibility, or, more precisely, the desirability of reversing the technology itself, raises these questions.

Marx treats the labor-and capital-saving technology of capitalist production as part of the historical legacy that a socialist society will inherit. For Marx, the importance of production in the self-formation of humans is destined for a decisive reversal. Moreover, notwithstanding his ascription of science and technology to the accumulated social knowledge and human powers over nature, there is little doubt that he regards the advent of modern industry as progressive in content, even if related intimately in origin to the purposes of capital in securing its own domination over society in general and over the productive process in particular. It is this radical distinction that allows orthodox Marxism to argue that scientific knowledge, especially experimental science, has an inherently neutral character, even if "objectively" subversive to the survival of the capitalist mode of production in its effects. Marx seems willing to employ the advances made by modern industry within the framework of a new mode of production. But his specific denial of Fourier's project to transform labor into creative work within the process of material production, except through the vehicle of experimental science and invention, lends credence to the claim of those who would transform Marx into a technological determinist, at least in regard to his later writings. (In this connection, it must be remembered that the *Grundrisse* constitutes but a series of elaborate notes for *Capital*, which must be viewed as the more mature and authoritative work.)

Nevertheless, there is evidence in *Capital* itself that while Marx tenaciously holds to his view that "large-scale industry tore aside the veil that concealed from men their own social process of production"[29] and rejects any intimation that technology is a force of nature, technology and its internal logic become revolutionary for Marx. The revolutionary character of technology consists in its release of self-activating forces within the productive process—forces which, through scientific means, transform the means of production and thereby cause changes "in the functions of labor within society and incessantly throw masses of capital and of workers from one branch of production to another."[30] In this sense, Marx certainly appears to hold with a historically created autonomous role for science and technology as such, which only in the "last instance" depends on the determination by capital. In this statement, Marx seems to ascribe a determining force to science and technology, in relation not only to labor but to capital as well.

It appears that Marx took the view that capital had created a golem in its incorporation of scientific and technological advances within the process of production. The golem was destined to turn against its masters. The social and political power of capital would be unable to contain the revolutionary effect of the sciences of chemistry and mechanics on production within existing social relations. It may be argued that it was not the modern proletariat which for the late Marx constituted the real gravedigger of capitalism, but the scientific/technological forces unleashed by capital itself. The subjective side of these forces, the "accumulated knowledge of society," has a specific repository in the social and technical division of labor: the scientist, engineer, and technician. To these may be added the managerial stratum directly responsible for superintending production.

These conclusions are certainly not present in Marx's writings as an explicit revision of the views contained in *The Communist Manifesto*. Since the closest Marx came to assigning to science and technology the main revolutionary impulse in the late capitalist society was contained in some of the *Grundrisse*'s statements and a few remarks in *Capital* itself, it is probable that he ultimately had reservations about the implications of the logic of his position. Yet, the nagging evidence has a cumulative power. In a letter to Engels, Marx refers to his discovery in *Capital* that "dialectic law" applies equally to both society and natural science: "You will see from one conclusion of my Chapter III, where the transformation of the handicraft master into a capitalist . . . as a result of merely quantitative changes is touched upon, that in that text I refer to Hegel's discovery . . . the law of merely quantitative changes turning into qualitative changes . . . as holding good alike in history and natural science."[31] Here Marx refers to his footnote in *Capital*, volume 1 regarding "the molecular theory of modern chemistry" being dialectical in the same way, specifically with regard to changes in the quantities of hydrogen and carbon and of hydrocarbons resulting in new compounds with different properties from each other.[32]

In these fragments, Marx seems to insist on the objective dialectical character of natural science, asserting the dialectic as a universal law applying to both society and nature. These suggestions lend substance to claims of positivist tendencies in Marx. That is, the notion of the objective material world as prior to human will, possessing an independent power which can be discovered through scientific investigation, specifically through experiment. These were the suggestions that motivated Lenin and Engels to argue that dialectical law inhered in nature and was reflected in human knowledge and social relations. Of course, the evidence of the positivist strain is far more fragmentary in Marx's work than in Engels's. It would be absurd to arrive at a position that Marx became either a technological determinist or a thoroughgoing "scientific" materialist on the basis of the evidence of his earlier writings. Much of it contradicts both his earlier work and other parts of his later works.

Yet, many of the statements themselves are fairly unambiguous. Moreover, there can be no doubt that he regarded modern industry as a qualitative leap both in the mastery of nature and in the role of the instruments of production as a motor force of historical change. The scientific-technological revolution cannot be permanently harnessed by capital because, for Marx, it has unintended consequences. The qualitative leap in the development of the productive forces engendered by capital's own subsumption of science and technology results in a vast accumulation of constant capital and the shrinking of the proportion of variable capital, that is, living labor. The reduction of the proportion of labor embodied in the production of commodities, even the means of production, results in a constant tendency for the rate of profit to fall. In turn, the capitalist's solution to counteract the depressing effect of technology on the rate of profit is to invest capital in even more technologically advanced machinery in order to increase the volume of surplus value.

This objective is accomplished by reducing the amount of socially necessary labor time required for the production of commodities, on the one hand, and the proportion of necessary labor to surplus labor within the working day, on the other. But this merely exacerbates the problem of overproduction since the number of laborers required for social production tends to be constantly reduced. Here, capital *appears* to produce more capital. Marx wants to argue that not only is labor necessary to motivate the machines, even under conditions of high automatic production, but that intellectual labor in the form of scientific and technological knowledge has become the decisive productive force of society. The crisis of capitalism in its last stages consists, for Marx, in its tendency to revolutionize the means of production against its own will and against its own interests in the long run.

This self-contradictory state of affairs has come about precisely because the aim of capitalist production is always the extraction of surplus value, which is achieved most efficiently, from the capitalist's standpoint, by intensifying labor and raising its productivity rather than lengthening the working day. The limits to raising workers' productivity within the framework of methods to intensify labor through speedup and stretchout are quickly reached as the human body refuses to go further. Even such wage incentives as the piecework system are ultimately bound by the physical organization of human beings. The introduction of machinery becomes the major means to overcome the falling rate of profit; it is in this sense that Marx ascribes a certain autonomy to the productive forces. The new scientifically motivated productive forces having been created by capitalism now become determinative of its dynamic development.

According to Marx, the epochal significance of the advent of modern industry consisted in the liberation of "man" from the realm of necessity, allowing for entrance into the realm of freedom.

The new self-activating productive forces generated by the application of the sciences of chemistry and mechanics to the labor process would provide the material basis for the end of human prehistory, that is, the domination of humans by nature and the domination of "man by man" that rested upon this relationship. Modern industry irrevocably undermined the material foundation of all social relations based on class domination because the hierarchical organization of society rested in the last instance on the capacity of ruling classes to maintain their power over the means of production and represent the highest possible development of the productive forces, in the face of the relative underdevelopment of human productive powers.

In capitalist society, the bourgeoisie is obliged to socialize the productive forces and unleash the social productive powers of humanity to an unprecedented degree. At a certain point, these productive forces become highly concentrated and centralized, and tend to reduce social distinctions to their starkest and simplest terms. On one side is the capitalist class, which owns and controls the means of production and establishes its ideological and social hegemony over all society; on the other, the mass of proletarians, recruited from agrarian and artisan populations who constitute the critical productive force in capitalist society. The contradiction between the social character of the productive forces and social relations based on private appropriation of them rents society and produces a revolutionary transformation that brings the social relations in conformity with the new productive forces.

From Marx's paradigm of revolution, it is clear that the productive forces which have become the motive force for historical change in the late stages of capitalism are largely autonomous, in substance, from the prevailing social relations and become powerful in proportion to the inability of the capitalist class to control either their configuration or their pace of development. Labor power, machinery, and raw material are the material embodiments of the "self-expansion" of capital, which is fettered but cannot be reversed by bourgeois social relations. In Marx's theory, as long as the forces of production were insufficient to liberate humanity as a whole from a life of toil under conditions of material scarcity, the domination of one class over another represented the relatively low level of society's domination of nature. Human freedom was predicated on the abolition of the material conditions that produced bondage; the insufficient development of the productive forces. In the 1840s and 1850s, Marx never tired of reminding the utopians that they could not achieve communism as an act of will alone. "No social order disappears," he admonished, "before all the productive forces for which there is room in it, have been developed; and the new higher relations of production never appear before the material conditions for their existence have matured in the womb of the old society."[33] British and French socialism at the turn of the nineteenth century were regarded by Marx as brilliant prophecies of the emergence of a society that was unable to come into existence as long as the capitalist mode of production represented the further development of the productive forces relatively unhampered by its social relations.

According to Marx, the "new higher relations of production," namely, socialism, had come into existence within the womb of Western European society by the latter half of the nineteenth century because capitalist social relations were no longer able to provide room for the automatic, self-activating machines that were created by "the accumulated knowledge of society," that is, science and its application to industry through technology.

The new technology, objectified as a huge mass of fixed capital, promised the emancipation of labor through the emancipation *from* labor. It promised nothing less than a historical break in two important respects. Not only did the enormous productive powers represented by scientific and technological progress present the possibility of transcending capitalism as a specific form of class society, and the creation of a society based on the self-management of the production forces by the "free association of producers." It almost promised the transcendence of the dialectic of labor, at least in the traditional sense of the term. If, in all hitherto existing societies, productive labor has been the means by which humans create both their social world and themselves, and if production has consisted in the subordination of nature in order to meet human needs, then no more powerful instrument for the achievement of basic human needs has been created than automatic machinery. Marx held that freedom rested upon the collective power of humanity to transcend the realm of necessity in the production of its means of subsistence. For the proletariat, the technologies already in existence, and those that Marx foresaw as a necessary development from existing means of production and the state of scientific and technical knowledge, were sufficient for the liberation of human beings from subordination to the routines of daily labor.

In fragmentary passages, Marx understood socialism to mean the possibility of creating vast quantities of time for the full development of the individual, now free to pursue intellectual, recreational, and social ends not determined by the productive process. The conquest of nature, or more exactly its domination by highly developed social labor, meant that the domination of humans by humans would disappear in the period of modern industry.

But there is a problem with this formulation. The ruling classes, by virtue of their ownership and control of the means by which nature is subjected to human needs, have also subjected the mass of humanity. The domination of nature has not been a neutral human project, independent of its relationship with the project of the domination of some humans by others. Marx understood this relationship well in his description of the transition from the artisan period of production to the manufacturing period in which a single capitalist established power over a labor force that became degraded through the fractional division of labor. Marx argues for the twofold character of management during this manufacturing period. On the one hand, he regards management as a productive force to the degree that it controls and coordinates the labor of many workers, none of whom can individually grasp the totality of the production process owing to the increasing rationalization of tasks. To this extent, the new, higher forces of production are represented by the progressive breakdown of traditional skills and the synthetic function of the capitalist as manager. On the other hand, Marx understands management as a form of domination of humans. It is the expression within the labor process of capitalist social relations insofar as labor becomes degraded and is transformed into a subordinate and dependent variable of capital. Here, the dialectic between the domination of nature and the domination of labor by capital is absolutely clear.

When it comes to modern industry, Marx seems to jettison his ascription to the social relations of production a determining role in the shaping of technology. The capitalist purpose is still the motive force for the entire production process: the extraction of surplus value and the accumulation of capital. The origin of modern industry results from the appropriation by capital of science and invention and their application to the labor process. But the new technologies can no longer remain in the control of capital. They assume a determining and independent relationship to the social relations that contain them. Science entails the logic of domination by stripping nature of its qualitative aspects, leaving only quantitative measurement, its compulsion to construe the natural world in terms of the control of objects, and its relentless instrumentalization of all human action. Science, as Christopher Caudwell shows,[34] is *bourgeois science* to the extent that it reduces nature to a machine and is fully comprehensible in terms of mathematical categories. Just as the lawfulness of human society is based on coercion and the establishment of rules which appear eternal and absolutely part of the natural order of things, so nature becomes self-acting as the outcome of domination.

Marx accepts this domination wholeheartedly, almost with awe. His view is that science marks the end of human bondage. By recognizing the lawful character of the natural world, humanity is now capable of subordinating nature to human requirements. But the domination of nature proceeds, as I have noted, by the sharpening of the division of intellectual and manual labor. Machine technology cannot be separated from the social relations that created it. The logic of domination remains embedded in the machine, which is an instrument for the perpetuation of social oppression and exploitation by virtue of not only its uses but its construction as well. The introduction of automatic machinery on a large scale in the cotton textile industry within the United States in the 1890s has been attributed by V. S. Clark more to economic than to technical conditions. The Northrup loom "embodied two fundamental improvements upon existing looms," remarked Clark: "A filling changing mechanism by which the time hitherto lost in replacing exhausted shuttles was wholly saved, a perfected warp stopping device that stopped the loom instantly when a single warp thread broke."[35] According to Clark, the importance of these innovations was that "within a decade, the number of looms a weaver could attend increased from eight to twenty four." Here, we can observe that new inventions were introduced in order to save labor time. Clark notes that the choice of technologies in the textile industry had other purposes. For example, the increasing use of the ring frame loom rather than the mule spinner after the eighteenth century was motivated by the desire of employers to eliminate English and Scottish craftsmen with long trade union and socialist traditions and replace them with women and children who were presumed less troublesome.

The concept of efficiency itself is by no means a self-evident category in the productive process. On the one hand, laborsaving technologies do not merely reduce the quantity of socially necessary labor time required for the production of the commodity. They often result in the degradation of labor and the sharper separation of intellectual and manual labor. The introduction of continuous flow technologies into the oil, chemical, and food industries transferred effective control of the labor process from skilled artisans to professional chemists. The skilled artisan was reduced to a watcher and a manual laborer, whose chief work, in addition to handling materials, became the monitoring of self-regulating machines designed by trained engineers and chemists working under the supervision of managers. On the other hand, the historic function of laborsaving technologies was to make it impossible for labor to successfully challenge the centralization of knowledge and control by employers and their professional managers. Labor saving is also labor degrading under the capitalist mode of production, since it almost always results in the polarization of the division of labor to an extreme.

The late nineteenth century witnessed the beginning of the self-conscious domination by capital of science and its application to the labor process.

The most important science employed by industry was that of chemistry. Andrew Carnegie, the leading capitalist of the iron and steel industry, was regarded by his competitors with some amusement when he hired the first full-time chemist in the Carnegie Steel Works in the 1880s. From the beginning, Carnegie regarded the employment of professional chemists and metallurgists as an extension of his strategy of achieving efficient management through measures that reduced the costs of producing steel. As stated by Oliver, "Carnegie declared that the nation that made the cheapest steel would, within a few years, have all nations at its feet."[36] Scientific personnel were employed as a specific measure to reduce the large number of skilled craftsmen upon whom much of the steelmaking process still depended as late as 1887.

Oliver: "In the early days, the making of steel was an empirical process, and determining its composition was a craft, not a science. Gradually, rule of the thumb practices were eliminated, and the skilled artisans relinquished their positions to research scientists in the laboratory."[37] His description of the transformation of the steelmaking process from artisanship to science is instructive: "The steel furnaces and converters are in reality huge test tubes where ingredients according to formula are transformed by chemical changes into a new metal product."[38]

This illustrates the determining role of social relations in the development of the productive forces. The gradual transfer of power over the steel production process from the crafts to the chemists is expressed in the process itself becoming a metaphor for the laboratory. Scientific practice in the early days of the Industrial Revolution remained isolated yet was supremely reflective of the ideology of domination insofar as the experimental laboratory is another instance of control, quantification, and instrumentalism that characterizes the bourgeois era. Its application to industry, in this case the steelmaking process, was facilitated by the desire of large capital to seize control of the labor process from the craftsmen.

But the ability of science to be coopted is an expression of the congruence of its methods with the aims of management—their sharing of a worldview. To be sure, scientists engaged in "pure" research proclaim the objectivity of their findings, that is, the correspondence of their theories with natural law. But the method of chemical science, in which the researcher combines natural elements according to formula in order to make a new product and thus reveal the properties of the elements themselves, is at the same time convenient to the industrial laboratory. The chemist, metallurgist, and engineer are considered part of the management's staff. In reality, they have become the most important new productive force and simultaneously the central instrumentality for reducing all other labor to dependence. The happy conjuncture of science, which was developed historically by investigators recruited from artisanship as well as from the middle and aristocratic classes (and who shared the general worldview of the bourgeoisie that the natural world could be made lawful in the image of capitalist rationality), and the development of scientific management that sought to eliminate once and for all the access of workers to the totality of the production process, is still the chief characteristic of the scientific-technological revolution. Far from being a force for liberating labor, it has become a means to dominate it.

The instance of the textile industry demonstrates similar processes. The discoveries of industrial chemistry originally important in the dyeing process resulted both in the abolition of the dyemaking crafts in the early years of the twentieth century and in the gradual replacement of textiles by synthetics in the United States. The development of the synthetic fibers industry was not undertaken by textile employers. It became the province of the new chemical combines, particularly DuPont, Union Carbide, and Allied Chemical, which established their preeminence within the chemical industry by the 1920s. DuPont accumulated much of its primary capital by the production of gunpowder during the Civil War and the Spanish American War, extending its tentacles into other industries owing to its vast accumulation of physical capital, but also its concentration of human capital, that is, of scientific and technical employees.

The development of the electrical industry illustrates the degree to which the choice of technologies depends on the competition between different capitals. Thomas Edison invented an electric power system that could give light by a central generator. But, by using direct current to transmit electrical energy, there was an inherent limit in the distance the electricity could travel. At most, generating systems were limited to a single town or large plant.

The invention of the transformer for alternating current from high to low voltages by George Westinghouse and his associates permitted the transmitting of electric power over long distances, the centralization of power stations, and the development of motors that supplied twenty times the power of the largest direct current apparatus.[39] Of primary importance, the new technology spurred the acceptance of electricity as a source of power. Electricity became a cheaper source of power than water or pulleys, and made it unnecessary to locate plants near water. The widespread use of electric power for industrial plants signaled the end of the concentration of industry in a few large urban areas. Although capital was to increase its concentration into ever fewer hands, the geographic centralization of fixed capital was eventually undermined. Yet, within each geographic area, electrical power supplies were extremely centralized as never before. The acceptance of the Westinghouse system of alternating current rather than the Edison system was guided by the increasing requirements of large-scale industry. The choice of technology was a social, not a scientific, consideration. Conceivably, in a system of decentralized, smaller-scale industry, the economies derived from alternating current would disappear. Thus, in this case at least, science and invention were directed by the requirements of capital, rather than the other way around. This conclusion is confirmed by British historians of technology. According to T. K. Derry and A. S. Williams,

> The earliest generators produced alternating current: that is,
> the direction of flow constantly reversed, with a frequency
> depending upon the speed at which the machine was turned.

This was looked upon as a most serious disadvantage—partly, at least, because the workers were accustomed to working with the direct current provided by batteries but toward the end of the 19th century, it was realized that for large scale use alternating current had decisive advantages over direct.[40]

It was large-scale industry, engendered by the concentration of capital, and the purposes of giant corporations that were emerging in the late nineteenth century in both Britain and the United States, that determined the use of one technique of electrical production rather than the other. "Technological superiority" was a function of capitalist competition between different electrical interests, on the one hand, and between giant and small capitals, on the other. Another interesting point to note is that the principle of alternating current was discovered in 1830 by Farraday but was not deemed practicable for industrial use until the advent of centralized, large-scale industry. In this instance, the concept of efficiency and practicality is coincident with the notion of economies of scale. It is probable that the use of direct current supplied by batteries enabled workers to assume greater control over the speed of work, whereas the use of alternating current facilitated a greater division of labor between those who supplied electricity and those who used it.

The prominence of electricity in industry, replacing gas and steam power, was no doubt enhanced by the development of technologies that employed Farraday's discovery of alternating current. Marx is right to point out that the systematic application of science to industry generates new industries, destroys others, facilitates the constant revolution in the instruments of production, and determines the fate of masses of workers. The motive force for these changes, however, remains closely intertwined with the concentration and centralization of capital into fewer hands, the subordination of science to the management strategies of degrading labor, and thus, the domination of the forces of production by social relations. There is no internal logic to the labor process itself that leads to the removal of control from the shop floor to the planning departments and the laboratories of the corporations. The evidence suggests quite the opposite. Corporate purposes determine the employment of scientific discoveries within industry in the late nineteenth and twentieth centuries. Science and invention are servants, not of industry in general, but of capital in particular. The history of the textile, steel, and chemical industries suggests that invention itself was motivated by the exigencies of the struggle for control over the labor process as much as the struggle within the marketplace among capitalists. In both cases, capital does not operate entirely as a blind force to determine the shape of technology but, in the era of monopoly capital, becomes self-conscious. Ruling ideas can no more exempt the configuration of science and technology than the shape of social ideologies. Caudwell:

> The point is not just a piece of nature. It is a piece of nature associated in bourgéois society with human beings who work the machine. It is the kernel of a social complex which gives it its shape and significance . . . The machine is a piece of humanized nature. It is composed of particles arranged according to a plan of a human desire. But the society which uses the machine is a naturalized society.[41]

Just as ideologies are often generated behind the backs of their makers but are nevertheless representations of a worldview that bears the unmistakable stamp of the dominant class, so science and technology, understood by practitioners as true reflections of the natural law and having the force of inevitability, make discoveries in accordance with imperatives of the domination of nature and capitalist rationality as if these were the only possible choices. Yet, the tendency of capital is to consciously determine the shape and significance of science and technology in accordance not only with the laws of the blind marketplace, but with the specific purposes of the employers. For example, the development of the self-acting loom was largely prompted by the spinner's strike of 1824 in which it was shown that skilled craftsmen in combination could effectively stop production. Richard Roberts, a machine manufacturer in Manchester, immediately set out to perfect the self-acting mule spinning machine. Conversely, the introduction and retention within the United States of assembly-line methods of automobile production are attributable not so much to an abstract concept of efficiency but to the conscious efforts of management to extend their control over the labor process. Assembly-line methods are inherited from the manufacturing period in the sense used by Marx. Even though the transfer mechanism has been made continuous, the "working up" of the material remains largely a hand operation. Tools have been specialized in accordance with the specialization of tasks. Here, the introduction of interchangeable parts, an engineering invention, facilitated the degradation of skills. Automobile parts plants operate on the principle of specialization as well. Stamping plants still utilize the most primitive methods of production. These correspond to the early stage of machine industry in which the machine imitates the hand operations, and only the mechanization of the so-called motive power, i.e., the substitution of electrical and mechanical energy, replaces human strength. In this case, capital has consciously fettered the development of productive forces that replace human labor as much as it has refused to consider other methods of production in which multivalenced laborers are in control of the pace of production, the assignment of tasks among the collective of laborers, and ultimately the techniques employed. The considerations motivating the recalcitrance of large auto monopolies may have been their absolute power over the market and the availability of large pools of labor which reduces the need for introducing laborsaving machinery. The refusal to mechanize or to consider alternative methods also corresponds to the "efficiency" of the assembly line in terms of retaining management control over the labor process.

Of course, in recent years, major United States auto makers have introduced computer mediated machine processes such as robotics and numerical controls both to the assembly line and to parts production. Robots are used primarily as tools for painting and handling materials, replacing one of the most onerous jobs on the assembly line: pushing auto frames on and off the line and moving equipment from one place to another by hand. Numerical controls are computer programs that regulate the speed and feed of machine tools and automatically set dimensions of parts production. In parts plants, robots carry metal stock to and from machines. Despite these innovations that improve productivity and eliminate a considerable amount of hand labor,[42] the general organization of the factory has been relatively untouched. Rationalization in the form of specialized routinized tasks still marks assembly-line work; and the traditional craft of the machinist is modified to a considerable extent by the degrading aspects of numerical control. Yet, the management system invented by Ford has not been significantly changed in assembly-line production. Few of the innovations introduced in Sweden, where small groups take responsibility for assembling engines and cars, have been applied in the United States because the employer's interest in control still dominates the configuration of technological transformation. To visit a parts or assembly plant today is an experience similar to that which existed thirty or fifty years ago. When machinery is introduced within the framework of the old organizational technology, its effect is to make matters worse from the perspective of the operator, even if new technology might permit organizational changes while maintaining or even increasing productivity.

What emerges from this discussion is the need for a radical critique of several significant tenets of Marxism as they derive from Marx's own work:

1. It seems to me that a careful reading of Marx's own concrete studies of the significance of social relations for determining the configuration of science and technology and, in turn, the productive forces, suggests a reversal of the formulation contained in the oft-quoted Preface to the *Critique*. Social relations, specifically the domination of capital, the hegemony of its ideology in all human pursuits within capitalist society, becomes the critical feature of the period in which scientific knowledge is the crucial productive force.

2. Ruling ideas, expressed as organizational technique, are themselves productive forces. Science, with its imperative of prediction and control, has determined, as much as reflected, the prevailing technologies of the capitalist mode of production.

3. Marx's argument that it is necessary to adopt the historical inheritance of the bourgeois productive forces addresses only one major precondition for the emergence of a society of emancipation: the potentiality for abolishing material scarcity. Central to the contradictions that inhere in the particular bourgeois solution to the question of material deprivation is that capital offers material plenty only on the basis of the surrender of skill, workers control, and craft or professional autonomy. This surrender is immanent in the structure of machine technology and the social organization of labor that is intimately connected with it. The technical division of labor under capitalism cannot be separated from the social divisions. There is no "purely" technical question that does not imply social hierarchy, especially sexual divisions and the divisions based upon the radical separation of intellectual from manual labor. The issue that remains is: can material scarcity be overcome without introducing the hierarchies intrinsic to the capitalist mode of production? More specifically, must we accept the continuous flow technologies of modern industry—does cybernation offer an alternative to centralist degraded labor?

4. A major conception implicit in the late writings of Marx is the separation of work from labor. Labor is defined as the task of producing the necessities of life which still must adhere to the most alienating forms of labor. Work, in Marx's conception, is lodged in the context of postindustrial society as activity only indirectly linked to production. Work is the self-actualizing activity, but it is clear that as long as we accept the imperatives of modern centralist technologies, the only solution is to let humankind devote a diminishing proportion of its

waking hours to "necessary labor," albeit of an alienating kind. Creativity is now to be achieved within the framework of what has been defined as *leisure*. The radical separation of labor from leisure raises sharply the question whether humans can tolerate the maintenance of the division of intellectual and manual labor, and the relegation of that which has been, at least in the past, the main self-actualizing activity— the production and reproduction of life—to a position in society outside the labor process. For, if society can achieve a nearly complete automated basis for the production of material necessities, does it follow that most labor can now be devoted to the provision of the services required for the social reproduction of life, namely, health, community services, and so on? Leaving aside (only for the moment) the world problem of food production, the lack of available fossil fuels, the material backwardness of many of the world's nations, the issue remains: can labor be transcended as a fundamental human "need"? Is the dialectic of labor destined for oblivion? If not, the clear implication is the necessity for new technologies that are radically different from those of the present. These would correspond to modes of life in which the division of labor itself undergoes basic transformation.

The notion of human interdependence that we have inherited from the division of labor particular to capitalist society would require transformation. Humans would continue their direct negotiation with nature, not as an increasingly peripheral activity, but as a central concern. This would not argue against trade, international relations, and so forth. But it would reject categorically a social division that would lead to the monopoly of production by those trained for intellectual labor. Some form of multivalenced laborer would reappear, not a romantic revival of crafts as they were practiced in the feudal epoch, but at least the idea of craft production as essential human activity would be recognized.

5. A massive examination of technological "imperatives" would result from the recognition that work as human activity would not be overcome in the new society. Those activities that perpetuated the divisions based on race, sex, and skill would be rejected. A new concept of the division of tasks based upon the specific activities rather than presumed levels of advanced versus backward skills would be introduced. It would not be necessary to state the problem as Marx and Marcuse would have done, in terms of necessary versus creative labor. On the contrary. A new look at Fourier's suggestion that work would be merged with play, and that both would correspond to "necessary" rather than "leisure" time, is called for.

The notion of leisure is a reflection of the distinction made within the capitalist mode of production. In a society not dominated by the labor process, such distinctions would disappear to the extent that work was transformed into play within the process of production of material necessities.

6. A mode of production in which the division of labor, based on the distinction of intellectual and manual tasks and the divisions based upon town and country, disappears can be achieved only within a decentralized social economy. Modern Utopians argued that cybernetic technologies lend themselves to decentralized economies. E. F. Schumacher and Murray Bookchin are among those who have taken the position that cybernation is, by its nature, an integrated rather than fragmented technology. Indeed, it has been difficult for capitalism to reconstruct fragmentation within computerization. The distinctions between programmers and systems analysts, on the one hand, and computer operators, on the other, are certainly as arbitrary from the point of view of production as those distinctions that appear on the assembly line.

Yet the problem remains: Are computers no more than a tool for the transmission of information? What are the qualitative implications of computerization of machine tool production or the processing of chemicals, oil, food products, and so on? Does standarization serve human interests? Or are the interests of any social hierarchy served by the quantification of social products, no matter what the ostensible social relations?

7. A question arises from the innovation in transportation and information: since much of the world's economy depends on these technological categories, what are the conditions for democratic society? Can a democratic, self-managed society exist when production is organized according to the efficiency of information systems? Did Marx abandon the visions of worker self-management when he accepted the supremacy of the "scientific-technological revolution"? If the answer is "yes," it may be that the Soviet Union *is* the wave of the Marxian future, and the entire corpus of the late writings is open to question. (See chapter 8.)

PART II

CHAPTER 4
ENGELS AND THE RETURN
TO EPISTEMOLOGY

For many who want to protect Marxism from a linkage with mechanistic thought, the slogan "back to Marx" intends to show the discontinuity between his thought and those of his immediate followers. The title of one writer's effort in this direction is *Engels Contra Marx*. Indeed, the project is already present as early as Lukács's *History and Class Consciousness*:

> To be clear about the function of theory is also to understand its own basis, i.e., the dialectical method. This point is absolutely crucial, and because it has been overlooked, much confusion has been introduced into the discussion of dialectics. Engels' argument in the *Anti-Duhring* decisively influenced the life of the theory. However, we regard them, whether we grant them classical status or whether we criticize them, deem them incomplete or even flawed, we must still agree that this aspect is nowhere treated in them.[1]

Lukács acknowledges that Engels stresses many of the significant features of dialectics, for example, interaction versus mechanical causality. "But he does not even mention the most vital interaction, namely, the dialectical relation between subject and object in the historical process, let alone give it the prominence it deserves."[2] Lukács contends that dialectics has no meaning other than this relation and that a dialectics of nature is therefore impossible. For to posit that nature embodies immanent dialectical laws falls into the trap of teleology or its antinomy, mechanism.

Yet, this is precisely what Engels tried to do. Beginning with his *Anti-Duhring* (1878), Engels proposed Marxism as a universal science whose propositions about the world held for society and nature. Dialectics became not so much a methodological principle for understanding history but were now conceived as laws of thought that reflected (approximately) the outer reality. Even though the concept of reflection was more complex for Engels than a mere "copy," the idea of correspondence between scientific knowledge and the "laws of motion" of matter and history were believed to be both objective and natural.

For Engels, the spontaneous development of natural science represented a vindication of the dialectical view of nature, even if natural scientists themselves were unaware of the contradictory character of observed phenomena. Engels's *Dialectics of Nature* may be understood as his most definitive statement of the view within the Marxian tradition that tried to eliminate the radical distinction between nature and society by establishing the universality of dialectical laws. The book was assembled from notes written during the last twenty years of Engels's life and was published posthumously. *Anti-Duhring* was published in 1878, during Engels's own lifetime and fully five years before Marx's own death. In fact, Marx himself wrote a chapter for the book so that it can be safely assumed that he not only was aware of the contents of the whole book, but approved of them.

In his Preface to the second edition, Engels states plainly his belief in the universality of the dialectic:

> It is, however, precisely the polar antagonisms put forward as irreconcilable and insoluble, the forcibly fixed lines of demarcation between classes, which have given modern theoretical natural science its restricted and metaphysical character. The recognition that these antagonisms and distinctions are in fact to be found in nature, but only with relative validity, and that on the other hand, their imagined rigidity and absoluteness have been introduced into nature by our minds—this recognition is the kernel of the dialectical conception of nature. It is possible to reach this standpoint because the accumulating fact of natural science compels us to do so . . . natural science has now advanced so far that it can no longer escape the dialectical synthesis.[3]

According to Engels, the role of philosophy is to help science to make conscious what its practice already reveals: that nature obeys dialectical laws and that these can be adduced from the actual discoveries of natural science. The first part of *Anti-Duhring* contains several chapters that attempt to extract from the findings of contemporary scientific knowledge the three laws of dialectics. If, later, Stalin was obliged to codify the law of the transformation of quantity into quality, the law of contradiction, and the negation of the negation as universal laws in nature and in history, he was merely taking his lead from the works of Engels. Following Engels, conventional Marxism has asserted the socially neutral character of scientific and technical knowledge. To be sure, science can be pressed into the service of the bourgeoisie, as in times of war when scientific discoveries are translated into techniques that advance the destructive powers of the new productive forces. Similarly, during periods of economic expansion, capitalists may, in their own interest, utilize the technologies available for the production of new commodities or the improvement of old ones. For example, the development of the chemical and electrical industries at the turn of the twentieth century was due, in no small measure, to capital's need for new investment outlets and new consumer markets. As Marx commented, the bourgeoisie constantly revolutionizes the means of production even as bourgeois social relations hold back their development. In sum, Marxists have never denied the subordination of science and technology to the scientific and technical revolutions that have occurred only under requirements of capital under specific historical conditions. In the latter instance, Marxist investigations in the history of science and technology have readily pointed out that scientific discoveries and technical inventions are not produced by great personages as much as they are the outcome of transformations in the material relations of production.

Samuel Lilley, writing from an orthodox perspective, states the case succinctly:

[Capitalism] provided a better economic mechanism for accumulating capital needed to use large scale machinery and similarly for supplying the wage labor to work it. Through its free competitive market, it could arrange for the commercial expansion that was necessary to make large scale production economic. Free competition forced less efficient producers out of business and thus put a premium on progressive and inventive methods. And, above all, capitalism radically altered the division between rulers and workers that had in the past held back the advances of technology.

A new flexibility came into economic and technological affairs. In the 18th century and much of the 19th, the ordinary workman with ingenuity and ambition could rise to become an industrial magnate. Capitalism did not abolish class divisions, it merely changed their nature. But these changes were such that for a time, most of the troubles which previously held back progress were abolished or were reduced to negligible proportions.[4]

Thus, according to Lilley, changes in social relations are the necessary conditions for advances in the degree of human mastery over natural forces manifested as progress in technology. But the sufficient conditions for explaining the configuration of science and production technology must be sought, according to this line of thought, in the realm of the internal progress of scientific ideas and material production. In other words, for Marxist orthodoxy, there is no such thing as "bourgeois" science or "bourgeois" technology as such. Science, in the orthodox tradition, has its own laws, and these laws are universally valid, considered historically. That is, the empiricist assumptions of English science can, to a great degree, be separated from the social conditions that produced the great discoveries of the sixteenth, seventeenth, and eighteenth centuries. The propositions of Newton's *Principia*, for example, may have been made possible by the advent of the bourgeois epoch, but their logic is simply the reflection of the external world, mediated by the stage of development of human knowledge.

In his effort to show that Marxism could be accorded the status of a science as much as a political creed, Lenin devoted an entire work, *Materialism and Empirio Criticism*, to a polemic against some positivist philosophers of the early twentieth century who, he believed, fell into the mire of subjective idealism. In Lenin's view, Marxism asserts the objectivity of the material world independent of human will and consciousness. In the process of producing their means of subsistence, humans discover the laws of motion of the material world which are true, relative to the stages of human knowledge of these laws. Not only is the natural world knowable, according to Lenin, but the social world obeys definite laws and can be discovered in the manner of natural science.

Except for his perennial qualification that scientific knowledge was relative to the material means in existence for the production of knowledge, Lenin's militant defense of scientific materialism is bereft of Marx's own dialectical understanding of the relationship between social being and social consciousness. Contrary to Lenin's interpretation, Marx posited that humans make themselves as well as remake the world. The mediations between both the natural and social world and human understanding are as complex in Marx as they are simple in Lenin. The reflection theory of knowledge—that is, the view that human consciousness is nothing other than a mirror view of the material world—was, for Marx, a reified view of the nature of reality and human comprehension of it. Marx held that practice not only mediated between consciousness and the external world but became a constituent of that world in reified form. Marx understood the development of modern industry, to some degree, as the objectification of the social domination of capital over the working class and society as a whole. Here, it is important to understand Lenin's philosophy not as an aberration from the general Weltanschauung of Marxism at the turn of the twentieth century, but as fully consistent with the commonly agreed upon interpretation of Marxist theory by the dominant leadership of the socialist movement after Marx's death until World War I. In the United States, a contemporary Marxist theoretician of science (and translator of *Capital*, volume 2), Ernest Untermann, held almost identical views to those of Lenin. Similarly, the positions of both Karl Kautsky and George Plekhanov were fully consistent with Lenin's "scientific materialism," and there is much evidence that he drew his inspiration directly from Plekhanov on these issues.

Until the recent past, there have been few tendencies dissenting from the dominant effort to endow Marxism with the status of a natural science, and natural science (most particularly, physics) with the mantle of universal truth. The first spectacular effort to argue that natural science was subject to the same mediations of class struggle that informed social theory was undertaken by the British Communist intellectual Christopher Caudwell. In his *Crisis in Physics*,[5] Caudwell attributed the conflict between wave and corpuscular theory in physics to the breakdown of capitalist society and culture. Written during the world economic crisis in the 1930s, Caudwell's book tried to demonstrate the congruence between the development of knowledge within bourgeois society and capitalism. To be sure, Caudwell argued that the early science of the capitalist epoch was "true" because it translated successfully into social and industrial practice, whereas late science is increasingly remote from production.

According to Caudwell, the breakdown of causality in nature generated by modern discoveries, such as Heisenberg's uncertainty principle, shattered the mechanistic physics of the earlier period. Since absolute determinism, or, to be more exact, scientific predeterminism, was an expression of the bourgeois worldview, the discoveries of modern physics represented the shattering of that view, a harbinger of the end of bourgeois hegemony. The incapacity of the physicists to integrate the new discoveries into coherent laws of nature reveals the persistence of the archaic worldview in the wake of experimental practice. Bourgeois science, for Caudwell, reflects the division of labor characteristic of bourgeois commodity production. The specialization of knowledge had become a barrier to physics' comprehension of the totality. On the other hand, the specialized knowledge generated by this division had yielded masses of new knowledge that simply burst the bounds of the old social relations. The crisis in physics consisted in the incapacity of the scientific community to arrive at a paradigm of scientific law that could be generally accepted by scientists themselves.

The second departure from Engels's version was Georg Lukács, who tried to solve the dilemma of his adherence to a strict interpretation of Marx's thought, according to which Marxism was a dialectical theory that was antiempiricist, and the identification of contemporary Marxism with the successes of natural science in transforming the material world through industrial practice. Lukács argued that dialectics, and therefore Marxism, was applicable only to the social world. Its attack against bourgeois social theory as ideology, i.e., false consciousness of the bourgeoisie was, for Lukács, evidently true. But Lukács insisted that Marxism had nothing to say about the natural world. In the human world, social practice transformed the nature of truth as well as the configuration of reality so that cognition, upon self-critical examination, could penetrate the essence of social development. For Lukács, history provided the key to social knowledge and yielded to the proletariat the capacity to grasp the totality of *social life*.

Lukács adhered to Marx's early statement that "the history of nature and the history of men are dependent on each other so long as men exist." The history of nature as such, according to Lukács, is of no concern to Marxism since Marxism deals specifically with the manner in which nature is transformed from a "thing-in-itself" into a thing for humans. The object world is always a reified form of the results of humanized nature, according to Lukács. What appears to the senses as an "external world independent of human consciousness" is nothing more than the thinglike form of social production, considered historically. Thus, the idea of a Marxist philosophy of nature is rejected. Instead, insofar as nature is considered as reified social relations made into an "object" of human knowledge by the exigencies of human practices within which are embedded the division of labor, social hierarchies, etc., historical knowledge provides the basis for understanding science as ideology:

But Engels' deepest misunderstanding consists in his belief that the behavior of industry and scientific experiment constitutes praxis in the dialectical, philosophical sense. In fact, scientific experiment is contemplation at its purest. The experimenter creates an artificial, abstract milieu in order to be able to *observe* undisturbed the untrammeled workings of the laws under examination, eliminating all irrational factors both of the subject and the object. He strives as far as possible to reduce the material substratum of his observation to a purely rational 'product,' to the 'intelligible matter' of mathematics. And when Engels speaks, in the context of industry, of the 'product' which is made to serve 'our purposes,' he seems to have forgotten for a moment the fundamental structure of capitalist society which he himself had once formulated so supremely well in his brilliant early essay. There, he had pointed out that capitalist society is based on 'a natural law that is founded on the unconsciousness of those involved in it.'[6]

As we shall see below, what remains undeveloped in Lukács becomes a fully elaborated theory of science and technology of the Frankfurt School. However, with these remarks, Lukács opens the door wide to a dialectical critique of science. His views are based not on the works of Marx and Engels but rather on Weber's theory of the tendency of industrial society toward technical rationality—making nature "intelligible" in terms determined by social arrangements that tend toward the domination of administration over all human relations. Lukács here understands scientific practice as a form of the administration of things, as social intervention rather than a process of discovery. Far from constituting a "praxis," a term which for Lukács always entails reflexive self-understanding and activity that understands the "other," or "object," as an aspect of the self, science denatures the objective and makes it something "for us," a *thing* distanced from social relations and certainly not derived from them. Here, Lukács comes closest to the view of nature as an artifice, constructed by social relations for human purposes—the purposes of domination.

The first essay of Lukács's *History and Class Consciousness* is an argument that "orthodox Marxism" consists in the rigorous application of the dialectical method to social theory. The first methodological principle of this argument is the concept of the *concrete totality*: the idea that all social phenomena are connected to each other and can only be understood historically. Lukács's starting point, in his effort to reconstruct Marxism on the basis of this principle, is drawn from Marx's *Poverty of Philosophy*, (1847), in which Marx makes explicit his view of the lawfulness of historical development: "the relations of production of every society form a whole." Lukács: "the objective forms of all social phenomena change constantly in the course of their ceaseless dialectical interactions with each other. The intelligibility of objects develops in proportion as we grasp their function in the totality in which they belong."[7]

Lukács portrays some tendencies of Marxism as having succumbed to the sin of fetishism when they attempt to apply the method of natural science to the study of society. When socialists, agreeing with the commonsense standpoint of bourgeois scientists, maintain the position that "the manner in which data immediately present themselves is an adequate foundation of scientific conceptualization and that the actual form of these data is the appropriate starting point for the formation of scientific concepts,"[8] they are applying the scientific method to the social fields which, in Lukács's opinion, may advance the progress of science but, "when applied to society, turns out to be an ideological weapon of the bourgeoisie."[9] Here, it can be observed that Lukács makes a radical departure from the late position of Engels and Marx that nature obeys dialectical laws, particularly the idea of the concrete totality and the laws of contradiction. He opposes this proposition, despite Marx and Engels's statements affirming the universality of the dialectic. But, for Lukács, there is no doubt that the dialectic can only be embedded in the structure of social reality. According to Lukács, reality is "a social process," not a series of discrete phenomena that can be observed and measured in the manner of natural science. It is the capitalist mode of production that "necessarily" produces the "fetishistic" forms of thought that obscure the inner core of capitalist social relations by asserting the essential reality of surface phenomena that appear as object relations rather than as social relations.

Lukács's reading of Marx has encountered much opposition since its publication in 1923. Particularly controversial is his idea that Marxism is essentially a social, not a universal, theory. Since Marx discovered that human activity is the unseen essence of the objective social worlds, it follows, according to Lukács, that the dialectic is meaningless apart from subject-object relations. Dialectical materialism describes the historical process by which the individual becomes able to transform his or her own conditions of life. Marxism is a class science and expresses the capacity of the proletariat to transcend its own particular existence and, by doing so, to transform society as a whole.

For Lukács, the process of the self-transformation of the proletariat represents a specific historical development in which the social conditions were created for a fully human praxis. Lukács's conception of praxis consists in the "emergence of consciousness as the decisive step which the historical process must take toward its proper end (an end constituted by the wills of men, but neither dependent on human whim, nor the product of human invention.)."[10] Lukács is no utopian insofar as he criticized the idea that humans could make their own history regardless of social limits. But he holds that, having brought to consciousness the social and changing character of their existence, humans could force "reality to strive toward thought," as Marx asserted. In other words, the interaction between the self-conscious acting subject and the social conditions produced by prior human activity constitutes the heart of the dialectic of history. Nature's laws were always true relative to the purposes of humans who discovered them. In his last work, the *Ontology*, Lukács makes a strict distinction between the concept of mechanical and dialectical causality. Lukács posits labor, which for him is always purposeful, as the central mediation between humans and nature. The transformation of the material world always presupposes the intentionality of collective human labor, so that causality is not separate from the process of production. Mechanical causality, on the other hand, depends for its validity on the notion of immanence of natural regularities and laws within an objective world that is not subject to the teleological determinations inherent in human action. Lukács suspends the question whether we can know the "real" laws of nature since our consciousness always mediates our relation to it:

> Teleology and causality are not mutually exclusive principles in the process, in the being and quality things, as had up to that time been deduced from every epistemological or logical analysis, but are principles which are certainly heterogeneous, which despite their contradictory characters, provide the ontological foundation of certain movements only when together, in inseparable dynamic co-existence only ontologically possible in the field of social existence.[11]

The *Ontology*, composed shortly before his death in 1971, is a vindication of the orientation of the earlier work. Nature's objective existence is never denied, but the question whether natural law can be separated from the process of production of use values is for Lukács a meaningless question for Marxism. In this later work, we can see all the old themes of the earlier position. Against the assertions of those Marxists who interpret Marx's call for the "abolition" of philosophy as a mandate to merge philosophy with science, Lukács insists that a Marxist ontology must be constructed because philosophy can only be abolished to the degree that society appropriates the natural and social world in accordance with its self-conscious telos.

Lukács wishes to retain the distance between Marxist philosophy and natural science because he opposes the reduction of the laws of the social world to natural-mechanical laws. The teleological projections of the higher levels of practice transform nature itself and do not merely obey its laws. Lukács labels as an idealistic fetish the effort to impute to socially created "laws of human practice" an independent existence instead of regarding them as mediated realities. Similarly, he launches an attack simultaneously upon the concept of economic determinism as an explanatory phenomena in the so-called superstructure and the idea of self-contained laws. Thus, even the "law" developed by Lukács regarding the presence in all objects of the results of teleological projects is not endowed with the status of universal truth regarding social development. Lukács offers a variant of Weber's concept of unintended consequences insofar as he attempts to establish the objective status of certain aspects of economic-social reality. That is, the economy, "despite its foundation in particular teleological projects of individuals, consists of spontaneously necessary casual chains, the phenomena which historically are always concretely necessary in them, may express the sharpest opposition between objective human progress, and therefore objective human progress—and its human consequences."[12] In this passage, we can observe Lukács's reluctance to substitute a new dogma for that which he rejects. Human beings are clearly responsible through their conscious acts for the social process as a whole, but some aspects of the process assume their own particular laws that are independent of the human will and may even determine it. Lukács discovers in the twin character of casuality a historical antagonism present in all epochs of history. Thus, humans live constantly in an ambiguous relationship with their own activity. On the one hand, they can never control the outcome of their intentional acts. The new circumstances created by the relation between humans and nature, as mediated through labor, obey their own laws and possess a dynamic that defies human determination.

Coming on the heels of the Bolshevik Revolution, which they saw as the triumph of self-consciousness which attempted to draw the object, society, closer to thought, the Marxism of Lukács and Karl Korsch may be said to represent the theoretical expression of the revolutionary era following World War I, just as Hegel's *Phenomenology of Spirit* represented the quintessential theoretical moment of bourgeois revolutions in the eighteenth and nineteenth centuries. In both cases, the blind forces of Second International Marxism appeared to give way to the triumph of reason. Whereas in Hegel, reason possesses a universal character not explicitly connected with a rising class, both Lukács and Korsch make clear their adherence to the proletariat as the embodiment of reason's historical development.

The weight of the postrevolutionary analysis of Lukács and Korsch was to emphasize the importance of consciousness as a critical aspect of the totality. Such was definitely not the case for the theorists of the Second International, whose leading spokesman was Karl Kautsky, the foremost theoretician of German social democracy. In his pamphlet, *The Socialist Republic*, Kautsky reflects the dominant Marxist formulation of the dialectic of historical development of the prerevolutionary era. It was also to serve as a majority statement following the ebb of the revolutionary tide, both for social democratic and communist thought in the 1920s. "Already in the forties," writes Kautsky, "Marx and Engels showed that, in the last analysis, the history of mankind is determined, not by the ideas of man, but by the economic development which proceeds irresistibly, obedient to certain underlying laws and not according to the whims and wishes of people."[13] Here, Kautsky, relying on the *Poverty of Philosophy* (1847), attempts to establish a direct, causal relationship between economic development and human action. This interpretation of Marx may be said to be the dominant paradigm of Marxist theory in the late nineteenth and early twentieth centuries.

Kautsky advances a conception of the meaning of economic development that ascribes to it the motor force of history. Economic development consists in the development of the productive forces. These productive forces, which, for Kautsky, almost always mean machinery, "developed in the lap of capitalist society, have become irreconcilable with the very system of property upon which it is built. Every further perfection in the powers of production increases the contradiction that exists between these and the present system of private property."[14] It is the task of socialism, according to Kautsky, to abolish labor as the central activity of human beings: "It is not the 'freedom of labor' but the freedom from labor such machinery will make possible in a socialist commonwealth that will bestow upon mankind freedom of life, freedom to engage in science and art, freedom to delight in the nobler pursuits."[15]

The revolutionary road, according to Kautsky, was being paved by capitalism itself. The development of modern industry within capitalist social relations constituted a fundamental antagonism to these relations. The profit motive was the spur behind the accumulation of capital in the form of highly centralized forces of production.

What Marx termed the "tendency of the rate of profit to fall" as the proportion of stored-up living labor increases in each unit of capital, became for Kautsky an iron law of capitalist development. The more the rate of surplus value fell, the more capitalists attempted to overcome the loss of the surplus by increasing the investment of capital in the form of machinery rather than living labor. In turn, this drive to preserve the mass of profit results in a falling rate of profit to the point where capital itself becomes a limit on its own further expansion. At a certain point, capitalist production no longer corresponds to the requirements that a certain rate of profit be maintained. The system comes, more or less, to a crashing halt. Huge masses of wage laborers are thrown out of work. Those remaining at work face sharp wage cuts, speedup, and other ways in which capitalists attempt to overcome the contradictions of the system. Then comes the era of social revolution.

This scenario was portrayed by Kautsky and other theoreticians of social democracy as true to Marx's own prognostications, especially in the third volume of *Capital*. To be sure, Marx himself never completed his crisis theory, but the implications of the law of the tendency for the rate of profit to fall seemed clear enough. The philosophical basis of the crisis theory advanced by Kautsky and others was the inexorability of "blind economic forces" of production, owing to the relatively independent development of natural science and its application to industry in capitalism's declining era. Thus, while economic necessity determined the use of science in industry, the advance of science itself engendered by capitalist social relations became the ultimate cause of the breakdown. History itself was propelled by the dialectic of human knowledge manifested in the domination of nature by humans. Having reached a certain historical stage in its capacity to subordinate nature to human needs, the collective powers of humankind were destined to overturn the private appropriation of these powers.

In this conception, the working class became an instrument of the unfolding of history's immanent laws. Even the activities of the working class, e.g., the formation of unions and political parties that contended in the legislative arena, were responses to the great social forces set in motion by capitalist social relations. But the real historical subject, in Kautsky's conception, was not the working class at all. It was the social theorists, now transformed into revolutionary intellectuals who had succeeded in mastering the science of society, who formed the real historical subject. Even though the proletariat was the *necessary condition* for social revolution, for most of the leaders of the Second International, it was the revolutionary party, armed with a true conception of the historical process, that was *sufficient* for the task of revolution-ary social transformation.

The proletariat was created by capitalism and would inevitably strug-gle against it because the social system would prove ultimately incapa-ble of meeting its needs. But the distinction between trade union and revolutionary consciousness, which became the celebrated Leninist contribution to the Marxist movement, was borrowed directly from Kautsky. Kautsky originally advanced the idea that socialism was always brought to the working class from outside, since the workers by their own efforts were incapable of generating their own theory although they were able to conduct militant combat against the oppression of capital.

It fell to Lukács, by his theory of reification, to provide an adequate theoretical basis for the Leninist conception of the party's role. The theory of reification proceeds from Marx's notion of the "fetish char-acter" of commodities in which social relations become relations between things because of the requirement of capitalist exchange that use values be divested of their qualitative aspect and be reduced to quantities capable of measurement. On the one hand, Lukács's idea of reification explains the ability of capital to dominate workers even in the wake of the breakdown of economic life. By introducing a category that shows that the everyday life of the marketplace and of material production mediates human perception, not only of nature but of soci-ety as well, Lukács offered a way of dealing with the counteracting ten-dencies to the inexorable laws that did not rely on automatic and objective economic categories. Lukács, often referred to derisively as a "cultural Marxist" who ignored the political economy of Marx, actually explained the concrete development of working-class consciousness by means of Marx's own category of fetishism of commodities. At the same time, however, he provided an epistemological theory of the Leninist conception of the party that grounded the role of the socialist intellectual in the actual process of production rather than in the empirical observation of the history of the working-class movement. That is, Lukács went beyond the collective experience of the working class's apparent inability to seize the opportunities presented by the capitalist crisis to argue for the centrality of the party in the class struggle.[16] He proceeded from dialectical thought, particularly the problem of the concrete totality. Capitalism generated a reified totality in the concrete form of the commodity that effectively mediated work-ing-class consciousness. Although capitalism's evolution has tended to "undermine the forms of reification . . . one might describe it as the cracking of the crust because of the inner emptiness"; on the other hand, "we find the quantitative increase of the forms of reification."[17] Thus, Lukács, while relying on the self-activity of the proletariat for its own liberation, held that the conditions of capitalist crisis do not provide the adequate basis for revolutionary activity. In the first place, according to Lukács, different sectors of the proletariat are not uniform in their degree of class consciousness. That is, sections of the working class remain under the "tutelage" of the bourgeoisie even in the wake of the powerful impact of the crisis upon their daily lives. Second, the proletariat can therefore engage only in "spontaneous mass actions" to alleviate the worst features of the crisis, if left to its own devices. For Lukács, the political character of the struggle cannot be deduced from blind economic laws, owing to the ideological hegemony of the bourgeoisie based on the persistence of commodity relations even within the crisis. Even in the midst of the crumbling of the economy, the proletariat, according to Lukács, "is still caught up in the old capitalist forms of thought and feeling."[18]

Neither the blind course of economic laws nor the spontaneous defensive response of the proletariat to the new conditions of exploitation and oppression constitute by any means a guarantee that workers will draw the correct conclusions from the new social environment. Thus, the party's role is not merely to provide a correct theory of the crisis or to conduct propaganda to spur the working class to mass action. As Rosa Luxemburg pointed out, and here Lukács agrees, the task of the party is to provide political leadership to the mass struggle, thereby transcending the purely economic struggle.

What for Lenin and Kautsky was merely a matter of belief based on observation of the limits of the workers' movement becomes for Lukács a theory—the working class by its own actions will transform society and abolish itself. But its consciousness cannot be inferred from the tendencies within the capitalist system toward breakdown. Lukács admonished those who transformed the "law of the tendency" to remember that Marx established only the *tendencies* of capitalism toward crisis, not its inevitability. Moreover, even the crisis does not generate mass revolutionary political activities without the party's leadership of the mass struggle. If the Second International failed to develop a theory of revolutionary organization, this omission reflected its belief in the ultimate power of the productive forces to transform society, or the ablility of the working class to lead the struggle for liberation. On the other side, Lenin and Lukács remained skeptical that those productive forces could be anything more than one important social determinant. But their articulation of the theory of political organization may be said to replace the "productive forces" as the sufficient condition for revolutionary transformation. Here, the party is conceived as congealed class consciousness of the proletariat which is not conferred with the status of preeminence in the revolutionary process.

The antinomies inherent in Lenin's thought can be ascribed to his adherence to the Marxism of the Second International on all theoretical questions, except those bearing on potential politics. And George Plekhanov, whom Lenin called the "father of Russian Marxism," was the chief transmitter of the theories of the Second International to the Russian Marxists.

Plekhanov invokes Marx to refute the view that teleological projects determined the actual development of the productive forces, even if these, in turn, influenced the course of social change.[19] Plekhanov denies the interactionist views of those who attempt to establish a dialectical relation between human activity and these productive forces. He invokes two arguments to refute the idea that the productive forces are in any way the outcome of a certain stage in the development of consciousness. The first, borrowed from Darwin and originated by Engels in his essay "The Part Played by Labor in the Transition from Ape to Man," argues that a certain development of physical capacities, particularly the perfection of the thumb, was a presupposition of the development of the intellect. The second, an argument much less intrinsic to the Marxist tradition, is the dependence of the development of the productive forces on certain geographical conditions, particularly the availability of certain materials, the proximity to means of communication, and travel and climatic conditions. These, according to Plekhanov, are the material determinants of the more or less rapid development of the productive forces and, in turn, the character of social relations considered as a reflex of the stage of humans' conquest of nature. Plekhanov takes literally Marx's statement in his famous *Preface to the Contribution to the Critique of Political Economy* that the superstructure of society, ideas, laws, and institutions arise on the economic foundation that is determined by the stage of the productive forces' development.

Perhaps the clearest statement of Plekhanov's Marxism is to be found in his description of "scientific socialism" in his important *Socialism and the Political Struggle*.[20] Once again, the achievement of Marx and Engels is compared, in the sphere of social theory, with the "amazingly simple yet strictly scientific theory of the origin of the species" advanced by Darwin. According to Plekhanov, the great principle of the theory of social organization is the struggle between the productive forces and the backward, "social conditions of production." The historical teaching of Marx and Engels is the genuine "algebra of revolution," in Plekhanov's view, and the scientific-mathematical analogy is meant as much more than a metaphor. Plekhanov believes that the founders of "scientific" socialism had discovered the inner laws of society in a strictly empirical manner, in concert with the methods of natural science.

In his lecture commemorating the sixtieth anniversary of Hegel's death in 1891, Plekhanov defines the dialectics of Marx in these terms:

> Modern dialectical materialism has made clear to itself
> incomparably better than idealism the truth that people make
> history unconsciously; from its standpoint, the course of
> history is determined in the final account not by man's will, but
> by the development of the material productive forces.[21]

It is not the absolute spirit that determines the development of society, in Plekhanov's view, but society that determines the spiritual life of humans.

What is Plekhanov's conception of dialectics? "It is a *theory* of cognition and, besides, a theory of a definite type, a materialist theory of cognition."[22] Plekhanov, following Marx and Engels, inverts Hegel's dialectics, but not merely into a materialist theory but a theory of knowledge, or, more exactly, a theory concerning the cognitive process by which knowledge is obtained. Plekhanov accepts the discursive ground of English empiricism and Kantian philosophy that there exists a problem of knowledge. We know the world through experience, says Plekhanov, and the world exists independent of our will. The object of our experience of nature is "matter," which, however, is only known gradually:

> There is not and cannot be any other knowledge of the object
> than that obtained by means of the impressions it makes on us.
> Therefore, if I recognized that matter is known to us only
> through the sensations which it arouses in us, this in no way
> implies that I regard matter as something unknown or
> "unknowable." On the contrary, it means, first, that matter is
> knowable and, secondly, that it has become known to man in
> the measure that he has succeeded in getting to know its
> properties through impressions received from it during the
> lengthy process of zoological and historical existence.[23]

In his polemic with the neo-Kantians, particularly Ernst Mach and the Russian Bolshevik A. Bogdanov (who had become a follower of Mach's "empiricomonism"), Plekhanov writes voluminously about the relation of the thing-in-itself and thinking. His conclusion, that thinking corresponds to things that exist *outside* experience, is, for Plekhanov, verified by science and technology. Materialism is the philosophy of science.

Yet, Plekhanov cannot avoid dispelling accusations from the neo-Kantians that he holds a naive copy theory of knowledge:

> The forms and relations of things-in-themselves cannot be what they *seem* to us, i.e., as they appear to us as 'translated' in our minds. Our representations of the forms and relations of things are no more than *hieroglyphics*; the latter designate exactly these forms and relations and this is enough for us to study how the things-in-themselves affect us, and in our turn, to exert influence upon them. I repeat: if no correct correspondence existed between objective relations and their subjective representations ('Translations') in our minds, *our very existence would become impossible.*[24]

The problem posed by the terms "translation" and "hieroglyphics" to describe representation cannot, for Plekhanov, be resolved by reason. These terms connote noncorrespondence but also render an account of perception of *appearances* contemplative. Plekhanov's solution to the problem of knowledge is to be found in Engels's adaptation of the English phrase "the proof of the pudding is in the eating," that is, the questions posed by the view that sensations can only reveal the object as appearance are resolved "by industry" and experimental science. In contemporary terms, our knowledge of the external world can only be verified by specific practices: "anticipating a given phenomenon means anticipating the effect that a thing-in-itself will have on our consciousness. It may now be asked whether we can anticipate certain phenomena. The answer is: of course we can. This is guaranteed by our science and technology."[25]

In his *Essays on the History of Materialism*, Plekhanov enters a detailed exposition of the dialectics of nature, once more echoing Engels but, since *Dialectics of Nature* was not yet known, the source is Hegel's *Science of Logic*, which Plekhanov summarizes in relation to scientific ideas.

Lenin and many leaders of the Russian revolutionary movement in the first decades of the twentieth century learned their Marxism through the teachings of Plekhanov and Kautsky. By the outbreak of World War I, Lenin was to have occasion to take sharp issue with both of his teachers on questions of the strategy and tactics of revolution. But it was not until much later, perhaps not until he read Hegel during exile in the years immediately following the outbreak of World War I, that Lenin's philosophical position began to mature.[26] Lenin reads Hegel epistemologically. In the first place, he reads Hegel from the perspective of *dialectical materialism*, that is, with Plekhanov and Engels in mind. The laws of dialectics are identical with the laws of social development and nature. Or, put another way, nature and society obey dialectical laws conceived in evolutionary terms, that is, as a process of development. Lenin interprets Hegel in terms of the problem of knowledge. How, for example, can we know anything? Where Hegel speaks of the forms of thought, he is concerned with "how they correspond to the truth in themselves." Lenin understands this phrase as a symptom of Hegel's preoccupation with the relation of thought forms to their idea. He wants to substitute the results of natural science for the internal agreement of thought with its own development. That is, he transforms Hegel's historical discourse into a problem for the theory of knowledge:

> Consequently, Hegel is much more profound than Kant and others, in tracing the reflection of the movement of the objective world in the movement of notions. Just as the simple form of value, the individual act of exchange of one given commodity for another, already includes in an undeveloped form all the main contradictions of capitalism . . . so the simplest generalization, the first and simplest formulation of notions . . . already denotes man's ever deeper cognition of the objective world. Here is where one should look for the true meaning, significance, and role of Hegel's logic.[27]

In other words, Lenin's judgment of Hegel's superiority to Kant is a means to return to the epistemological issues raised in the earlier *Materialism and Empirio-Criticism*, issues which were deeply implicated in the problem of how do we know, what do we know, when do we know, etc., in short, the Kantian questions. That Lenin answers these questions in dialectical materialist terms does not bring him closer to Hegel, who wishes to abolish these questions, or to Marx, for whom history (and the critique of political economy) would replace epistemology. True to Plekhanov's conception, Lenin contrasts materialist dialectics with Hegel's account of mechanical and chemical science:

> The laws of the external world, of nature, which are divided into mechanical and chemical, are the bases of man's purposive

activity. In his practical activity, man is confronted with the objective world, is dependent on it, and determines his activity by it.[28]

Lenin goes on to argue that there are two forms of the *objective* process: "nature (mechanical and chemical) and the purposive activity of man . . . the mutual relation of these forms . . . at the beginning, man's ends appear foreign ('other') in relation to nature. Human consciousness, science, reflects the essence, the substance of nature, but at the same time this consciousness is something external in relation to nature (not immediately simply coinciding with it)."[29]

These passages are commentaries on Hegel's claim that science is dependent on a telos, even though it appears to "stand opposed" to nature. Hegel's point is to show that the duality of humans and the external world is only illusory, but both can be subsumed by the category of Notion. Lenin inverts this claim by, on the one hand, placing science in the context of natural history, providing an evolutionary materialist way of resolving the dualism and then reasserting the duality of thought and its object. On the other hand, Lenin asserts that "in actual fact, man's ends are engendered by the objective world and presuppose it—they find it as something given, present."[30]

Having once more asserted the materialist outlook, Lenin returns to the dialectic and reformulates what, for Hegel and Marx, are historical questions, including the history of "nature" as an epistemological issue: "The splitting of a single whole and the *cognition* of its contradictory parts is the *essence* of dialectics. . . . The correctness of this aspect of dialectics must be tested by the history of science" [my emphasis].[31] There follows Lenin's attempt to equate the "laws" of dialectics with some axioms of mathematical and physical, chemical and social science. The result is to equate a dialectics of nature with a dialectics of social life as a continuum of forms of self-movement valid for "all phenomena and processes of nature, including mind and society."[32]

Lenin's philosophical thought, even in this late amendment to the earlier *Materialism and Empirio-Criticism*, is still firmly rooted in Engels's Hegelian objectivism, even as he praises the sources of dialectics in philosophical idealism. And, it is from philosophical idealism that he is to rediscover the problem of knowledge, yet he holds with Plekhanov that the question raised by the distinction between things and their representations must be answered by the "history of science," just as science must verify the hypothesis of nature as a dialectical unity of self-movement. Therefore, processes of cognition must be dialectical in order to attain cognitive "approximations" of natural processes.

Despite his growing respect for Hegel, born in the process of reading and making extensive comments on the *Science of Logic*, Lenin reinterprets him in accordance with materialist and objectivist criteria. "Causality, as usually understood by us, is only a small part of universal interconnection, but a materialist extension, a particle not of the subjective, but of the objectively real interconnection."[33] There is no concept here of Marx's admonition that the old materialism has failed to grasp things *subjectively*, or of Lukács's insistence that dialectics implies the relation of subject and object and that their identity results from the historical process, from human praxis. In Lenin's later reflections on science and philosophy, we still see a dialectics of nature and history considered as forms of matter, or, more precisely, of "universal connection" of all things. In the last analysis, for Engels, Lenin, and Plekhanov, the problem of knowledge is resolved ontologically by positing the equivalence of thinking and being whose substratum is matter. Nature and society appear, in this conception, as instances of the same dialectical evolution of matter in motion whose highest expression is thought, understood here as a material object.

Of course, the identity of thinking and being, Plekhanov's formulation, can apply only to scientific thought. The problem arises: how to explain the proliferation of incorrect ideas, that is, ideas that do not correspond to the external world, that are nonidentical with it? The Marxist answer is to refer to the class struggle as mediation between thought and its object, a mediation suffused with interests. This conception of ideology is still grounded in an objectivist account of history, in which bourgeois and other forms of ruling-class domination distort understanding on behalf of interests that are rooted in the social infrastructure. In the social world, the task of science is to separate knowledge from those interests that distort thinking. Of course, for Marxism, knowledge cannot really be separated from interests. But the orthodox tradition, of which, in this regard, Marx himself is a part, holds out only two sources of truth: the proletariat, for which truth is an *interest*; and natural science, which throughout its history has been obliged to struggle against mysticism in order to establish the legitimacy of its findings. If, as Lenin says, the history of science confirms both dialectical and materialist propositions, it may be said to correspond "objectively" to the proletarian interest. We might reformulate one task of orthodox Marxism as the forging of an alliance between science, representing the new forces of production, and the working class, representing the new productive relations. This might explain the considerable literature, in turn-of-the-century Marxism, on the relation of science to materialism and dialectics, as well as its struggle against neo-Kantians for whom truth was always relative (although not, as Marxists claimed, entirely subjective).

Stalinism took the argument a step further in three areas: psychoanalysis, genetics, and linguistics. As an inheritor of some of the traditions of Marxist orthodoxy, it mounted an attack on psychoanalysis on the grounds that it was an ideology, rather than a science, in the service of the capitalist class in its epoch of decline. The attack was twofold: on the one hand, psychoanalytic practice was criticized for its effort to adjust the individual to the bourgeois world rather than trying to transform it. The critique of psychoanalysis as a form of conformist ideology was undertaken with special vigor during the 1930s and the early years after World War II, when all but collective solutions to individual psychic pain were eschewed as a diversion from the revolutionary project. Freudian practice was viewed as a kind of mental repair shop for the petty bourgeoisie. On the other hand, the theory of the unconscious was labeled a mythological, idealistic effort to deny the material base of all human faculties. The most philistine of the Stalinist arguments against the unconscious was the charge that no empirical evidence was available for its existence and that all Freudian "proofs" relied, in the last instance, upon the inferences derived from the interpretation of dreams and everyday behavior by means of categories that were no more than arbitrary value judgments. Thus, the Stalinist critique invoked traditional scientific methodologies to prove its bourgeois character.[34]

But it was not sufficient to dismiss psychoanalysis altogether. In this connection, it is instructive to compare the views of Karl Popper with those of Soviet science during the Stalin era. Popper makes a similar critique of psychoanalysis. Psychoanalysis has much to say "of considerable importance and may well play its part one day in a psychological science which is testable."[35] But it is not science precisely because its "clinical observations" are not refutable. As we shall see in chapter 9, Popper holds that no discourse may be considered a science if it cannot be subjected to repeatable, falsifiable tests, regardless of the value of its theories. Psychoanalysis is guilty of a "dogmatic attitude" in comparison with genuine science, which "is ready to modify its tenets, which admits and demands tests."[36] This is the critical attitude and is always characterized by a "weaker belief," whereas dogmatists hold strong beliefs.

An alternative had to be found that, at once, embodied a materialist theory of psychological phenomena, one that agreed with the views of Engels and Lenin, had to view mental questions as an extension of physiological and environmental determinations, and, in addition, could be subjected to "tests" requiring none of the mythic metaphors that permeate Freudian psychoanalysis. The solution was provided by I. P. Pavlov, who, though not a product of the revolution, found his theories compatible with the philosophical orientation of the Soviet leadership.

Pavlov reduced all psychic phenomena to physiological causes. The theory of conditioned reflexes preserved the Stalinist attack against psychoanalysis by attributing all neurotic and psychotic behavior to lesions of the brain and the nervous system, which could be identified in a purely empirical manner. These lesions manifested themselves as the temporary destruction of cells, which could be restored by means of sleep therapy and other paramedical techniques. Soviet psychology was merged with physical medicine, and the roots of psychic disturbances were traced to the influence of the environment upon the organism.[37]

According to the theory of environmental causes of mental illness, the only alternative to the organismic approach to treatment was to change the environment in which the individual lived. Patients exhibiting symptoms of emotional disturbance could not be treated by finding the source of the problem within a nonexistent unconscious, according to Pavlov.

Pavlovian psychology located the emotions in various portions of the brain: the cerebellum (the sense organs), the cerebral cortex (responsible for the "higher" or intellectual functions), and the thalamic region (which was the repository of memory and corresponded, roughly, to the unconscious). Although there is no independent psychic *structure* in Pavlovian psychology, the brain is the repository of psychic *functions*. For the Marxists, conditioned reflex therapy and its scientific basis removed mysticism from psychology and made it a natural science. Pavlov's contribution to Soviet science was that he placed psychology on an experimental basis and provided an explanation of mental illness that linked somatic and genetic features of behavior. Accordingly, the interaction of humans with their natural and social environment was held responsible for some features of mental disorders, whereas purely physiological impairments accounted for others. In any case, *experience* was significantly linked to physical pathologies, and changes in experience could remove such problems.

My concern here is not to enter into a detailed discussion of Pavlov's science, except to elucidate its relation to Marxism. Loren Graham argues that "from the standpoint of the history and philosophy of science, the greatest significance of Pavlov derives from his success in bringing psychological activity within the realm of phenomena to be studied and explained by the normal methods of natural science."[38] To have explained "man's psychic activity on a physiological foundation"[39] does sharply contrast with the "introspective approach" of turn-of-the-century investigators into psychic phenomena, as Graham notes. The "revolution" carried on by Pavlov and his school was embraced by Russian materialists in much the same way as Engels was embraced by Plekhanov and Lenin. As Graham shows, the identification of materialism and socialism was deeply rooted in Russian nineteenth-century history when czarist censors opposed the attempts of Pavlov's predecessor, Ivan Sechenov, to found psychology as a natural science. In this context, Soviet Marxism's fierce attack on Kantianism becomes much more than a scientific or philosophical dispute: it has profound political significance.

The Soviet rejection of Freud and Lukács takes on a new coloration, for these theorists, each in his own way, could be viewed as counter-revolutionary insofar as they challenge the application of "normal methods of natural science" to history and, especially, to the study of psychic activity. The case against Freud is less clear-cut than that which the Communists mounted against Lukács, for several reasons. Lukács proposes to violate the cardinal rule of Marxist orthodoxy emanating from Engels, a rule which views the human sciences as an extension of natural science since nature and history are linked by the dialectical character of matter. If the human sciences obey "laws" not commensurable with those of natural science, because the interaction of subject and object constitutes the social world, then the "normal" methods of science are useless or, worse, yield results that do not correspond to the criteria of critical science but are "an ideological weapon of the bourgeoisie." This claim is intolerable from the point of view of a Marxism that wishes to achieve the status of "objective, natural science." Here, it must be noted that Graham's suberb study of Soviet science is, on the whole, friendly, precisely because he (correctly) grasps the history of that Soviet science as the story of its tendency to normalize science, as opposed to the ideological and critical aspects of Marxism that inhibit such developments. Movements within Marxism such as critical theory, phenomenology, and more recent French developments such as post-structuralism (but not structuralism), can be viewed as impediments to normalizing science. On the other hand, the tendencies within Soviet philosophy and science after Stalin's death to draw closer to systems theory, cybernetics, and other sciences that owe much to Kantian themes and their development by Wittgenstein may have been caused by the attempt of Stalin's successors to break with his dogmatism. Even more, it may have been caused by a tendency to be drawn toward that side of the Marxist tradition that privileges experimental science and mathematics as the "guarantee" of the truth of its claims.

As is well known, Freud begins as a medical scientist, a neurologist whose initial commitment is to the normal methods of science. Intentions notwithstanding, Freud discovered categories of psychic activity that could only be "tested" symptomatically, through interpretation. As his *Interpretation of Dreams* amply displays, theories of psychoanalysis require a different conception than those of experimental science in order to establish itself. In this respect, Paul Ricoeur's attempt to link psychoanalysis with phenomenology, particularly its hermeneutic methodology, is entirely justified. On the one hand, the dream work, because of the resistance of the unconscious, is always encoded, virtually a semiotic field full of traps and dead ends.[40] One "reads" the dreams in terms of the psychoanalytic categories that are presumed to enjoy the status of transcendent judgments. And, whether these specific categories are accepted or not, any investigator is obliged to rely on some theoretical paradigm of the unconscious as the ground of interpretation unless the dream work is dismissed as an unreliable informant concerning psychological activity, literally regarded as "poetry" in the way orthodox empirical science regards all unmeasurable information.

However, if Freud's paradigm of human action is accepted, one would need to accept its description of the psychic structure as axiomatic for the system as a whole. Further, the requirement that experiments be repeatable was mediated by Freud's insistence that he work with human subjects, a practice that foreshadowed a second difficulty in regarding his work within the rules of natural science: in a nonbehaviorist model, the person and his or her multiplicity of relations are unique. Even if one can posit a common psychic structure, the individual's family, experiences and mode of life are subject to wide variations.[41] Although the relation of these "environmental" influences to the psyche varies with every individual, Freud insists that his work is scientific because he is able to establish categories that are trans-individual. The foundation of psychoanalytic theory is the posited psychic structure, the ineluctibility of sexuality as the key determinant of emotional life, and so on.

These categories are constituted as axioms without which any discourse claiming the status of science cannot function. Freud's theory satisfies another criterion imputed to genuine science: reduction of all observed phenomena to determinate relations that can be deduced from its axiomatic structure. In fact, Popper's claim that psychoanalysis is infected with the dogmatic attitude is simply not true. In his later years, Freud felt constrained to amend his metatheory to make room for a new drive or instinct, that of death. This represented a fundamental alteration of the dyadic axiom that the struggle between the pleasure principle and the reality principle constituted the dynamic of human development. To this was added thanatos, or death, which he now saw as the dominant principle of human existence. Where, in the old theory, human aggressiveness was explained as the consequence of obliged repression of the pleasure principle in the service of civilized activity (work being the most important), now aggressivity was proposed as directly linked to the death instinct.

Yet, Popper is right to claim that psychoanalysis is incommensurable with the normal experimental and quantitative practices of natural science. Its method corresponds more closely to hermeneutics, a discipline that presupposes the object is constituted intersubjectively, or, in Freud's own version, that human agents are constructed by the interaction of biologically given characteristics and the mode of life, which are mediated by other human agents such as parents, the work imperative, and so on.

The attempt of Stalinism to refute the findings of classical genetics provided the needed "objective" science from which to stage the bases of Stalinist scientific critique. The Soviet geneticist Trofim Lysenko aimed his fire at the theory of the relative autonomy of the human genetic structure and, in consequence, offered a revived version of the theory of the inheritance of acquired characteristics. The Lysenko theory was consistent with the views of Pavlov that environmental influences penetrated the living tissue of organisms. But Lysenko argued that, once having been penetrated, the new characteristics were carried over into future generations. Thus, the "new" society of the Soviet Union not only would produce the "new man" in the poetic sense, but would produce a physically superior human being. Of course, the experimental basis of the Lysenko claims were established more modestly with varieties of wheat rather than persons. But the implications were hardly lost on Soviet ideologists, who were quick to point out that, if true, the Lysenko experiments had provided a strictly scientific-empirical proof that the previously discredited theories of Lamarck were more consistent with Marxist environmentalist ideology. Mendel-Morganism was nothing more than the application of metaphysics to the realm of science, because it posited an immutable gene subject only to contingency, i.e., mutations. If Lysenko's claim to have bred certain qualities into plants intergenerationally was proven accurate, the whole basis of modern biology, according to which genes function relatively independent of the external environment and change only through mutation, would be refuted.

The findings purported to have resulted from Lysenko's work were published widely by the Soviets and produced a furor both among "bourgeois" scientists and among those who considered their work as a part of the Marxist tradition. Biologists such as J. B. S. Haldane refused to declare themselves in favor of the new science, despite their strong adherence to dialectical materialist philosophy. Others within the Soviet Union were ordered to adhere to the new doctrine or forfeit their scientific and material privileges. However, it would be a mistake to attribute the demise of Lysenkoism either to Stalin's death or to the counterfactual arguments made by its detractors. What must be explained is why Soviet science and, more pertinently, the Communist leadership of the Soviet Union became vulnerable to such arguments. In a word, how did Western criteria of what counts as scientific evidence and the truth of scientific propositions become normatively accepted in Soviet science? We shall return to this in chapter 8.

As a Marxist linguist, N. Y. Marr hypothesized that language was derivative of social relations and changed both form and substance with changes in class structure and the development of the productive forces. Stalin argued that language belonged neither to the base nor to the superstructure but transcended them.[42] In effect, his argument paralleled that of the structuralists, who attempted to assign to language and its forms an invariant structure within human capacity that resists historical changes except by mutation. In other words, Stalin seemed to endorse the notion that language does not reflect changes within human society but rather obeys a separate set of "natural" laws.

It is interesting to compare Stalin's occasionally class-based conception of science with the Leninist view of technology. Although Lenin was aware that such capitalist production techniques as the assembly line were degrading to labor, he appeared to view them as the inevitable price the working class was obliged to pay in order to achieve industrialization. The Bolshevik leadership in Russia were fully prepared to adopt Western production technologies.

Marx and Engels were themselves well aware of the social costs the working class had to pay for scientific and industrial progress, and there was no question of adopting the position that in the transition

from capitalism to socialism, it would be possible to transform the productive process overnight. On the basis of this reasoning, Lenin and
the majority of Bolshevik leaders rejected the program of the so-called
workers opposition (1921), which asserted that workers' control within
the factories be immediately instituted and that, since factory managers represented the worst features of the old mode of production,
workers should institute self-management instead. The attack against
the Left within the Bolshevik party was conducted on the specific political objection that it was impossible to implement self-management
under conditions of capitalist military and economic encirclement.
According to Lenin, such a rapid transition to new production relations
would risk collapse of the whole revolution. The members of the workers' opposition met this objection by contending that self-management
would permit wider support for the revolution among workers,
whereas a policy that rescinded the promise of workers' power would
inevitably result in defection from the revolution.[43]

Although the Soviet Communist party leadership adopted material
incentives, rationalization of the production process along Western
capitalist lines, and canons of labor discipline resembling other industrial countries, these were justified as the inevitable and ineluctable
features of technology. Unfortunately, the Soviet leadership has contended that the crisis is permanent, as long as capitalism survives. Stalinism was also ensconced within the canons of orthodoxy with respect
to the achievements of science and technology in the bourgeois
epoch. Its criticism of certain aspects of science was confined to the
contradictions that appeared in the late capitalist period: psychoanalysis, relativity and quantum mechanics, and genetics were by no
means accomplishments of the golden age of capitalist development
and could be assigned to the ideological realm rather than to science.
They were all developments that occurred at the turn of the twentieth
century, when the bourgeoisie, according to conventional Marxist
analysis, had already begun its steep descent. The results of modern
scientific discovery could be disputed precisely because they were
generated under conditions that some Marxists said were no longer
hospitable to the forward march of the productive forces. According to
Caudwell and others, the metaphysical character of modern science
was due almost entirely to the exhaustion of the ruling-class morality
and the beginning of the disintegration of its cultural hegemony.

Caudwell traces the development of modern science to Newton's
achievement, especially his "atomistic scheme [which] gives a basis for
deleting God from the Universe as a causal influence once it is
treated."[44] Newton still had recourse to God by separating "nature"

from reality, which took on a supersensible aura. The crisis begins when experimental physics admits that natural phenomena are unknowable and Einstein recognizes that these unknowabilities are in fact important principles of knowledge about nature. The theory of relativity "states that absolute length, mass, energy, space, time and motion do not exist." Einstein makes these limits "real descriptions of the nature of matter in relation to us."[45]

The old physics portrayed nature in the image of bourgeois society—as a self-motivated machine that was subordinate to the bourgeoisie, as a piece of property. But this worldview was "shattered," Caudwell argues, when Einstein and others placed human activity back into nature and turned from the idea of "a self-sufficient nature." Newtonian physics reproduced in its laws the bourgeois marketplace—atomistic, self-motivated, and blind. But it was a stable world that was pictured in these laws because, argues Caudwell, bourgeois society was still in its prerevolutionary phase marked by a compromise with the aristocracy.

Having said all this, Caudwell stoutly maintains the objectivity of physical knowledge:

> Newtonian physics is not a reflection of bourgeois society; if it were, it would not be knowledge about reality, and its practical success indicates its real content of positive knowledge. Physics is necessarily the science of the most objective components of phenomena. It is the most generalized and formal aspect of matter.[46]

At the same time, physics possesses a "special distortion" owing to its close affiliation with bourgeois society—its abstraction of quantity from quality, whose other side comprises the developments in biology and geology.[47]

According to Caudwell, Einstein's solution, positing the space-time continuum within the context of the process of knowing, simply inverted the error of absolute space and time. "Einstein's world is contradictory because it is still bourgeois and mechanistic—it is not a world of complete relativity."[48]

Caudwell argues from the position of the dialectic between continuity and discontinuity, but Einstein established the relativity principle merely to reestablish continuity. For Caudwell, Einstein's theory was the last possible solution to the multiple contradictions generated by the bourgeois worldview. Only dialectics' unity of opposites, transformation of quantity into quality, etc., can resolve the crisis brought about by the shattering of the old physics. A new physics would have to understand the world as the mutually contradictory and determining relations of subject and object, which are the basis of a dynamic process. Not only is the world in mechanical motion, it is in a process of flux in which its stable moments are temporary moments in a larger picture of discontinuity between past and future. Further, consonant with the Marxist tradition, Caudwell takes the point of view of the totality that posits the interconnectedness of everything as well as its difference.

In terms of the history of the Marxist philosophy of science, Caudwell demarcates his own views from that of Marxist epistemological objectivism associated with Lenin and Engels. In Caudwell's view, modern physics having abandoned its "mechanistic interpretations . . . has not as yet, however, found any substitute for the categories which its own research has revolutionized. In abandoning, therefore, the categories of mechanism, it attempts to use the categories of subjectivism, both bourgeois"; or, as with Heisenberg and Dirac, it attempts "to do without categories altogether."[49]

Caudwell holds out no hope that physics can resolve its agonizing contradictions until social relations are transformed. Therefore, his dialectical program is a weapon of criticism rather than an alternative. Yet, Caudwell accepts an epistemology that understands the development of science to correspond, albeit indirectly, with movements in the political economy of bourgeois society; and in this sense, however objectively representative of human knowledge as such, science is at the same time class science. He anticipates the work of Kuhn by arguing for the discontinuity of scientific discovery based on the appearance of anomalies that cannot be resolved within the Newtonian paradigm. However, unlike Kuhn, Caudwell argues that these "contradictions" make their appearance in consequence of both the changes in the course of experimental science and the transformations of the economic and political relations that produce a social world racked by uncertainty and disorder. "Worldviews" mediate the relation between changes in the social world and science, but the chain of causality begins with the class struggle.

I want to argue that Caudwell's account of contemporary physics marks a break in the Marxist philosophy of science. Published in 1939, two years after Caudwell was killed in Spain but at the height of Stalin's political and ideological influence on the world Left (especially its Communist parties, of which Cauldwell was a member), *The Crisis in Physics* is a remarkable document. In many ways, Caudwell's position parallels that of Lukács, e.g., the dialectic as a category of the relation of humans to nature, and the refusal to reify nature as external "object." Further, his account of the crisis in physics begins with a critique of the mechanical world picture that emerges from seventeenth-century science, and it tries to understand the whole development of modern science in terms of efforts to resolve contradictions between this worldview and experimental discoveries that render old theories obsolete without altering their fundamental character. Finally, Caudwell traces the crisis to the vicissitudes of the commodity form and the relations of production, especially what Engels described as its most telling feature, human loss of control over social relationships. Thus, indeterminacy and discontinuity appear to the bourgeois not as a creative process that gives rise to a new synthesis "at a higher level," but as chaos. Freedom, for the bourgeois, is "freedom from necessity" in proportion as large-scale industry and giant machinery cut humans off from nature. Caudwell's description purports not only to explain the emergence of the crisis, but to show the social constitution of scientific knowledge that is to link "external" and "internal" accounts in a single discourse.

To be sure, his efforts are marked by considerable reductionism. The bourgeois and proletarian worldviews are described more as antinomies than as mutually determined opposites. Further, given Caudwell's analysis, it is difficult to sustain the sense in which theoretical physics remains "objective" truth rather than ideology. Nevertheless, his is the first Marxist work to posit both the *relative* autonomy of scientific knowledge from its social relations, and the determination, in the last instance, of physical laws by the social context of their production.

CHAPTER 5
THE FRANKFURT SCHOOL:
Science and Technology as Ideology

Despite the political gulf that separated socialists after the Bolshevik Revolution, the leading tendency within Marxism concerning science and technology was deeply influenced by the position of Engels as articulated by Plekhanov and Lenin. Even those anti-Leninists, like Kautsky, adhered to a relatively uncritical view of science and technology, critical only of the uses to which they were put by capital. However, the preponderance of Marxist theorists greeted the "age of science" with virtual silence, which signified that they accepted the scientific and technological "revolution" as a progressive aspect of industrialization. Science and technology became part of the "givens" of the social world, requiring no further analysis. For most Marxist intellectuals in the twentieth century, Marxism was identical to historical materialism; the dialectic of nature stemming from Engels and the late Marx either was greeted with embarrassment or was explicitly denied.[1]

From the perspective of historical materialism, science and the scientific method became a cultural ideal worthy of emulation. This worship of natural science and strictly empirical methods of social investigation was, of course, always mediated by a "dialectics" which, in principle at least, was strongly anti-positivist. But Marx's injunction that the propositions of historical materialism could be verified by strictly empirical methods was taken seriously by most of his followers, as was the maxim that Marxism signified the end of metaphysics, speculative philosophy, and, in some quarters, philosophy itself.[2]

Yet, as the twentieth century wore on, it became increasingly difficult for some theorists to ignore the long shadow that science and
technology cast on all discourse; beyond industrial practice, it was
becoming clear that the scientific/technological worldview had
become a distinctive ideology and had penetrated culture to its foundations. The idea that science was an aspect of social life that could be
placed alongside others was questioned on several grounds: the
development of communications technology (telephone, radio, and
then television) raised sharply the question of the massification of culture and, concomitantly, the decline of working-class culture and other
autonomous popular forms. On this basis, a significant tendency in
Marxist theory, associated with the Frankfurt School, challenged the
possibility of maintaining a political intellectual opposition that could
imagine an alternative to rampant technologies.

Of course, the Frankfurt School, writing about mass culture from
the Left, was by no means the only quarter from which criticism emanated. José Ortega y Gasset condemned mass society from a deeply
conservative perspective. For Ortega, there was no question of culture
emanating from the "masses." On the contrary, he argued that mass
democracy was the enemy of culture.[3] Such emanations were echoed
in the Critical Theory of the Frankfurt School, whose attacks against the
emerging technological society were often accompanied by a passionate defense of high bourgeois art and appeared to confirm what conservative cultural critics had known all along—that the pursuit of equality had unintended consequences in the decline of taste and civilized
sensibility.

However, it was not until the period after World War II that a broad
account of science and technology becomes important in the Marxist
debate. Perhaps the core of new interest in what constitutes a science
among Marxist intellectuals in the West resulted from the conjuncture
of the Western capitalist economic miracle in the 1950s and 1960s,
which seemed to postpone the revolution in advanced capitalist societies and the deepening economic, political, and ideological crisis of
Soviet and Eastern European societies after Stalin's death in 1953. The
sharp contrast between relative prosperity and political stability in
much of the West and the upheavals in the really existing socialist
world was disconfirming for a historical materialism that had predicted
postwar upsurge in those countries that had been victimized by fascism and betrayed by liberals. Instead, with the notable exception of
countries such as Greece, Yugoslavia, and Albania, whose antifascist
liberation movements were led by communists, the preponderant
countries of Europe, especially in the West, fell into line of the deals

made by Roosevelt and Stalin at Tehran and Yalta which divided Europe into spheres of influence. For the Left, until the mid-1960s, the postwar era was one of defeat and marginalization, the exceptions being the socialists in Scandinavia.

These political and economic reversals triggered a crisis in Marxism, especially in France and Italy where communist and Left-socialist participation in the resistance had produced a powerful left-wing challenge to the legitimacy of post-war governments of the Right and center. While it would be excessive to trace a direct relationship between the political climate and the return to a kind of 'normalcy,' the dissatisfaction of two generations of Marxist intellectuals with these arrangements fed into an unprecedented outpouring of philosophic and literary ruminations on the status of Marxism as a science and on the entire question of historical agency. For the postwar period was marked by what Marcuse termed the "integration" of the working class, its trade unions, and Labor and Socialist parties into a new corporative relationship with capital.[4] That militant strikes, bitterly contested elections, and the maintenance of the anticapitalist ethos on the Left accompanied this new political environment does not detract from the validity of this characterization. In retrospect, the period of insurgency on the Left (1965–72), although in some respects a serious repudiation of the character of Left accommodation in the entire postwar period, must be seen as a lengthy *episode* in an otherwise general crisis of Left and Marxist theory and practice.

When the period is viewed in this way, it is not unexpected that those who responded by providing a discourse on science and technology engaged these issues with the intent of contributing to the renovation of Marxism from a position to the left of Stalinist orthodoxy. Needless to say, the range of areas covered in the entire debate—the critique of the status of the working class as historical agent, questions of economic theory, the history of socialism and the labor movement, and the cultural contradictions of capitalism—extends beyond the discourse on science. But the significance of this particular debate is that it addresses the metatheoretical level—the question whether Marxism is science or "ideology." For, in the process of exploring the nature of science as a form of knowledge as well as that which distinguishes it from nonscience, the claims that dialectical and historical materialism constitute the most general sciences of nature and society become the underlying problematic of Marxist philosophy. Even when the object of inquiry is the social function or the purely "methodological" distinctions between science and nonscience, the implicit issue must be the question of Marxism itself. As we shall discover, most writers in this metaphilosophical debate understand that this is so.

One of the most important problems for a critical theory (critical in the general sense) is the ground from which that critique can proceed. In contrast to almost all versions of Marxist orthodoxy, postwar Marxisms felt obliged to draw elements of their respective axiomatic structures from other traditions, even though these had been condemned by the leading Second and Third International theorists. Many of these graftings were consonant with national intellectual cultural traditions. Since the postwar resuscitation of Gramsci's *Prison Notebooks* by Italian Communist leader Palmiero Togliatti, Italian Marxism was marked by a discourse on ideology and political hegemony, but these categories were mediated by the peculiarly Crocean reading of Hegel that dominated Italian philosophy at the turn of the twentieth century. In France, the Communist party and its theoretical and cultural apparatus stood firmly beside the Stalinist appropriation of Hegel, mediated by the work of Engels and Lenin. Although Gramsci could not be overturned, postwar Marxists concerned with responding to the breakdown of the moral order of their respective national communisms, especially after 1956, could not easily break from the Communist party because it still enjoyed substantial working-class support and had successfully claimed the inheritance of the radical side of the democratic revolutions in their respective countries. Consequently, the debate on science takes on the character of the debate between Kant and Hegel, which becomes a code for the struggle between freedom and determinism. Everywhere, except in Eastern Europe, Kant and the practices of natural science signify the liberation of Marxism from the grip of "world-historical" Hegelian predestination that coincidentally culminates in the primacy of the Soviet Union and its ideology over all possible Marxisms. The affirmation of science, and Marxism as a science, becomes a declaration of anti-Stalinist intellectual war. If Marxism only can articulate its own metatheoretical principles with the worldview of modern theoretical physics, and, for some, Freudian psychoanalysis, it can regain its place as the science of social relations.[5]

For Sartre—who may be, in this reprise of the dominating tendencies of postwar Marxism, a kind of anti-Christ because he retains a deep respect for the speculative Kantianism of German phenomenology—this means, nevertheless, the need for a non-Freudian psychology. Others, however, identify the non-Communist Sartre with Stalinism, or, to be more precise, with speculative reason that has reduced Marxism to an ideology. Louis Althusser, Etienne Balibar, and Jacques Rancière in France, Galvano Della Volpe, Lucio Colletti, and Sebastian

Timpanaro in Italy, become the manifestation within Marxism of the tendency in all Western philosophy to "return to Kant" and, by doing so, to repair the breach between empirical science and philosophy that was created by Hegel.

Thus, it is not surprising that these Marxists should have returned to Lenin and Engels, notwithstanding their substantial reservations concerning the generality of dialectics claimed by earlier theorists. Rather, they are attracted by Engels's *will to scientificity*,[6] his insistence upon a Marxist discourse about science and technology that had been ignored by orthodox Marxism during the interwar period, when science and technology had been discussed only in terms of their practical economic applications, specifically their key role in the processes of socialist construction or capitalist breakdown. But the theoretical debate seems to have ended with the death of Lenin, or, more precisely, with Stalin's consolidation of his hegemony over the communist movement, which, by the mid-1930s, had become official Marxism for all intents and purposes. From the early 1930s until the late 1950s, Marxism appeared a closed universe, hermetically sealed by an epigony that excluded as much as it included in the heritage.

Marxist structuralism is constituted as a dialogue with "humanism," the doctrine according to which "man" stands at the pinnacle of evolution and has powers and prerogatives superior to those, not only of the animal and plant spheres, but also of "lower" forms of the human species, namely, those civilizations that had not entered the Western ideological and technological orbit. It was not difficult for structuralist anthropology to deconstruct humanism and to show it as the servant of colonialism, particularly at the time of the French intervention into Vietnam in the 1950s and the more politically explosive Algerian war. Thus, structuralism presents itself as an alternative to the essentialism inherent in evolutionary thinking, its imperialistic political implications, and its racist social ideology. Claude Lévi-Strauss, in his essay "The Science of the Concrete,"[7] leads the attack on evolutionary humanism by striking at the claim of science to be a unique discourse that presupposes the development from historically lower to higher social and mental forms. He shows that the so-called higher functions of what we call science—classification, observation, and experiment— are already present among so-called savages. Ethnographic evidence abounds in this respect, and Lévi-Strauss tacitly infers that the sharp break between science and magic is false, that their relationship is one of continuity as well as discontinuity. To be sure, science and myth are not to be confused. Both are systems of knowledge, possess theories of causality, and so on. But the distinction between them rests on two points. Science is systematic not only with respect to its ends, a feature shared by magic, but also with respect to its means. Here, Lévi-Strauss introduces his famous metaphor of the magician as a *bricoleur*, roughly translated as a handy worker whose tools are collected haphazardly from the surrounding environment, whereas the tools of science are as one with its general system. Thus, the technology of science is literally *scientific* whereas the tools of magic are identical with the refuse of everyday life. The second distinction, however, is the opposite: despite its informal methods, magic postulates a complete and all-embracing determinism. Science, on the other hand, is based on a distinction between levels, only some of which admit forms of determinism whereas, on others, the same forms of determinism are held not to apply. Having said this, Lévi-Strauss is quick to add that the passion for ordering, classification, and inferential thought is common to science and art. Magic may "hit" on the right explanation for particular phenomena but, at a certain point, lacks the anticipation, effect, and ability to use the evidence of the senses to correct false conceptions on the basis of what is known. Thus, it is science's systematic open-endedness that marks it off from magic.

This theme, the ability of science to revise theory in the light of evidence and therefore separate itself from the determinisms of "ideology," and structuralism's attack against evolutionary essentialism form the two legs of the structuralist attack on Marxist humanism. This debate dominates postwar intellectual circles in France and Italy and, as we shall see, has its reverberations in the German tradition as well.

In this chapter, I shall explore the critique of science and technology emanating from both the Frankfurt School and its successors, not so much because they are chronologically prior to structuralism but because it is precisely against the attempts of Horkheimer, Adorno, and Marcuse to criticize science as a form of domination that the new structuralist defense of science is directed. As we shall see, what remains objectionable about the Frankfurt critique of science, from the point of view of Althusser and Colletti, is its implied judgment of science, including Marxism, as ideology. Beyond this remains the question whether we can offer secure, reliable foundations for knowledge. For while Horkheimer, for example, rejects the return to metaphysics and while Adorno's attack against postwar phenomenology as a "jargon of authenticity" is infused with a Marxological presentation of antiessentialism, there nevertheless is a genuine nostalgia for philosophy in the old sense, that is, for the possibility that speculation can yield more than just that—the Frankfurt equivalent of philosophy of science that merely exposes the methodological problems in empirical or historical inquiry. What lurks beneath the Frankfurt School's critique of science is a yearning for the integrated civilization that Lukács and Ernst Bloch posit as a basis for the critique of the present. But this time it is not evoked as a heuristic device alone, but as a genuine alternative perspective against which the modern world must be judged. As I shall argue, this alternative remains incomplete, is not a real alternative, and so what inevitably recurs is the Frankfurt School's reconciliation with technology and the discourse of high modernism with which it is connected. The unfinished character of the alternative may also be connected to another conundrum of Critical Theory: the fear that on the other side of the critique lies unreason, the space of the religious attack against evolution, of the antiscience of fascism itself, whose return to mythology the Frankfurt School (along with Wilhelm Reich) exposed. Following this discussion, I shall explore the contributions of Sohn Rethel, who linked the Frankfurt critique of the scientific enlightenment with Lukács's reading of the significance of the commodity form for understanding the character of knowledge.

Despite its considerable debt to Marx, Critical Theory stood outside the orthodox tradition to make a fundamental critique of modern science and technology. Unlike Caudwell, the Frankfurt School did not critique science as a symptom of the decline of bourgeois social relations, but in terms of early bourgeois social development and especially its ideological presuppositions. Horkheimer, Adorno, Marcuse, and others associated with the Frankfurt Institute for Social Research, the leading center of Critical Theory, did not content themselves with a critique of reification as the camera obscura of the commodity form, or with linking developments in modern science to historically specific conditions such as the market. Instead, they made an immanent critique of modern science and technology in the capitalist epoch, beginning with the foundation of the scientific method itself. They attempted to show that the worldview of the bourgeoisie had its concomitant in the method of investigation as well as in the results of modern science and technology, which were judged a "historically congealed renunciation" of human pleasure.

The crucial works on science and technology of the Frankfurt Institute—*Dialectic of the Enlightenment* by Horkheimer and Adorno, *Eclipse of Reason* by Horkheimer, and *One Dimensional Man*[8] by Marcuse—were inspired, in part, by the apparent success of Western capitalism not only to survive the ravages of World War II but to emerge stronger than ever, at least within its own sphere. Second, Horkheimer, Adorno, and Marcuse witnessed the collapse of the working-class movement in the wake of the onslaught of fascism in the 1930s only to discover that the postwar era promised no more in relation to the conditions for revolutionary change. The working class seemed safely integrated into the capitalist order, its consciousness befogged by the battering of mass culture and its material demands met by capitalism's ability to "deliver the goods" owing to its tremendous technical and economic resources. The disappearance of a historical agent capable of transforming society was the concomitant of the growth of totalitarian forms of capitalist rule. In the 1940s, the Frankfurt theorists undertook a study of the historical development of the ideology and culture of capitalism in order to explain the emergence of these new features of capitalist power. In undertaking the study, Critical Theory made a radical break from one of the sacred canons of orthodox Marxism. Instead of only celebrating capitalism as a stage of social development that advanced humans' mastery over nature on the way to liberation, Critical Theory also discovered the roots of the crisis of contemporary humanity in this domination itself. The drive toward "mastery" over nature, according to Critical Theory, was prompted by the human fear of nature, particularly its capacity for destruction, that is, to enslave human beings to its vicissitudes. People created myths in order to vitiate the destructive capacity of nature by explaining and, in spirit, conquering it, a goal that the existing level of technology could not achieve.

The function of natural law in primitive societies was to create order out of chaos, but primitive science was no more than a speculation since it reflected a very tenuous hold on the natural world. The achievement of the Enlightenment was the demythologizing, or, in Weber's terms, the disenchantment, of the natural world by generating scientific theory and technologies that represented a qualitative leap in human domination of nature. The means by which this mastery was accomplished was the quantification of all natural phenomena, and the rationalization of nature through the instrumentalization of reason. In other words, mastery became domination when reason lost its distance from industrial technology. In essence, technical rationality meant the subordination of all natural problems to questions of social control. The machine became the chief mediation between human beings and nature by extending their productive powers to transform. "What men want to learn from nature is how to use it in order to wholly dominate it and other men . . . however, the only kind of thinking that is sufficiently hard to shatter myths is ultimately self-destructive."[9] Critical Theory found the sufficient condition for the domination of nature in the capacity of humans to gain domination over themselves. The development of scientific reason was predicated on the repression of the pleasure instinct and the separation of humans from nature so that nature could become pure "otherness."

Instead of regarding science and technology and its rules as one mode of thinking among others, the Enlightenment and its aftermath imposed the rigors of scientific thought upon the whole of society. Instrumental reason became the only possible mode of thinking, and all other modes were relegated to the realm of myth. In the wake of the triumph of science on the basis of the criteria of utility and its capacity to subordinate nature to human will, it became reified. "The more the machinery of thought subjects existence to itself, the more blind its resignation in reproducing existence."[10]

The key concepts of technological rationality were: subjection of individuals to the productive apparatus, the elevation of the productive apparatus to the status of the source of values and authority, and the collapsing of all distance between thought and technical means (Marcuse's concept of one-dimensionality).

The pervasive logic of technical reason—that is, the idea of domination as the motive force of human activity, the stripping of the world of all qualities so that it can become the object of mathematical calculation and industrial rationalization, and the subordination of all decisions to the criterion of efficiency conceived in terms of productivity—was inevitably extended to the social world. In Marcuse's words:

> The technological a priori is a political a priori inasmuch as the transformation of nature involves that of man and inasmuch as man-made creations issue from and re-enter a social ensemble. One may still insist that the machinery of the technological universe is "as such" indifferent toward political ends—it can revolutionize or retard society. . . . However, when the technique becomes the universal of material production, it circumscribes an entire culture: it projects a historical totality—a "world."[11]

So comprehensive was the victory of science and technology in the bourgeois epoch that machine technology became a part of nature. The logic of industrial production, that is, the division of labor organized hierarchically and by means of coercion, was now an unquestioned aspect of the natural order. It was no longer regarded as a human product that had evolved historically under certain social conditions. It became the only possible way of dealing with the external world. "The scientist knows things insofar as he can manipulate them."[12] The criterion of all became whether a given idea or technique worked in the supreme task of dominating nature in order to meet human needs.

Critical Theory rejected the reification of human domination over nature and attempted to demythologize science and technology. Against the prevalent view that in its essence "thinking is an automatic self-activating process . . . an impersonation of the machine,"[13] Critical Theory argued that mechanistic thinking created the machine, not the other way around, and that the invention of machines was an objectification of the split within society and within human beings themselves. It appears that machines have established their priority over humans so as to subordinate the human will to the requirements of the productive apparatus. Critical Theory argues ambivalently for distancing thought from its object to show that the unity of thought and object under capitalism serves specific ends: the domination of human by human through the mediation of the domination of nature. "Even the deductive forms of science reflect hierarchy and control. Just as the first categories represented the organized tribe and its power over the individual, so the logical order, dependency, connection, progression and unity of concepts is grounded in the corresponding condition of society—that is, of the division of labor."[14]

In this perspective, the so-called forces of production are the objectification of domination within human society as well as the measure of the degree of dominance over nature. The forms of science and technology are mediated by social divisions, and "the loss of memory is the transcendental condition for science." A sense of history is inimical to the project of domination because it would generate questions that cannot be answered instrumentally. Since technological society presents itself as a closed system, indeed the only possible way of seeing, "history is bunk," in the words of Henry Ford. Even if the renunciation of the self is the precondition for the domination of nature and becomes problematic under circumstances where machine technology has the capacity to produce sufficient quantities of goods to satisfy elementary human physical needs, the system demands its own perpetuation and the suppression of the historical rationale for coercion and hierarchy. This is the reason for the suppression of the memory and the end of transcendence under late capitalism.

The members of the Frankfurt Institute ultimately despaired about the chances of reversing the established society because they became convinced that the historical process by which capitalism achieved its victory over feudalism was simultaneously a process by which it transformed the genetic makeup of humans.[15] For capitalism is not merely a system of production according to rules that privilege production for profit. The private appropriation of the means of production had become a secondary feature of the system as it developed in the nineteenth and twentieth centuries. The subsumption of science under capital meant that technology became the new a priori of all social relations, not so much the commodity form as Lukács had argued. Technology provided the means by which needs were satisfied. However, rather than producing new needs that could overtake the capacity of either productive relations or the productive forces to satisfy, these needs were ensconced in the system of social relations which were, themselves, determined by technology. Just as the commodity penetrated all corners of the world in the nineteenth century (Lukács), the new condition for social life is *administration*, a totality of relations that subjects all interaction to the mediation of instrumental rationality; everything becomes means, or, to be more precise, means and ends are entirely conflated.[16] In the process, the individual, often touted as the crowning achievement of the bourgeois epoch, suffers apparently irreversible decline. Individuality is forced to submit to the imperatives dictated by industrialized social and economic administration. "The individual no longer has a personal history," writes Horkheimer. "Though everything changes, nothing moves. He needs neither a Zeno nor Cocteau, neither an Eleatic dialectician nor Parisian surrealist, to tell the Queen in *Through the Looking Glass* when she says, 'It takes all the running you can do to stay in the same place.' "[17]

Horkheimer acknowledges that the individual "despite everything does not entirely disappear in the new impersonal institutions" but reminds us that "his life seems to fit any questionnaire he is asked to fill out" and that "his intellectual existence is exhausted in the public opinion polls."[18] Moreover, "every instrumentality of mass culture serves to reinforce the social pressures upon individuality, precluding all possibility that the individual will somehow preserve himself in the atomized machinery of modern society."[19] Thus, the cultural apparatus parallels the economic and political apparatuses to produce mass society, one in which its participants can no longer conceive alternatives to the present social order, much less devise an effective politics to oppose it. The absence of alternatives leaves no recourse to those who would oppose the prevailing setup. Individuality cannot express itself in the positive act of participation in a public sphere of social and political discourse, because what passes for public activity is really determined by the apparatus, the machinery of conformity. What's left is the assertion of sheer individuality—refusal. But, as Paul Willis found when he studied a British comprehensive high school in the 1970s, even refusal results in defining, not individuality, but merely the terms for the induction of rebels into the industrial order. Willis shows that working-class kids get working-class jobs by refusing the curriculum, which leaves them with no alternative but to perform unskilled or semi-skilled factory labor.[20] In short, refusal may become a form of deprivation when it results in illiteracy. Those who will not imbibe what counts as socially approved knowledge are consigned to definite and subordinate places in the social order, places which entail performing labor that denies individuality.

The critical analysis of the Frankfurt School raises sharply the question of scientific neutrality. Its flat conclusion that all science and techniques are mediated by social relations did not mean that the Critical Theorists were prepared to advocate a return to preindustrial society. On the contrary, their theories were rooted in the liberatory potential of modern technology, notwithstanding its destructiveness to human life and spirit. Like Marx, they accepted the inevitability of the capacity of modern industry to relieve humans of backbreaking labor. Their vision of the humane use of technology found its apogee in Marcuse's projection of the automatic factory that would require people to labor only a few hours a day or week, leaving the remainder of their time for creative play, the arts, crafts, human communication, and other activities considered "leisure" under contemporary repressive industrial conditions.[21] The real "work" would be that which is now considered trivial by our culture, and necessary labor under highly automated technologies would be reduced to the absolute minimum.

Thus, Critical Theory stops short of a truly comprehensive critique of technology and, consequently, of industrial society. Its fatalism about the chances of a reversal of technical rationality's control over both science and society is revealed by its belief that humans must ultimately submit to the machine because, after all, progress in alleviating the most onerous aspects of human existence is attributable to the capacity of science to provide the tools for the domination of nature by humans. Critical Theory advocates an end to the hegemonic status of instrumental reason over social life. Recognizing that critical reason has no power within the realm of the productive apparatus, including the working class which seems to have been overcome with awe by that apparatus and its achievements, the Frankfurt Institute's theorists solved the dilemma by a compromise with the machine. They generated a vision that offers a compromise to technical rationality, guaranteeing it a role in future society but one sharply circumscribed by self-conscious social control.

The outcome of the Frankfurt critique of science and technology is an unwitting accommodation to the presuppositions that have informed both Marxist and bourgeois social theory—the inevitability of self-controlled renewal of industrial society. The polemic against hierarchy and domination, against coercion and a division of labor that produces a reified object world that appears to control human action and consciousness from without, turns into despair. Critical Theory reifies the reified by asserting its irreversibility.

Marx celebrates the advances of science and technology even as he ruthlessly exposes their repressive aspects within capitalist social relations. In this sense, both the Critical Theorists and the orthodox Marxists truly represent the master and find justification in his writings. Marcuse testifies eloquently to the end of critical theory:

> Here technological rationality stripped of its exploitative features, is the sole standard and guide in planning and developing the available resources for all. Self-determination in the production and distribution of vital goods and services would be wasteful. The job is a technical one and as a truly technical job it makes for the reduction of physical and mental toil. In this realm, centralized control is rational if it establishes the preconditions for meaningful self-determination. The latter becomes effective in its own realm—in the decisions which involve the production and distribution of economic surplus, and in the individual existence.[22]

Thus, Marcuse deals with the power of technological rationality, especially its historical hegemony over human activity, by assigning it a special role in future socialist society. It is to be accorded reign in the sphere of production of necessities, whereas self-determination will reign in distribution and that portion of production designated as "economic surplus." From this, we can assume that the power of technology as developed within bourgeois social relations cannot be radically transformed at its core, only at its periphery.

I shall take up these questions concerning the social organization of work in the next volume of this work. The core of Marcuse's theory of science is his statement that "the science of nature develops under the technological a priori which projects nature as instrumentality, stuff of control and organization."[23] At the same time, he acknowledges that "pure science is not applied science; it retains its identity and validity apart from its utilization."[24] But his point is exactly the reverse. Approvingly, he cites the following remark by von Weizsacker: "And what is matter? In atomic physics, matter is defined by its possible reactions to human experiments and by mathematical—that is, intellectual—laws it obeys. We are defining matter as a possible object of human manipulation." Marcuse concludes, that "if this is the case, then science itself has become technological."[25] Here, Marcuse redefines the "neutrality of science" by showing that this understanding of the science of nature is possible only if we grant the presuppositions of technological rationality, which are to universalize the domination of nature by placing it within the realm of universal, rather than specific, interests. If we do not grant the validity of this claim, the entire apparatus of scientific truth collapses. That is, technological rationality operates in the interest of domination, and science is an aspect of it (even if its language suggests autonomy). Marcuse argues that by suspending metaphysical questions such as "What is?" in favor of the functional "How?" the science of nature actually loses this autonomy from the rationality that increasingly propels social life.

If this judgment is right, human sciences that wish to emulate the science of nature also wish to become part of the same rationally organized administrative universe that has integrated natural science. Of course, this is exactly what contemporary social science, including its Marxist variant, wants. One way of effecting this outcome is to emulate the methods of natural science, to transform all experience into information, as Walter Benjamin ruefully observes.

Experience cum information becomes raw material when it is processed into a form appropriate for scientific analysis. Processed information must be purely quantitative in order to be commensurable. When data has been produced (no longer "raw," that is, in preanalytic form), it can be compared according to certain criteria of "scientific" method. Social science works with averages, means, medians, and frequencies—the standard statistical categories. The individual loses identity except insofar as her/his experience may be rationalized in relation to these methods and the purposes for which a study has been made. Information that does not fit within the algorithm of a given study cannot be "used" and often disappears. Recall von Weisacker's complaint that the "matter" of experience is defined as an object insofar as it can be manipulated.

The starting point of social scientific manipulation is the same as that of physics: framing investigation in terms of human purposes and determining whether data can be fitted into an a priori methodological scheme. Thus, Critical Theory insists that no hypothesis is innocent of the telos of design as well as of method. Quantitative methods permit measurement by means of averaging and assume what Durkheim contended was the starting point of all social science—that these averages are representations of the social, which must be treated as an irreducible fact. The particular datum must be treated as an instance ($+$ or $-$) of the averaged data. Although Durkheim's conception of the social was never merely the aggregation of individual instances, the sum of datums, much of contemporary social science accepts this definition of the social.

Data are collected from human subjects by means of surveys, the contents of which should permit quantification. This entails as many determinate replies as possible, leaving little or no room for ambiguities or contradictory answers. This requirement dictates the nature of the questionnaire. It is not an open-ended system. Therefore, much of social science research is similar to marketing or voter polling surveys. The results are likely to be useful, say, to a political candidate seeking to reduce public issues to voter preferences or to test whether her/his own positions prove acceptable to a given constituency. Similarly, a corporation seeking to sell a new product can test a sample of potential users to determine whether the product should be placed on the market.

Of course, the typical survey sheet is not limited to product preferences. The designer of the survey wants to know the sex, race, ethnicity, and economic status of the target voter or consumer. In this respect, social science is needed to evaluate the relative weight of each of these factors for explaining the result. Scientific polling must advance a hypothesis on whether a given candidate or candidate's program will "fly" with a rationalized constituency. Mathematical methods are devised to implement this hypothetical segmentation. And, as Engels said, "the proof of the pudding is in the eating." Post-tests are developed to determine whether votes or purchases or a given candidate/product correspond to the "profile" of the client indicated in the survey design. "Scientific" social research organizes itself around the principle of falsification: the survey should be a predictor of human behavior, given the framework within which such behavior must take place—public life in which democratic participation is bounded by voting and buying. Implicit in survey research is the assumption that attitudes translate into action; action is defined as behavior that responds to stimulae.

Like modern physics, quantitative social research is an *intervention* conditioned by the purposes for which the research is conducted. Increasingly, such research is "sponsored," not only by political parties and/or their candidates, or by those who wish to sell something to consumers, but also by the state. The scale of research is often linked to its necessary collective methods of work, which typically involve the efforts of many persons. Research funds are awarded by public and private sources for the purpose of assisting those in power to design social, military, and other aspects of public policy. With few exceptions, the social scientist becomes an instrument of public policy even when he/she proudly refuses to hire out as a "gun" for soap companies or political consulting firms. The history of quantitative social research is, of course, intimately tied to both its commercial uses and its public policy uses, and is relatively recent. Methods of survey research were developed concurrently with three simultaneous developments in American economic and political life: (1) the rise of the national market for consumer goods, which engendered and was accelerated by advertising, but also marketing technology, (2) the emergence of the "human relations school" in labor and industrial policy, which, incidentally, coincided with the growth of large corporations as the typical business institution in economic life; and (3) the proliferation of mass communications as an instrument of both commercial and political propaganda. The development and dissemination of electronic media as the field for market and political action changed the face of politics and culture. For the messages transmitted by radio and television sold not only "products," but a way of life as well.

The discovery by social scientists of their value to this new world has been determined by developments within the social sciences and new currents in social policy, chiefly the rise of the welfare state in the 1930s and 1940s as well as the broad changes described above. Like Marxism, social science has been treated by the sciences of nature as imprecise, at best, and in the most pejorative characterizations, as a form of astrology, since, as we have shown, science has been virtually identical with mathematicization even more than observation. Because much empirical social investigation remained qualitative throughout the first half of the twentieth century, its standing as science has always been in doubt. But the advent of "mass society" brought with it a new respect for statistics and other quantitative methods in social sciences. Developments in social theory stemming from Max Weber conjoined with the will of the social sciences to respectability. Recall that Weber characterizes industrial societies by, among other things, the development of a new system of rationality, principally what he calls *instrumental rationality* in which all forms of action tend to be oriented toward limited ends such as profit and the priorities of public and corporate policy. Weber says that social relationships ranging from "friendship" to the "state" tend to become rational in this sense, whether or not by mutual consent.

Later writers coined the term "mass society" which extends the scope of instrumental rational relationships in proportion as other types of relationships weaken and even vanish. The "individual" becomes a problematic concept because the nexus of our relations is bound by the limits imposed by instrumental action. I am not claiming that Weber's theory gave social science "permission" to treat individual experience within the norms of rational calculation, but that it did provide an empirical theoretical underpinning for aggregating social action.

A second element of mass society theory is, of course, the corollary assumption of Weber's idea of capitalist rationality, the theory of bureaucracy. Modern bureaucracy tends to dominate the state, civil society ("business relations"), as well as voluntary associations such as trade unions, fraternal organizations, and social movements—in short, the whole of society. Bureaucratization becomes the norm of all forms of social organization. In this context, bureaucracy is merely the social form of instrumental rationality. Yet, its norms govern more than organizations: the rules of personal behavior merge with those of public action, and the concept of the individual as an autonomous being becomes problematic. Further, the *community* as an autonomous cultural as much as a geographical site of social action is increasingly weakened by the various strands of massification, especially the centralized state and the equally centralized institutions of mass communications. The possibility of private life, installed by the bourgeois revolutions of the eighteenth century as both political program and putative social norm, is challenged.

The power of this thesis is not in the least affected by the movement, in our own times, away from conscious public life, so evident among all sections of the North American and Western European populations. That the tendency to privatization has become stronger is a corollary of consumer society, the widespread perception that modern politics entails media and other modes of manipulation and that all social problems are subject to purely technical solutions—at least this is the prevailing ideology. The question is whether private life is possible. Even if an individual chooses to withdraw in all but the most minimal activities from public participation, not only in elections but in social institutions that make up the notion of "community," the weight of technological domination, much of which is centralist in orientation and application, prevents such a choice. In these societies in which late capitalist social relations prevail, an ersatz public sphere is created by television; information about missing persons adorns milk cartons; and what counts as "news" inevitably focuses on personalities. Most subversive to the idea of the "private," a welter of products appear on the market that address the symptoms of what C. Wright Mills calls "private troubles"—headaches, insomnia, emotional and physical stress, for instance—while the mass media explore the substance of these troubles, i.e., divorce, sexually communicable diseases, child abuse, incest, and so on. Unemployment is investigated as a private trouble as experience is simulated in the "news." In short, we have the conflation of private and public, their mutual determination even as the illusion of the autonomous individual remains the dominant American ideology.

Of course, Weber regarded these phenomena as *tendencies* that inhere in processes of industrialization and modernization. Writing a half-century later, Marcuse saw massification and instrumental rationality as empirical descriptions of the prevailing situation. His concept of one-dimensionality merely encapsulates these twin forms that seem to have overtaken the indeterminate elements of capitalist social relations. But the implication of Marcuse's theory is curiously resonant with mainstream sociological will to scientificity. Even though his perspective remains critical throughout, he accepts the premise upon which contemporary social science strove for scientific legitimacy—the decline of the individual and the idea that aggregation was a correct methodological procedure in the light of a society mediated by mass technology.

It may be argued that the other social sciences—psychology, history, sociology, and political science—were merely imitating the already mature innovations of scientific economics that emanated from the Vienna school of Menger and Bohm-Bawerk, the English neoclassical economics of Marshall, Jevons, and Keynes which had, by the late nineteenth century, transformed the categories of political economy into a positive science, replacing eighteenth century "economic philosophy." This transformation was based, paradoxically, on the assumption that economic behavior in a market economy was a matter of rational choice in the interest of maximizing individual advantage. Later, Mancur Olson was to codify these assumptions as a "logic" governing collective as well as individual action in what John Elster calls the assumption of methodological individualism.[26] Yet, the difference between this method and survey research is that the former builds models based upon an ideal, self-interested rationality which forms the options for mathematical calculation. This deductive method contrasts with the inductive procedures of surveys in which hypothesis remains largely implicit, that is, is revealed only by a "reading" of the questions to discover the tacit purposes of the inquiry.

Obviously, these are merely variants of positivism, although model building corresponds more to the methods of physics. In each case, the hypothetic structure is validated, according to its protagonists, by verification. In many cases, social investigators of the inductive school are not aware of their own assumptions, adopting survey procedures without apparent hypotheses. Nevertheless, one can detect these purposes in the definition of the problem under investigation as well as in the methods of inquiry. For example, efforts to find explanations for "deviant" behavior such as crime have varied according to the prevailing political and ideological climate. The individual is presumed, in one mode, to be the product of his/her family, school, and broad, generally economic, environment. Modern liberals tend to define "deviance" as a rational response to these circumstances. Improving the life chances of individuals becomes the primary aim of social policies that accept this paradigm of social behavior. This entails such measures as improving educational opportunity, providing job training for "offenders," and, where possible, jobs. On the other side, if the crime is ascribed to innate aggressivity or to subnormal intelligence, rehabilitation is futile and the best hope for thwarting crime is to use retributive measures such as long prison sentences or capital punishment. Surely social scientists for whom corroborative evidence is requested by government authorities are "free" to refuse to collaborate with those who adopt tacit positions that are inimical to their own moral strictures. The consequence of such a decision may, of course, exclude these social investigators from consideration when funds are dispensed. In effect, the ethical issues in social science research cannot be avoided no matter how much one wishes to separate value from fact. For the facts of the social world are enmeshed in ideology, but, even more broadly, social science is an integral part of the prevailing social order. In this respect, it strives for parity with the sciences of nature which have been incorporated massively by industry and the state, particularly research (which has military applications).

The critical theory of science proposed by the Frankfurt School has been marginalized by philosophy and social studies of science precisely because of its insistence that science is social relations. At the same time, scientific Marxism excoriates the Frankfurt School's critique of the Enlightenment. Listen to Lucio Colletti, a major figure in the Marxist-structuralist school, whose teacher Galvano Della Volpe may be considered the leading Italian Marxist philosopher of science: "Together with Marcuse, they [Horkheimer and Adorno] are the most conspicuous example of the extreme confusion that can be reached by mistaking the romantic critique of intellect and science for a socio-historical critique of capitalism."[27] Colletti accuses Critical Theory of a "nihilistic negation of the highest achievements of human thought."[28] Citing Lukács's *History and Class Consciousness*, Colletti reveals his own purpose in criticizing Lukács and the Frankfurt School—to distance Marxism from Lukács's judgment that "nature is a social category" and "is a development out of the economic structures of capitalism." Colletti identifies this point of view with that of Sartre's existential Roquentin, for whom "the scandal of alienation is that a natural world should exist." From here, Critical Theory is merged with Bergson's "spiritualism" and other late nineteenth century romantics such as the neo-Kantians of the historical school, particularly Rickert and Tonnies. Now the influence on Lukács of Weber's idea that capitalist rationality signifies the disenchantment of the world is undeniable. The concept that nature is construed socially can be interpreted from premises that derive from another major neo-Kantian tendency, phenomenology.[29] Colletti's criticism amounts to a demonstration that these are not consistent with materialist premises that posit the object of scientific knowledge independent of human will, but does not address the validity of the claims themselves. Instead, he resorts to what Peirce calls the *method of authority* for establishing truth, showing that Lukács's views do not correspond to those of Marx, or, more precisely, that Lukács misunderstands Marx when he extends the critique of reification to positive knowledge. In other words, Colletti, following Della Volpe, wishes to make a radical distinction between social life and science.[30] Naturally, like many others, Colletti holds Marx faultless; he is portrayed as victim of his own epigones. Further, natural science, especially the prescriptions of Galileo for transforming science from the grip of its Ptolemic prison, is invoked but remains unexamined. For Colletti, in contrast to Lukács and the Frankfurt School, nature is "given" as the object to which scientific knowledge—laws and propositions—corresponds. Accordingly, to argue that nature is socially constituted, and therefore subject to economic, political, and especially ideological mediations, nullifies science and is little more than a return to mysticism, romanticism, and the other enemies of positive knowledge. Thus, any effort to link the content or methods of scientific knowledge with its social/historical preconditions must be denied. Colletti must, therefore, maintain that philosophy is a critical and even ideological discourse, as his critique of Lukács clearly indicates. But Marxism is a science precisely to the degree that it transforms the critical categories into positivities that enable precise scientific investigation. The indeterminate abstractions of the early Marx, "civil society," the "state," etc., which remain logical, speculative categories for both Hegel and Critical Theory, must be replaced by concepts such as the "relations of production," whose content is capital, surplus value, profit, and other categories of political economy.

Despite the elegance of Colletti's polemical strategy—its deft use of Kant to restore scientific Marxism, for example—the intention is clearly to rescue Marx from the attempts led by Lukács in the aftermath of World War I to treat Marxism as ideology insofar as all discourses are ideological. Lukács was trying to propose a new interpretation of *Capital* as a critique of scientism, which for the theorists of the Second International as much as for Lenin established historical materialism on a strictly scientific basis. To be sure, Lukács's appropriation of Marx was always framed in the rhetoric of "orthodox" Marxism. But his orthodoxy meant the restoration of the critical meaning of Marx's categories. This strategy was made necessary by the restricted intellectual environment that had been produced not only by the emergence of Bolshevik hegemony over the world communist movement which extended into philosophy as much as it did to tactics of political struggle, but by the dominance of religious motifs in postwar Marxism itself. Wearing the mantle of orthodoxy also corresponded to the significance of the epithet "revisionism" within Marxist circles following the famous Bernstein/Kautsky/Luxemburg debates at the turn of the century, and to Lenin's excoriation of the onetime gatekeeper of orthodoxy, Kautsky.

Colletti's trenchant criticism identifies Lukács's major sin: if his critique of science as social relations is right, Marxism must also be socially constituted not only with respect to its origins but, more scandalous, with respect to its categories and its description of the social and natural worlds. Of course, this is already implied in Lukács's idea that dialectics can only refer to subject/object relations and that Marxism's effort to found a nature dialectic is misdirected, in any case, because it holds to a view of nature as not only ontologically independent but possessing intrinsic regularities that are not subject to the intervention of the investigator.

Ironically, Lukács's concept of scientific law as socially dependent parallels developments in modern physics that acknowledge that our knowledge of the object as both theory and "subject" dependent. On the other hand, the various brands of Marxist objectivism, including that of structuralists such as Colletti and Althusser, acknowledge only that theory may identify the social context within which science develops, but refuse the next step, which is to ascribe both the resultant theories and their methodological assumptions to social relations. Thus, Engels taught that advances in technique are the catalyst for scientific discoveries, and that these advances are socially determined. But the discoveries themselves are quite autonomous of their social and historical presuppositions. Similarly, Marxism in its pristine form arises indirectly from ideological and even metaphysical sources, especially Hegel, utopian socialism, and classical political economy (Lenin), and the social precondition for Marxism is the entrance of the proletariat on the world historical stage in the early nineteenth century. But its scientific status is independent of these influences. As Althusser and Colletti argue, Marxism becomes a science when it frees itself from the yoke of its metaphysical origins and the actually existing working-class ideologies which are limited by trade union, i.e., craft and industrial, demands. Put succinctly, Marxism achieves scientificity by means of criticism, but its theoretical discoveries are not critical in substance. How this is possible is spelled out by Della Volpe, for whom science is constituted by a series of negations, particularly of rationalism, teleology, and dogmatism. Against rationalism, he proposes to substitute "a mode of reasoning" that entails experiment and observation but is ultimately formed by a materialist logic. I shall return to this problem in Chapter 7; but before doing so, I want to examine the work of perhaps the singular instance of a Marxist critique of science that relies primarily on Marx's own categories, rather than adopting Weberian concepts, as the Frankfurt School felt obliged to do. I refer to the work of Alfred Sohn-Rethel, to whose *Intellectual and Manual Labor* I now turn.

Sohn-Rethel's fundamental thesis is that the formal analysis of the commodity holds the key, not only to the critique of political economy, but also to the historical explanation of the abstract conceptual mode of thinking and of the division of intellectual and manual labor that came into existence with it. Knowledge is derivative of specific social relations, the commodity and exchange abstractions. For Sohn-Rethel, these abstractions are real processes, not thought modes: "A derivative of consciousness from social being presupposes a process of abstraction which is part of this being."[31] Thus, economic relations, principally commodity production and exchange, are the sources of the characteristic scientific categories: nature becomes "pure object world" divested of the subject; it is configured as pure quantity just like the commodity, which, in exchange, has only quantitative value, having been shorn of its particular qualities. Sohn-Rethel's nonepistemological theory assumes the social constitution of knowledge but provides this concept both a temporal and spatial specificity:

> By its own physicality in terms of spatio-temporal action the
> abstraction from natural physicality, which exchange enforces
> by its separation from use, establishes itself as a physicality in

the abstract or a kind of *abstract nature*. It is devoid of all sense reality and admits of only quantitative differentiation. . . . This real abstraction is the arsenal from which intellectual labor throughout the eras of commodity exchange draws its conceptual resources. It was the historical matrix of Greek philosophy and it is still the matrix to the conceptual paradigms of science as we know it.[32]

Basic changes occurring in these paradigms indicate major changes in this matrix, and vice versa because the socially necessary forms of cognition in any epoch have no source from which they can originate other than the prevailing functionalism of the social synthesis. The separation of head and hand which originates in Greece can be traced, according to Sohn-Rethel, in the development of money, itself a synthesis of the commodity/exchange abstractions:

What defines the character of intellectual labor in its full-fledged division from all manual labor is the use of the non-empirical form-abstractions which may be represented by nothing other than non-empiricist 'pure concepts' whose material base is the real abstraction, commodity exchange. From there comes the social category of the independent intellect who is self-directed and 'self-alienating' and possesses its normative sense of 'logic' without material reference, especially to the exchange abstraction, money representing value.[33]

Under this regime, the intellect becomes impervious to the social formation to which it conforms. Thus, science as a further development of this alienating process imagines itself governed by an internal logic that discovers the "basic categories of nature as object world in anti-thetic contrast to man's own social world."[34] Sohn-Rethel insists that these "pure" concepts are derived initially from abstractions from manual labor for which socially constituted time and space is the norm. Thus, abstract time and space, causality, matter, abstract motion, that is, the nature which is the object of scientific investigation, are categories of a "second nature." Nature is nothing but the product of this abstraction and that is itself social. Scientific law, which the "independent" intellect imagines to be some kind of representation of an unmediated object, refers to that social nature, particularly the emergence of mathematics as the foundation of all science.

Sohn-Rethel renders a historical account of the development of science as dependent, not so much on technique, although he draws on the work of Lynn White, who identifies the invention of the stirrup and harness as characteristic technologies of medieval social relations.[35] Science outstrips its dependence on technique and appears to free itself from manual labor by the seventeenth-century postulate of nature as a machine. From this, technology develops from science, and the historical relation is reversed. Galilean science established "a clear-cut division between head and hand," with intellectual labor taking the dominant role in their mutual relations. After Galileo, knowledge of mathematics, which initially took its categories from such activities as estate accounting, international trade (commercial activities of merchant capital), as well as the artisanal mode of production, appears independent of its origins. Mathematics, which Sohn-Rethel defines as "the logic of socialized thought," is now denied to artisans whose work had no historical reliance on a symbolic language. But social production based upon machinery requires this language, which appears as the controlling instrument over production. This "cleft between a context of thought and human action" is linked to the domination of labor by capital, which by the nineteenth century has seized ownership and control over all elements of production. Just as capital takes control of production, so science takes control of the object by isolating it from its "uncontrolled environmental influences."

Manual labor is one of the "uncontrolled" elements of production that technology attempts to exclude or marginalize from the process. Concomitantly, Galileo provides the basis for excluding the subject from nature, an unwanted environmental influence, by positing the theory of inertial motion, which Bertrand Russell says "led to the possibility of regarding the physical world as a causally self-contained system."[36]

That Sohn-Rethel's theory of science is basically indebted to Lukács's theory of reification is obvious from his similar analysis of the results of the penetration of the commodity form in the social world, as well as the division of labor which derives from it. But the theory goes beyond that of *History and Class Consciousness* and the science critique of the Frankfurt School. Its advance is to make the critique concrete by examining scientific knowledge itself as well as its material or intellectual presuppositions.

Yet, like the Frankfurt School, Sohn-Rethel retreats to a defense of the results of science as well as its methods: "Let us be quite clear: methodologically classical physics has nothing to do with the exploitation of labor by capital. Its findings are valid irrespective of any particular production relations." This statement hangs on a single epistomological supposition: that exchange relations and the real abstraction embodied by them "have substantial identity with corresponding elements of real nature."[37]

Even as Sohn-Rethel acknowledges that science has been subsumed under technology, which in turn "hinges on the controlling power of capital over production, cuts up nature piecemeal by isolating its objects of study from the context in which they occur . . . the pattern of exact science is still that of classical physics,"[38] even though recent developments in science have altered the earlier determinisms of natural law. Therefore, we must conclude that, according to Sohn-Rethel, science is ideology to the degree that it imagines its practices to be free of social preconditions, but remains valid, that is, independent of the social context that produces it, because of the correspondence of its material foundations, commodity/exchange abstraction with nature mediated (presumably) by manual labor.

As I argue in the next chapter, the retreat from a critical conception of the nature of scientific discovery and the laws deduced from them may be connected to the difficulty the social theory of science has encountered in imagining an alternative to science which does not reproduce mysticism or romantic anti-scientific naturalism. That a realist theory of knowledge recurs in Sohn-Rethel's discourse testifies to the difficulty of sustaining a new critical conception of science and its historical development in the wake of its hegemony over the normatively construed canon of knowledge, within which physical science occupies a privileged place. This difficulty shows that even Critical Theory has its own second nature: the nagging doubt that the critical theory of science is not merely the mirror image of Marxist and mainstream scientism. This doubt will be replayed in post-Frankfurt School and structuralist Marxisms of the 1960s. For these readings of both Marxism and the intellectual hegemony of science and technology, there is no longer any question of a critical theory *of* science: theory must become scientific or risk oblivion.

CHAPTER 6
HABERMAS:
The Retreat from the Critique

The distinction between science and ideology, indeed, the concept of ideology as such, depends on a conception of science as a self-critical, self-correcting inquiry. The received wisdom of Western thought is that science is constituted by value-free knowledge of the external world. In this conception, the scientist, in Max Weber's invocation, understands that he/she approaches nature and society as objects of investigation unburdened by personal or political interests. For Weber, the task of the investigator is to purge his/herself of such interests both before and in the course of the research.[1]

Although there have been many versions of the concept of ideology and, at times, social theorists and philosophers of science have engaged in fierce debates regarding definitions, there is no end in sight. The reason for the persistence of ambiguity with respect to its meaning does not reside in some slipperiness of social thought, but in the dependent nature of the concept itself. Ideology is construed in the image of the epistemological and historical system within which it functions. As opposed to the optimism of some nineteenth-century thinkers, who employed the term to connote "false consciousness," "worldview," or "values and beliefs," all of which signify interests that produce partial totalizations whose half-truths are not true at all, twentieth-century Marxism has come to hold ideology to be an ineluctable feature of social life. From this insight follows a range of positions on the historical function of ideology in modern discourses. In contrast to orthodox Marxism, ideology is no longer situated in the superstructure, if, indeed, the distinction between base and superstructure retains any validity after Lukács's reading of *Capital*. Instead, for both Lukács and the Frankfurt School, ideology inheres in the structure of capitalist social relations, particularly in the "mysterious" permutations of the commodity form in the labor process, and indeed in the entire system of capitalist and scientific rationality. As we shall see, Althusser, reading *Capital* with very different lenses, attributes superstructures such as the legal system, one of the crucial "ideological apparatuses," to their correspondence with commodity exchange relations and the capitalist mode of production which is its basis. Of course, Althusser also uses ideology in other ways, for example, most relevant to our inquiry, as coextensive with nonscience, a formulation which places him close to the views of Karl Popper.

In contrast, commentators on the concept of ideology who are concerned with the role of intellectuals, such as Antonio Gramsci and Karl Mannheim, retain the link between ideology and interest inscribed not only in worldviews but also in forms of social practice. Mannheim, in particular, has emphasized that any group that emerges as a social category, such as the intelligentsia, places itself at the center of its worldview. All groups that grope for a social orientation first attempt a self-enhancing interpretation of society, and this bias corrects itself only on a higher level of reflexivity—a level which we approach through the sociology of knowledge. But the process of self-orientation occurs only by the group situating itself in relation to the worldviews of dominant classes or, in the case of the modern intelligentsia, in relation to the proletariat.[2] Alvin Gouldner, in his studies of the rise of intellectuals, defines the adoption of a specific *culture of critical discourse* as the marker signifying the rise of the intelligentsia to bid for class power on its own.[3]

This discourse is, of course, closely identified with the distinction between science and common sense, which, in the era of later industrialized societies, appears as a particular attribute of intellectuals to the degree that the division between intellectual and manual labor ruptures the link between craft culture and scientific and technical knowledge. This rupture is accompanied by the introduction of special languages of science and technology which seem to possess only logical necessity, particularly the notations of mathematics and adaptations of symbolic logic. Further, the common belief that these languages can be acquired exclusively by schooling completes the circuit: knowledge acquired in accredited institutions (most importantly of "higher" education), in which ordinary language, its syntax and characteristic

modes of utterance give way to mathematics and specialized discourses, counts as scientific knowledge. In contrast, the human sciences, which, with some notable exceptions, still employ ordinary language and speculative reason with its characteristic essay form, may
yield valuable insights from the perspective of science; but they lack
rigor, precision of results, and, most important, cannot be subjected to
repeatable experiment or be falsified, and are therefore types of ideology. In terms of the rise of the scientific intellectual, ideology takes on
a relational meaning: it is any discourse that uses the language of daily
life, except what is counted as "art." Popper called this the problem of
demarcation, refusing to discuss the issue of meaning precisely because of the metaphysical root of the term. Science is not concerned
with "meaning," only with truth, which is construed as the conformity
of a proposition to the outcome of a procedure of scientific experiment.

For the purposes of our discussion, I want to adopt, provisionally, a
concept of ideology that presupposes its difference from science. As I
shall try to show, this difference does not derive from the correspondence of a given proposition or set of theories and assumptions now
conventionally termed a "paradigm" with "reality." Rather, science is a
type of discourse with special languages, rules of investigation, and
forms of inquiry that determine the form of the result. Together, these
constitute elements of an ideology that is accepted by the scientific
community and, to the extent this ideology becomes hegemonic in the
larger social context, that is accepted as "truth." Other discourses
become poetry, religion, metaphysics, or whatever, but are zealously
marginalized from what signifies science by those who constitute the
scientific community.

For those not armed with Weber's weapons of criticism, the experimental method is said to provide an almost automatic corrective to the
tendency of scientists to permit personal, social class, or bureaucratic
interests to mediate and deform knowledge. For most scientists, the
experimental method is believed to be "ideology-proof" and constitutes a device to ensure the neutrality of inquiry with respect to interests that contaminate the results. Weber directed his attention to the
social and cultural sciences and advocated reflexivity as a supplement
to the experimental method. In the Weberian view, the human sciences limited the chances of value-free inquiry without ideology-cri-
tique, since the objects of investigation were humans themselves.
Although ultimately pessimistic about the possiblity of expunging
interest from the process of the acquisition of knowledge about
humans and thus from the results of investigation, Weber was imbued

with the spirit of the Enlightenment in his belief that a priori values could be expunged from scientific inquiry was possible and had been realized in the natural sciences.

We shall return to the problem of freeing social science from its ideological premises below. The starting point of this inquiry must be the problems of freeing science and technology as such from ideology, since natural sciences are commonly believed to be, at least in their objective possibilities, unmediated by social relations. The philosophers and scientists of the Enlightenment distinguished themselves from medieval theologians and Ptolemic scientists by attacking the latter's efforts to understand nature as not only factually incorrect but grounded in suppositions that were, in essence, mythological. The revolutions in astronomy and physics started by Copernicus, Galileo, and Kepler, and later by Newton, were based on the privileging of the senses, mathematical calculation, and experimental verification as sources of knowledge that were free from the mystifying power of religious belief to distort observation. The view that modern science represents the return to the things themselves by means that could dissolve ideological mediation has animated four hundred years of science and invention.

Of course, social research in the past ten years has begun to challenge the notion that technology could free itself of social relations, including systems of belief that served the interests of domination. The discovery that the *applications* of both scientific and craft-wrought knowledge to the development of machinery, to new methods of work organization, and to scientific knowledge itself were permeated by presuppositions that embodied social interest has now become a fairly respectable, if not completely accepted, position among many, including non-Marxists. We have learned that "scientific" management and its methods reorganize the labor process so as to degrade and deskill labor and deprive workers of their autonomy in their daily work; that industrial psychology, which purports to be a new science, is little more than a means to "motivate" workers to produce more, to identify their interests with those of the company so that they may cooperate in keeping industrial peace; that "motivation research" is an application of Freudian psychological principles to advertising in order to persuade people to purchase commodities. Some investigators are persuaded that the history of all these "sciences" is rooted in the problems of late capitalism: how to keep the rate of profit from plummeting, how to transform producers into consumers, and, in the case of the technologies of "administrative" sciences, how to plan and control the relations between the state and the corporations on a national and international level. They have argued that even culture has become an affirmative technology of domination.[4]

These conclusions, discovered for the most part by those influenced by the Marxist tradition, have cast a long shadow over the neutrality of technology and many human sciences that have been shown to be closely linked, both directly in their administrative aspect and indirectly through the work of academic social science, to the project of the social reproduction of capitalism. To extricate scientific management from its class determinations is to destroy the methodological presuppositions of the inquiry. Once we have discovered that workers may produce higher quality and as many goods by sharing rather than segmenting tasks, by controlling the total production process themselves—including the object to be produced, the methods of production, the speed of labor and its uses, and by becoming multivalenced rather than single valenced and fungible—the concept of technology as ideology no longer appears strange. However, many who have come to recognize the intimate relation of technology to domination retain their belief that the content of science is somehow, at least in principle, autonomous from social relations.

Of course, it is not hard to understand why most of us retain our faith in science's emancipatory potential. The overwhelming experience of moral and political corruption within capitalist society has relieved us of the notion that the state or civil society can stand apart from the competing worldviews, much less interests, based on class, caste, race, or sex. Years before Marx's *Critique of Hegel's Philosophy of the State*, James Madison pointed to the close relations of class and state.[5] The bourgeoisie of the revolutionary period alternated between its own ideology of the neutrality of the state (a mediator between classes, the autonomy of the electorate) and a keen understanding that private property formed the basis of both civil and political society. For the most part, Adam Smith's hidden hand of God became the last resort for those who realized that they acted in accordance with their own interests in both politics and business. To some extent, "science" became God's hidden hand when liberalism was hard pressed to abandon its faith in the business ethic. It was the proof that one realm of human existence remained inured from the "materialism" of the marketplace. Science guarded its prestige through nature, which technological "progress" had brutally transformed into an arm of domination; in Marxist terms, nature had become humanized, just as humans had become naturalized. Science appeared to retain nature as an object, not for domination, but as a field for disinterested inquiry. For Newton, scientific investigation possessed even erotic dimensions.

The Marxist conception of the machine as an object form of the relation of human activity to nature, rather than as a thing separate from all social relations, helped to endow science with the sole mantle of truth. Critical reflexivity, for Marxist science, and experimental method for natural science, were relied upon to yield reliable knowledge, that is, knowledge that could not be reduced to the interests of those who acquired it. The "logic of scientific discovery," whether dialectical or not, is held sacrosanct by most philosophers of science, Marxist theorists, scientific workers, as well as the public.[6] It may be extreme to claim that science has replaced religion as that activity in which the divine is apprehended; yet, the fierce defense of the neutrality of science, from social determination both at the level of its choice of inquiry and at the level of its methods and results, attests to the deep-rooted concept of science as our collective hope of redemption. Science has become our passport to the earthly paradise.

I shall reserve for Part III a discussion of the way in which the unreflexive belief in science expresses itself in the mainstream of the scientific and philosophical communities. I shall first review the relation of Marxism to science, since Marxism, as the critical theory of capitalism and that theoretical perspective which has insisted on the primacy of social relations as well as the concept of ideology as a crucial mediation in social discourse, has been the most ambivalent on the question of science.

Let it be stipulated at once that no major tendency of Marxism denies the relevance of the relations of production—that is, the division of society into classes—to the processes of scientific investigation. Indeed, Marxist sociologists and historians of science have insisted on the relationship between the rise of the bourgeoisie to political and social power and the freeing of science from the chains of illusion. For Marxists,[7] the rise of science accompanied the rise of the bourgeoisie for two reasons. First, science was the primary antagonist to religion at the ideological level; its own methods of observation and experimentation were derived as a critique of the a priorism of religion. The derivation of truth and hypothesis from observation, the privileging of sense perception as the ultimate source of knowledge, and the scientific attitude of skepticism are all understood by Marxist historians as inextricably linked to the bourgeoisie's revolt against the closed system of Church, and aristocracy, within the feudal system. The persecution of Galileo by the Church was the ideological displacement of the struggle of the older order against the rising merchant class. Second, even though not directly related to industry in the seventeenth century, the years during which modern scientific theory was formed (except in navigation, a useful adjunct of commercial exploration), scientific thought legitimated the critical attitude that constituted the basis of the bourgeois critique of the feudal system. In Protestant theological doctrine, God's law mandated that humans be held accountable for their own actions. Nature was no longer an untouchable artifact, but had to be conquered, ostensibly to relieve humans of backbreaking labor, to secure their collective safety from the onslaught of natural forces that threatened human life, or, in the more Protestant versions, to save them from damnation through the dignity of labor. As Horkheimer and Adorno have pointed out,[8] the conquest of nature was linked to the fear of nature. Only by reducing the external world to a "tool" of humans, by making it an instrument, could the destructive powers of nature be controlled. The passing of the organicist view of nature, in which the world is believed to possess its own integrity and beauty which can only be apprehended contemplatively, and the transformation of nature into an antagonist, achieved the necessary distance the bourgeoisie needed in order to dominate it. Now, "organic" conceptions of the natural environment were reserved for poets and philosophers, a manifestation of the huge divide introduced by the Enlightenment between the material and the spiritual. Science became the cutting edge of natural domination, albeit rationalized in terms of the quest for human freedom. As human control over nature was extended, the domination of humans became easier since "human nature," now the internalized beast, required harnessing.

The social relations of science derive, in part, from the subordination of science to the needs of industry in the mid-nineteenth century and thereby the inscription of scientific theory into the forces of production. It is not difficult to demonstrate this aspect of the relation of science to society; the command by the state over scientific resources to achieve more powerful forces of destruction inscribed in the arms industry constitutes elementary evidence that science is hardly an autonomous enterprise in the twentieth century. And even if we may not link science to industry, since in their respective origins the so-called scientific revolution preceded the Industrial Revolution by centuries, the scientific enterprise *does* correspond to the rise of the bourgeoisie to social and political power. It is not one that caused the other in any kind of simple, linear succession. Rather, the "mechanization of the world picture,"[9] a leading theme of the scientific revolutions of the sixteenth and seventeenth centuries, that is, the positing of the machinelike character of nature, was consistent with the entire "epistēmē" of the period.[10] In this particular period, "natural order" was conceived as a system of regularities, subject to quantification and regulation, and, if extrapolated from the multitude of relations of the external environment, could be controlled by a series of reproducible interventions called experiments. The reproducibility of reactions of nature to human interventions were confirmations, it was believed, of the orderliness of nature, its obedience to what was called natural law. For some, the subordination of nature to human will, understood as an outcome of the ability of science to reveal the properties of things and particularly the regularities of nature, was proof of God's sovereignty over all things. For, if the hidden hand of God was not immanent in all things, how could the chaos that was apparent in the external world ultimately obey laws? Even if an individual phenomenon had no apparent connection to any other, gravitation, the atomic structure of matter, and other discoveries of science showed the ultimate unity of all things.

The homology of this view of science to Hobbes's view of human society is not difficult to see. According to Hobbes, individuals vied for pecuniary advantage in the marketplace, buying and selling, cheating and fighting, engaging in the "war of all against all."[11] At this level of human intercourse, peace and harmony, law, was impossible. Hobbes feared the anarchy of production and exchange would result in the destruction of the human community unless the state is constituted as the higher body that saves humanity from itself by imposing a system of higher morality inscribed in laws. Social physics, no less than the natural physics, is necessary to maintain human survival.

In the past forty years, Marxist studies in the history of science have progressed from a view of simple determination of the economic over the scientific. In recent work, the relation of the economic base to scientific and ideological development is understood as a *mediated* relation.[12] Science is conceived as a relatively autonomous sphere of human activity under capitalism. On the one hand, the tendency of capital to subsume science under its drive for accumulation results in the subordination of science to the requirements of both the state and large corporations, at least in the twentieth century. There is a politics of pure science because in nearly all countries, the state is more than a patron of science—it awards contracts to produce goal-oriented products. Of course, not all assistance is rendered on condition that the results can be put to industrial or military uses. As I shall try to show in chapter 12, however, the form of the results of scientific knowledge leads to its instrumentalization. Of course, it does not follow that the *content* of all scientific work is determined by the funding source; "he who pays the piper" does not call the tune in a crude manner. But, as

Dan Greenberg has shown,[13] the predicament of "pure" science is that, since it is supported by public funds for the most part, the piper's demands usually entail, at least implicitly, the conversion of its results into forms of technique. While the scientific community adheres to its craft, and regards science as a kind of religion requiring no other legitimation than that it results in contributing to human knowledge in the abstract, the choice of fields of scientific inquiry is no less influenced by the requirements of capital than is applied science.

It is functionally desirable for the individual project or scientific worker to regard the work as free of political or economic considerations. Once the project is funded, the requirement of autonomy in relation to government, industrial, or administrative interference is usually observed in most advanced industrial countries. It may be argued that even if most science has been subsumed by practical applications to industrial and military uses, the existence of a "pure" science sector safeguards the objective possibility of the search for knowledge unmediated by the requirements of social relations. For the most part, Marxism shares the belief that, despite the tendency of social demands to direct the character of scientific research as much as technical development (an intervention which infects science with ideological elements), it may *not* be said that all science is ideological. The social relations within which scientific labor is conducted mediate, but do not determine, the scientificity of science. At bottom, Marxism shares with other perspectives the belief that ideology may be transcended, that truth may be known or, in V. I. Lenin's terms, "approximated."[14]

The standard shibboleth of Marxism is that it is not theoretical suppositions themselves that constitute an ideology, or, to be more exact, one cannot subsume Marx's science under ideological discourse. Some Marxists have acknowledged that Marxism is, among other things, the ideology of the working-class movement, just as individualism was the ideology of the rising bourgeoisie and capitalism's own ideological break with the feudal past. In this formulation, ideology conceived as a worldview functions to rationalize or legitimate the claims of a social movement, class, or even a class-in-formation to seek or maintain its social power. Of course, this does not imply that ideology is *merely* a kind of distortion of the "real relations of humans and things." These distortions, in Louis Althusser's terms, are themselves real insofar as they constitute themselves as material practices and as they are inscribed in institutions or political life. The relation of the imaginary to the "real" guides human conduct and experience as it is actually lived.[15]

Thus, the version of "socialism" practiced in Eastern European countries may depart from either utopian models or the implied conceptions of several varieties of Marxism. In such countries, Marxism is universally adapted to the requirements of legitimating the prevalent forms of social rule. Rudolph Barho's phrase to describe these countries—"socialism as it actually exists"[16]—is close to the conception of ideology as a kind of lived experience that constitutes itself, in one aspect, as the relation of the imaginary to the real which, nevertheless, has its objective side in the social institutions of these countries.

When Karl Mannheim accused Marxism of being merely an ideology, he took this feature into account and made of it a *partial totalization*.[17] For Mannheim, Marxism could never be a science precisely because it was linked with a particular class in capitalist society. Marxism was nothing but the ideology of a class seeking social, economic, and political power, whose claims to universality were no more persuasive than those of the bourgeoisie in its historical struggle against the feudal order. For Mannheim, every class seeking power makes claims to possessing universal truth, or scientific validity. But these claims lack verification, especially in the light of the failure of the world revolution to materialize in the twentieth century.

Mannheim argued that no science could proceed except on the basis of a community of disinterested observers. Weber's invocation to science to separate fact from value was clearly the inspiration behind Mannheim's discovery that only intellectuals could provide scientific truth, precisely because of their marginal status within class society. When intellectuals identified themselves with one or another participants in the class struggle, their work was fated to become ideological, that is, their theories would inevitably end up as versions of "false consciousness."

Marxism traditionally answers these objections with a historical as well as theoretical argument. The uses to which science is put are never neutral, and its development cannot stand apart from the class struggle and the interests of the dominant class to which it responds. Yet, under certain conditions, its identification—more exactly, its production by some social classes—opens the way for the production of objective knowledge that is *relatively* true (relative, that is, to the limits imposed on human knowledge by the stage of development of the productive forces, including the means available to penetrate the secrets of nature). During the period of the transition from feudalism to capitalism, the bourgeoisie's interests forced it to encourage the advance of human knowledge. Since the old feudal relations, including feudal science, regarded most questions of human knowledge to be subject to settlement by religious doctrine, the search for scientific truth, the exploration anew of nature and even of "human nature" corresponded to the requirements of the new social order brought into existence by the bourgeoisie.

Again, the logic of capital, its requirement for continuous accumulation at the most rapid pace possible, brought the Industrial Revolution into existence, wherein the results of science were subsumed by technology and human productive powers were drastically enhanced. In conventional Marxist explanations, the new capitalist relations of production became a fetter on the further advance of science and technology in the latter half of the nineteenth century. Henceforth, science would be subsumed under capital and directed exclusively to means of destruction or means of production that were internally contradictory to capitalist social relations. Scientific discovery would be slowed down, its instrumental character become more evident.

The proletariat needs science, according to conventional Marxism, in order to free itself from the burdens of the past. Marxism shared with the bourgeois tradition the belief in the emancipatory potential of science and technology, if only freed from the constraints of capital. Bourgeois social and psychological science, subordinated to the needs of capital, produces results that are immediately employed as an instrument of the domination of humans, just as natural science has become a servant of political and economic power. Thus, although it is in the *particular* interest of workers to liberate science, such liberation would be universally beneficial. Humankind as such would be emancipated from pestilence, war, and exploitation by the freeing of science from capitalism and the founding of a socialist society.

Georg Lukács argued that the bourgeoisie could not penetrate the reality of capitalist social relations because its particular interest remained intertwined with domination.[18] To be sure, the proletariat could not free itself from the false consciousness imposed by commodity fetishism. Marxism, as the theory of the proletarian revolution, represents the consciousness of the proletariat freed from the chains of bourgeois illusions, able to reveal the underlying laws of capitalist development, its inevitable decay, and the revolutionary transformation that arises from the internal contradiction.

Even though Lukács's version of the evolution of Marxism into the scientific theory of capitalism and its downfall remain controversial, in Marxist terms as well as among social theorists in general, there is a striking similarity here with the traditional claims of science. Marxism seeks to establish its own relative freedom from ideology to the degree that it wishes to assert that reliable knowledge of the social world is possible, that the tendency of bourgeois philosophy and social science toward agonisticism is a sign of the decay of the social order. Those tendencies of Marxism that fall prey to the seductions of ideological discourse are regarded by orthodoxy as a deviation from a true course. The truth is not determined exclusively by verification, since social science does not permit of the same type of procedures as natural science, except in its most positivist manifestations where all social relations are reduced to their behavioral and quantitative aspect. As with the natural scientific institutions, Marxist science is validated by the scientific community grouped around its flag, its traditions, and its accepted discourses. In some countries, these institutions have the force of the state or the party which acts as arbiter of scientific truth. Even in countries like France and Italy, where the Marxist movements do not hold state power, the parties committed to Marxism have enormous influence both in universities and in scientific journals, so that conceptions of official Marxism have, in the recent past, been no less powerful than those of normal science have been among physicists, chemists, and biological scientists. In some Western countries where left-wing parties are dominated by Marxist theories, the establishment of "normal" Marxist science is as pervasive; those who challenge some basic axioms of Marxism, e.g., the determination by the economic even in the last instance of law, education, culture, ideology, etc., are consigned to the role of pariah. The Marxist theoretical community achieves its solidarity by means that are similar to those of the natural or social scientific communities, and its sanctions are equally harsh for those who wander.

Despite the weight of Marxism in the West, it remains a minority tendency in social science, even if many French intellectuals, whose relation to the Marxist tradition is ambiguous, once proclaimed that "nous sommes tous Marxistes." Those who hold Marxism to be a science agree that the disintegrative effects of bourgeois ideas and evolving social reality must always be combatted in order to retain the integrity of the science. The methods of combat are understandable for those who are convinced that Marxism is *the* theory of capitalism, whose scientific status is indisputable and whose alternatives have been proven bankrupt. When exponents of tendencies within Marxism have observed that some features of late capitalism are anomalous with respect to the theory, official Marxism has been prone to invent categories that integrate these features into the theory (for example, relative autonomy to explain the disjuncture between economic base and elements of the superstructure), rather than challenge the underlying assumptions of Marxism itself.[19]

Thus, Marxism's defense of the concept of an ideology-free science is an aspect of its self-defense. If adjustments must be made in particular features of the theory, these are held to be a response to new developments within the capitalist order; or a result of new knowledge produced by the experience of the revolutionary movement; or even the results of more penetrating work either in the natural sciences, by bourgeois social scientists who are acknowledged to be capable of good results in specific areas but not in theory, or by Marxist research.

In Thomas Kuhn's terms, the paradigm, that is, the general framework of the theory, remains intact.[20] Marxism holds that its self-critique must be limited to particular historically transcended aspects of the theory, for, as long as capitalism exists, there can be no other *general* theory of capitalism.[21] Of course, these limits are no less severe than those imposed by Newtonian mechanics upon physical or chemical research. For nearly three centuries, Newtonian physics remained hegemonic, even if relativity was introduced to correct some of its obvious deficiencies. (Most scientists, however, would argue the reverse thesis: Newtonian science is made a special case by the more general theory — relativity physics.) But this obtains only at a high level of abstraction. The general methodological as well as theoretical *framework* of classical mechanics — its way of looking at the world — has been little disturbed in "normal science," no more than has Marxism's adherence to the fundamental category of the dialectic of labor as the explanatory framework for understanding the configuration of social relations. To challenge the mode of production of material life as the determining instance of all social life is to depart from the essential presupposition without which Marxism is not possible.

Thus, to challenge the necessity of demarcating science from ideology at all levels is to challenge Marxism itself, since it partakes in the assumptions of all science since the Enlightenment that it is the only possible, true way of looking at the (social) world. For this reason, contemporary Marxism cannot look at the social relations of science from the *inside*; that is, it has never been able to locate the scientific worldview, the mechanization of the world picture, the experimental method, the form of its results, and the legitimating weight of the scientific community in sanctioning that which is called science, within the framework of its critique of ideology. The duality of science and ideology in Marxist debate is the next subject I shall address.

It is by now well known that challenges to the orthodox Marxisms of the past now abound from many directions. Perhaps the central times when these challenges appeared in the twentieth century were during

the First World War and its aftermath, and following the famous reve-
lations about the crimes of Stalin revealed by Nikita Khrushchev at the
Twentieth Congress of the Communist Party of the Soviet Union. These
events prompted not only a reexamination of the political practice of
the communist movement, but also a searching probe into the theoretical presuppositions of Marxism itself. Not the least of these efforts
was the movement back to Marx in search of sources of the deformations that had given rise to Stalinism; alternatively, many tried to rescue
Marx from those who had distorted his ideas even while acting in the
name of Marxism.[22]

The critique of Marxist practice inevitably led to a critique of theory.
I want to examine two varieties of this critique. First, I shall render a
reading of the relation of Herbert Marcuse's critique of science and
technology to Jürgen Habermas's effort to simultaneously preserve its
implied critique of Marxism while trying to reformulate and transcend
the categories of Marxism. Second, I shall consider Louis Althusser's
defense of Marxism against those who tried to extract from the early
Marx paths to the renovation of social theory. In both Habermas, the
critic of Marx, and Althusser, the apparently staunch defender of Marxist orthodoxy, we shall find an almost identical reaffirmation of the traditional conception that science is incommensurable with ideology,
however their routes to this conclusion differ. I shall contend that
unless Marxism is able to understand itself as both a theory of the
actual development of capitalism *and* a type of ideology that both
expresses the particular interests of workers and other subordinate
classes and strata within capitalist society and contains ideological elements within its own theoretical formulations, it cannot make a significant critique of science as social relations. The consequence of this
refusal has been to ossify Marxism and to give rise to antimonic "post-
Marxist" critiques. I have chosen, first, to discuss Habermas's theory of
science and technology because he exemplifies one such post-Marxist
formulation of the problem.

The title of Habermas's essay "Technology and Science as 'Ideology' " already contains the clue to its meaning: ideology is put between inverted commas because Habermas intends to show that this
formulation contains an ambiguity.[23] Habermas deals with Herbert
Marcuse's attempt to show that, contrary to the pervasive belief that
technology and science are neutral aspects of the forces of production
and may be regarded as part of the legacy of a new socialist society,
they are, in fact, repositories of domination.

According to Marcuse, "domination perpetuates and extends itself
not only through technology, but *as* technology, and the latter provides

the great legitimation of the expanding political power which absorbs all spheres of culture." Moreover, "science, by virtue of its own methods and concepts, has projected and promoted a universe in which the domination of nature has remained linked to the domination of man—a link which tends to be fatal to this universe as a whole. Nature, scientifically comprehended and mastered, reappears in the technical apparatus of production and destruction which sustains and improves the life of individuals while subordinating them to the masters of the apparatus. Thus, the rational hierarchy merges with the social one."[24]

The implication of Marcuse's judgment is that what Max Weber calls "rationality" is a self-contradictory phenomenon within later industrial capitalism. That which is called rational contains the progressive organization and subsumption of all human action by criteria that subordinate action to the purposes of the organization of domination—corporate, state, and ideological. The penetration to all areas of the social by criteria of rational decision increases production, makes the functions of the state appear less arbitrary and increasingly subject to bureaucratic rules, and through its institutionalization harnesses science and technology to economic and political purposes.

Habermas accepts Marcuse's reading of Weber's theory of rationalization but rejects his interpretation of the significance of the subordination of science and technique to economic development and political domination. For Marcuse, humans have been forced to cut a deal in order to gain material comfort. They have been obliged to surrender their individual and collective control over their own destiny. Or, to be more accurate, the promise of freedom made by the old bourgeoisie has now been refurbished by the new apparatus of capitalist domination to mean the *freedom to consume*. Marcuse finds that this apparatus of domination appears rational, that is, the apparatus appears in the form of science and technology, and thus becomes virtually immune from attack.

According to Habermas, the central error in Marcuse's formulation of the problem is that he has retained a concept of ideology appropriate to an era long surpassed by the new capitalist apparatus. In the old concept, the ideological was linked to the worldview of a definite social class. Ideology represents not only a self-justification of its legitimacy within bourgeois society, but also the class striving for a new society. That is, the old model of ideology always contains a utopian element as wish fulfillment, delayed gratification, repressed desires. Ideology was not only a form of false consciousness, in Habermas's reprise, but a political weapon in the guise of utopia which, nevertheless, yielded practical results.

Technology and science are no longer types of ideology in this meaning of the term because, in modern society, the apparatus lacks a normative basis for decision making. The normative belongs to societies where classes vie for political power over a centralized state, where myths and religion are legitimating ideologies, and where the distribution of economic rewards has a class basis. Under these anterior conditions, classes were faced with *practical* problems, defined here as problems that are solved according to a priori normative rules. Habermas contends that these characteristics no longer obtain in modern society. Classes have not disappeared in contemporary society, but they have been irreversibly integrated into the apparatus. Thus, although the class struggle has not disappeared, it is suppressed and appears only as a factor in the technical, rational procedures of the apparatus. Technology and science, as aspects of economic growth and the instrumental rationalization of all social processes, destroy ideology as such, at least in the old sense, since myth and religion no longer have a material basis in class struggle. Instead, economic and many political problems become purely *technical* in character. As a result, ideology as system of normative-oriented action and belief can no longer exist. Its self-legitimation, still connected to political and economic power, no longer evokes a utopian side but consists merely in quantitative increases in comfort, grounded in the efficient handling of all social and political questions as technical problems to be solved through rational means.

Thus, technology and science cannot be called "ideology" in the old sense of the word, for they are neither mythic nor religious in the way that science was regarded in the seventeenth century. They have been fully subsumed and instrumentalized by a system of rational-purposive action which, in Habermas's words, "makes permanent the extension of subsystems of purposive rational action and thereby calls into question the traditional forms of legitimation of political power,"[25] i.e., ideology. The traditional distinction between means and ends has been reversed: means are no longer determined by and subordinated to ends, as they were in traditional societies. Ends have become means and means ends. The collapse of the two, in Habermas's view, is irreversible since the old normative systems that obeyed different logical principles have been permanently laid to rest.

Habermas proposes nothing less than the reformulation of the theoretical framework of Marxism, in which he finds Marcuse still caught. Marcuse's attempt to show technology and science as ideology is predicated on the assumption that people still obey the rules of *"interaction"*rather than the rules of what Habermas calls *"rational-purposive* action." The trouble with this assumption, according to Habermas, is that history has surpassed traditional societies. The old distinction between forces and relations of production, on the one hand, and the base and superstructure in which ideology functioned, on the other, must be reassessed.

Rational-purposive action and interaction are the two new a prioris of Habermas's theoretical system. Human societies in all epochs require labor to master nature in order to survive. But traditional societies rigorously separated *work* from other activities having to do with intersubjective communication, relations that are mediated by language, social norms, the construction of institutional frameworks that embody the moral development of children, and the maintenance of conventional authority.

Late capitalism, with its integration of the state with the economy, its nearly complete subsumption of science under technology, and its suppression of interaction, has attempted, in Habermas's description, to inform institutional frameworks for human action with the standards of technological and scientific rationality. For Habermas, Marxism has been suppressed precisely because it has occluded interaction as a separate category from its discourse: the hegemony of technocratic consciousness may not be overcome—it must be taken for granted.

For Habermas, Marxism's program of overcoming the problem of domination is anachronistic since the establishment of new "relations of production" to unleash the suppressed forces of production has already occurred within the framework of domination. Technology and science are at once the new hegemonic forces of production *and* the institutional framework for social life. Emancipatory practice must focus on the restoration of that which has been suppressed by late capitalism: the richness of intersubjective communication informed by social norms—in short, the moral life.

When Habermas calls for the abandonment of the critique of technology and science as ideology, it is not a sign of his celebration of their ascendancy; he recognizes that they have become forms of domination as well as emancipation from deprivation and arduous labor. However, since the entire framework of economic and political power is intertwined with the very canon of technical rationality, Habermas sees Marcuse's critique of rationality as a futile, romantic effort that is grounded in the confusion between purposive rational action bound up in *work* and the ethical life bound up in *interaction*.

By this formulation, we are asked to accept, for all intents and purposes, the technical-scientific structure of social relations as eternal and impenetrable. Habermas conflates work and science/technology in a single term—"rational-purposive action." Thus, the ideological disappears from their processes. Science and technology have become merely the historically evolved *form* of the relations of human to nature and to that part of social relations having to do with work. The task, for Habermas, is to extricate from the province of rational-purposive action that other part of social relations which may be subject to normative rules.

Habermas proposes to deny the dialectic of labor its internally contradictory character. Only labor's binary "interaction" may be the scene of ideological analysis and debate. Horkheimer's "instrumental reason" has so enveloped the world of work that the project of its transformation into emancipatory activity, which in Marx depended on the contradictory unity of work and interaction, now becomes impossible when rationality is entirely subsumed by the apparatus. What are the consequences of taking science and technology for granted, that is, of returning to the Weberian critique of rationality in order to shed the critique? Habermas has reintroduced the Kantian distinction between reason and judgment and tried to make of politics a new aesthetic. At the same time, the sundering of the unity of human activity paves the way for political quietism in science or, to be more precise, advances a theory that ends by asserting the neutrality of science and technology "in the last instance," despite its acceptance of Marcuse's linking of these with domination. For Habermas, domination is merely the price one must pay for material comfort and its presuppositions, economic growth and technical control. Since class struggle no longer constitutes the dynamic of historical change but has been replaced by rational-purposive action appearing as technology and science, any effort to argue their ideological character, which must always be predicated on the existence of social classes as significant social actors, is doomed to failure.

Habermas has exposed one of the crucial logical difficulties in Marcuse's position. In view of Marcuse's argument that there is a universal application for Weber's theory of instrumental rationality, in which science and technology are vital components of the productive and administrative apparatuses of capitalism, and also that this rationality bears the stamp of consensus in all but a small and diminishing intellectual avant garde, the status of Marcuse's appeal for a new science free of its obligations to the prevailing order is extremely unclear. More anomalous is Marcuse's lingering Marxism, now construed as an act of "pure refusal." For, if the predicate of the totally administered society is correct, the program of a politics that would intervene at its core—work, science, and technology—is utopian in the bad sense.

Habermas is also right to point out that Marcuse stopped short of carrying his critique to a logical conclusion: Marcuse fails to posit alternatives to technology and science, even hypothetically. Marcuse's refusal is grounded, not ostensibly in some intellectual acceptance of the new technocratic framework of modern society as desirable (although he has a tendency toward the view that it could be desirable)—a conclusion that must be drawn as well from Habermas's position, despite his gesture to the theory of domination—but in his belief that the working class as historical subject, the only material basis for ideology critique, had failed, at least for the time being. Moreover, Marcuse has a deep-seated Marxist suspicion of efforts to construct alternatives as logical exercises before the material conditions for their realization are manifest.

Marcuse's fealty to the concept of ideology is based upon his adherence to the foundations of historical materialism. Habermas's rejection of these foundations leaves science and technology as discourses, unexamined. Nevertheless, behind his own back, despite his stated intentions, Habermas has reinstalled the forces of production as an autonomous and sovereign historical power. For he has ended up in a technological determinism from which Marcuse has been saved by his ideology critique of science and technology.

Unless the categories of social class and ideology are retained, or replaced on the same critical plane upon which they were proposed, theory has no place to go but to conventional dualism. Habermas is caught in a series of binary oppositions—binary because they are autonomous, neither mutually determining nor subject to internal transformation. Moreover, they are posited as structural features of human relations, transhistorical phenomena whose form is altered by social development but whose content remains a property of what it means to be human as such.

Perhaps the most blatant and archaic duality retained by Habermas is the mind-body split implied by the distinction between work and interaction. For Habermas wishes to separate communications and normative judgments from the labor process and presents this conclusion as an empirically given proposition. The confusion of his thought results from the assumption that the exploitation of nature for human survival, in terms of both material subsistence and protection against natural forces, requires the repression of communication free of domination. The labor process, then, is defined as hostile to reflective social practice, which may be mediated by historical development. Its particular character is different in different periods. But, for Habermas, it can never be part of an emancipatory praxis because it already contains domination in two ways: it entails the domination of nature as well as domination of internal nature by the repression of interaction which could be free from the institutional framework ensuring work norms in opposition to communications norms.

Habermas is prepared to grant technical reason, that is, the domination of nature, its sphere of autonomy but insists that communications be accorded a space in the conscious rules that govern social life. But, since Habermas genuinely believes he can separate the labor process from communications, his theory is doomed to collapse. For the very argument inherited from Critical Theory—that science and technology have become modes of social domination—augurs badly for the project of human emancipation unless work and interaction are seen as moments internal to each other within a social totality. Thus, seeing technology and science as ideology is a step in the process by which emancipatory interaction may be achieved, not by permitting us to define a new sovereignty for communications, language, moral development, etc., but by insisting that work itself be transformed by restoring interaction as a normative principle.

The idea of work as interaction has been a utopian idea since the dawn of capitalism. For the status of labor as instrumental activity, that is, action that is subordinated to domination of nature and of humans, produced the notion that emancipation may be achieved only outside the workplace. The historical struggle of the working class for the shorter workday represented the most reasonable protest against the extraction of surplus value, and also a partial solution to the progressive degradation of labor introduced by the subsumption of technology under capital. The routinization of leisure as the sphere where public discourse, moral development, and undistorted communication could be carried on assumed that the labor process could be nothing but alien to human emancipation.

On the one side, Habermas's argument for the separation of work and interaction corresponds to the specific historical subsumption of work as it was understood traditionally—self-directed activity—under labor. Their unequal merger since the rise of industrialization makes reasonable the demand for an autonomous public sphere in which speech remains unconditionally free from domination. Habermas here ratifies the current setup but refuses its totalizing practice. This compromise parallels Marcuse's own assessment that the results of the scientific and technological revolution since the seventeenth century have been unexpected from the perspective of emancipation. In this sense, Habermas retains some of the subversive content of Critical Theory. We must seek the space for interaction, but not at the center.

At the same time, the politics of marginality no longer seeks an alternative totalizing practice, but struggles for spaces within a social order that appears to deliver the goods. "Communicative action" is not merely a supplement to purposive action, although in Habermas's account it functions in an extrarationalizing way—it becomes an alternative to the extent that it remains utopian despite its agreement to compromise. For "one-dimensionality" is nothing if not a totality of identity. The assertion of plurality of practices in the midst of rational-purposive activity undermines the technological imperative.

Still, while Habermas's desire for utopia through communicative action situates itself within at least one tendency in modern socialism, notwithstanding its distance from Marxism, I believe he has surrendered a major terrain of contestation for the restoration of human interaction he so fervently seeks. It is not so much that his refusal to address the question of work and its relation to interaction abandons the Marxist dream of the whole individual that is at issue. To be skeptical of this possibility merely echoes what has become a cardinal certainty in contemporary philosophy and social theory: the ineluctability of difference, the incommensurability of spheres of existence.[26] More significant is the effort, in the light of this judgment, to establish that utterances may enjoy a distortion-free environment without unpacking the ideological content of the idea of rational-purposive activity. Not only is this conundrum of Habermas's thought the consequence of the normative centrality of work for our culture, even as its historical necessity continues to diminish; it results as well from the role of work in the formation of moral identity in individuals and the extent to which work still defines moral horizons for most of us.

In renouncing the possibility of an intervention into the sphere of labor, Habermas has ratified the technical division of labor by producing a new ideology of human nature in which these divisions are elevated to a structural concept. The "mind" is always rational and directed toward natural and human domination, whereas the body is the sensuous, affective side of that species called human. In this metaphoric construction, the body is surrendered to the mind in the labor process or, more exactly, indulges its desire for freedom either through the fantasy life or in a distorted manner within the private sphere. Yet, in late capitalism, this life is colonized by technologically mediated culture, which now regards the problem of alienated labor as a technical issue, one subject to administrative manipulation, scientific investigation, and negotiation. Therefore, Habermas wishes to remove the body from the mind (and the mind from itself), steps made necessary by the totalizing force of technical reason in the labor process, and made possible by the end of material scarcity, except for marginal groups in late capitalist societies. To the extent that Habermas has a political program, his aim is to reconstitute political and social life on the basis of communicative actions rather than economic production. In this construction, science and technology are relegated to marginal activities in public life.

But this is no political program at all because politics is no longer free to contravene technological and scientific domination in Habermas's own conception of the repressive totality of late capitalism. To argue for a noninstrumental system of interaction advanced and legitimated by an institutional framework whose force derives from technological domination is to admit that what Habermas calls the "socio-cultural" phase of human development has no material affectivity. Having ceded to technological consciousness the entire sphere of rational-purposive action, Habermas is left with a constituency of university, college, and high school students whose political force derives precisely from their disjoined relation to the labor process—since they are not workers, technicians, or scientists (yet). Habermas believes that the legitimating processes of late capitalism, i.e., its ability to "deliver the goods" by a thorough mobilization of science and technology, are not operative among students because "their protest is directed against the very category of reward itself."

Habermas's optimism of the will has proven unable to cope with the changes in Western capitalism in the 1970s and the 1980s, in part because the material conditions he posits as the presuppositions for the struggle for a political sphere unburdened of distorted communication that is mediated by technical reason no longer obtain. The postulate of the end of material scarcity is now only a partial description of the social and political environment confronting the apparatus. Its failure to sustain growth beyond the Vietnam War and the early 1970s in Western Europe has challenged the legitimacy of the apparatus of domination and produced new problems for the political directorate, opening the spacing for a critique of technology and science as ideology. Since we have learned that the apparatus can fail to deliver the goods, the theoretical possibility of creating an opposition has returned even among the workers. Recent studies of the labor process, showing its race-, sex-, and class-based character, have exposed the extent to which "rational-purposive" activity has a specific history. Its character has been revealed to be temporal and structural only in terms of a particular regime of production.[27]

The protest against the degradation of labor that appeared throughout the capitalist West between 1969 and 1975 offered a moment of

reflection for workers. One of the questions raised by these movements against the "technical" organization of labor in order to degrade
and deskill it was whether technology and science could any longer be
regarded as neutral actors in the social process, "things" to be used by
alternative social systems in quite different ways. Although Habermas
does not claim that technology and science are neutral in their origins,
he makes a historical argument that implies that their reified appearance as immutable "things," as instruments, must now be accepted as
"natural" facts. For those who follow Habermas in insisting upon the
absolute transparency of rationality, its relegation to the labor process,
and the treatment of work as instrumental activity, the only solution to
the disappearance of classes and class struggle as relevant conceptual
categories of social analysis is to insert interaction as an autonomous
category of the social process.

CHAPTER 7
MARXISM AS A
POSITIVE SCIENCE

The virtue of Habermas's work is to remind us of the indissolubility of the Marxist framework. It is not possible to dissociate the theory of ideology from classes and class struggle, any more than science and technology can be regarded as either historically or logically independent of social relations. But that is exactly what Louis Althusser and his school have attempted to do.[1] Their assertion that Marxism is a science is specifically linked to the concept that in order to become a science, its theoretical system or discourse must separate itself from ideology. Althusser regards the critique of ideology as the first and crucial step in the development of science and claims that the early Marx may be partitioned from the late Marx on the basis of his critique of idealism. Althusser distinguishes science from ideology in three distinct features: (1) the object of knowledge is different in the two. The scientific object of knowledge, while different from the "real" world, is no longer informed by religious, abstract essences. In elaborating his claim that the early and late Marx may be differentiated on the basis of an epistemological break from the idealism of Hegel, Althusser argues that the condition for Marx's "theoretical revolution" was his constitution of a new object of knowledge. Marx moves from philosophical speculation to scientific practice when he discovers that society as a structure is the proper object of investigation, not the essence of humans.[2] (2) But it cannot be said that science and ideology are separated by a Chinese wall; science emerges out of its critique of ideology and constitutes itself in and through this critique. For Althusser, scientific knowledge is marked by its mode of production of knowledges (the plural here refers to Althusser's insistence that the knowledge modes of production, e.g., chemistry, physics, Marxism, are distinct practices that make up science: the use of the singular is ideological because it connotes a totality that is more than the ensemble of material practices). This mode of production, according to Althusser, may be likened to the labor process. The specific object of knowledge for each science is the raw materials from which the theoretical means of production (the methods of science including theory, experiments, technique, etc.) derives a result. Althusser acknowledges that the historical relations ("both theoretical, ideological and social") form the context of scientific labor. But these historical relations are accorded no determinative weight in the mode of production of scientific knowledge. Althusser holds that scientific knowledge is "concerned with the real world through its specific mode of appropriation of the real world . . . the mechanism that insures it."[3] This mechanism, the process of the production of knowledge, enables us "the grasp of the concept." For, even though the object of knowledge and the real world are distinct, science can, in Althusser's view, appropriate the real object or the real world through both a critique of the ideological object and its ability to form a mechanism of knowledge.

At this point, it is necessary to caution against the apparent homology between Althusser's concern with the process of the production of knowledges, its theoretical conditions, so to speak, and traditional concerns of epistemological inquiry. For he distinguishes his approach not only from the school of Marxist humanism, according to which (in his account, at least) the a prioris of appearance and essence and of the abstract totality climax in an idealistic ideology that must be banished from Marxist science, but also from Husserl and his school which proceeds from the problem of whether knowledge is possible.[4] Althusser implicitly agrees with the view that knowledge and its mode of acquisition is the object of philosophical inquiry, but not the question of how a subject can know the object. For Althusser, this is the wrong question since it ignores the real progress of science, a mode of production that has already established the possibility of science free of ideological determinations by its grasp of the real world through its mechanism and its means of production.

(3) Althusser places himself within one tendency within Marxist thought, a Marxism that makes a decisive break with reflection theory, according to which knowledge is "reflection" of the real world where ideas correspond to material processes as a matter of simple causality.

Althusser recognizes that the mechanism of scientific practices yields only a "knowledge effect," from which theory must make inferences. Althusser's theory of truth retains the traditional notion of the existence of an external world "independent of the process of knowledge" but makes no claim for the correspondence of ideas derived from scientific practice with this world as some kind of reflection:

> We can say, then, that the mechanism of production of the knowledge effect lies in the mechanism which underlies in the action of the forms of order of the scientific discourse of the proof . . . in the fact these forms of order only show themselves as forms of the order of appearance of concepts in scientific discourse as a function of other forms, which without themselves being forms of order, are nevertheless the absent principle of the latter. . . . the forms of order (forms of proof in scientific discourse) are the *diachrony* of a basic *synchrony*. . . . Synchrony represents the organizational structure of the concepts in thought—totality or *system* (or, as Marx puts it, 'synthesis'); diachrony the movement of succession of the concepts in the ordered discourse of proof.[5]

But proof is "not in the eating," in Engels's alimentary metaphor. The proof is found in the internal structure of science. The famous criterion of practice as the verification of a theory means the distinct scientific practice to which any specific discourse refers. For Althusser, the theoretical "practice" is self-contained by its own structural unity, a logical order that grasps the real "in thought" (although thought includes, even subsumes, the theory and technique of a specific scientific practice). There are no "guarantees" for scientific truth, except the norms of theoretical validity established by the scientific community.

Thus, we arrive at a convergence between the work of Thomas Kuhn, Charles Peirce, and Althusser. Kuhn locates scientific revolutions, defined as the replacement of an old paradigm by a new one, in the contradiction between the old paradigm and the anomalies of its experimental practice.[6] A shift to the new paradigm takes place, according to this theory, only when that paradigm is able to explain phenomena considered anomalous by the older one. The new paradigm that changes the substance of science may also entail new relationships both between science and nature, and within science itself. Yet, the norms of theoretical validity remain those accepted by the scientific community. Change is a process occurring within science; it takes place on the basis of the willingness of those whose work is "normal" with respect to existing scientific practices to accept the validity of the new paradigm. The scientist is already equipped with a series of concepts, a theoretical framework capable of grasping the real world in thought. Knowledge is not derived from observation but is only confirmed by it: there are no self-evident "facts." In the last analysis, Althusser implicitly follows Leibniz in his belief that the predicate of any true affirmative proposition lies, implicitly or explicitly, in its subject. This relation between subject and predicate resides in the "grounded connection" between the two or, in Althusser's terms, in their unity in structure. For Althusser has evolved a theory in which the forms of thought, the correspondence among the various elements, and the logical principle of order constitute the proof of theory. Here verification through practice, labeled by Althusser as rank empiricism, is subsumed by the logical principle of order. The "absent" link in Althusser is the relation of thought processes to nature.

This link is provided by the historical godfather of structuralism, Emile Durkheim. For Durkheim's major contribution to the legitimation of the social sciences was his insistence that they were continuous in their epistemology and methods with those of any other science: society as much as nature is "a structured and rational order, whose phenomena obey invariant laws and are determinate."[7] The critique of Durkheim's reliance on experience for obtaining truth from an Althusserian perspective refers back to the idea that theory requires no empirical basis for its propositions, but these propositions are true because they obey the same logical order as nature or society.

The notion of empirically grounded knowledge is always found in the realm of ideology. Thus politics, indeed, all the sciences whose subject is "man," must be ideological. According to Althusser, the subject of social science is social structure (the synchronic) and its ordered discourse which is prior to verification. Here, the reconstitution of the object of knowledge from "man" to "society" as a social fact irreducible to individuals, their subjectivity, their ideological relations to each other, and the social structure within which they live combines with the ordered discourse to constitute science itself. Thus, Althusser says that Marx constitutes the science of society as an object of knowledge through his critique of abstract "man," the object of the so-called human sciences, derived from Kantian premises that problematized the possibility of knowing anything outside human interaction.[8]

Althusser introduces a dualism in the study of the social world. There can be no science of social relations unless these are treated as a determinate ordered discourse, obeying definite laws already specified by the Althusserian canon that makes structures the true object of any science. Nevertheless, people do study interaction, social norms, and record their experiences of the social world unmediated by structural analysis. These studies are called ideology by Althusser and are defined pejoratively, although accorded the status of legitimate, ideological discourse—but not science. This is the sphere of "lived" experience or the "imaginary," which has an indirect relation to the real, imaginary because these inquiries are not theoretically grounded; they lack the apparatus of true knowledge and cannot grasp the real except as ideology.

Hence, class struggle, at the level of either trade union practice or practical revolutionary politics, can only be ideological since these practices arise from the lived experiences of workers. Science is somehow separate from the class struggle, even though class and class struggle may be the object of knowledge of scientific investigation provided they are viewed from the mechanism of structuralist analysis. In this mode, a priori, there are a finite set of categories, derived from the ideology critique from which the science has arisen, which form a grid through which reality is grasped, or appropriated. The progress of the scientific study of social structure, called the mode of production by Althusserians, takes place within the social context of lived experience, but this experience cannot have a decisive influence on the configuration of the science of society because it obeys a different set of rules.

In the end, Althusser claims that Marxist science, like any other science, can be value-neutral. It has overcome the "iron cage" of the imaginary within which all ideological discourse is imprisoned. Its language machine is capable of assimilating any raw material, chopping it up into discrete objects, ordering it according to logical principles, naming and mapping in advance, and "producing" knowledges that take on the aspect of a predicate of which the mechanism itself is the subject.

Althusser's metaphor of production and of the machine is not arbitrary. His Marxism turns out to be an almost conscious adaptation to the age of mechanical reproducibility, one in which the machine is both form and content, or, to be more exact, the form implies the content, which, however alien at the beginning, is produced as a reified object with no history. Althusser's attack on the search for origins as ideological is an attack against the effort to insist on the validity of knowledge before the Enlightenment, just as Copernican science criticized its forebears as mysticism. For, at base, Althusser is a rationalist: anything that refuses his mechanical idea grinder is labeled irrational, or ideology. A science of politics or of art is possible only on condition that these are treated as ideological discourses since they are premechanical.

Among other problems, the Althusserian theory of science seems incapable of finding the new features of social, economic, and political development since, in his own metaphoric analogy, these are part of the social unconscious, and he views the unconscious as the seat of the irrational, the structural root of all ideology. Following the metaphor, the process of acquiring scientific knowledge may be compared to the process by which the patient makes conscious irrational, unconscious desires and needs in order to control them. The congruence of Althusser's conception of science and the instrumentalization of reason that has been integrated harnesses the unconscious and makes it part of the conscious life, transforms it from alien nature to raw materials for the theory machine.

I do not believe that this comparison of Althusser's separation of science and ideology to the unbridgeable parts of the psychic structure in Freudian theory is farfetched.[9] Just as Freudian psychology has a side that seeks to subordinate anarchic, irrational human nature to purposive-rational action, so Althusser wishes to restrict ideology to certain spheres of human activity or, if possible, to progressively subordinate them to Marxist science, considered here sovereign because untrammeled by lived experience,

John Mepham has likened the relation of science and ideology to two diffferent generative sets having a different matrix of internal relations.[10] Both are considered structured discourses that may be understood as separate languages. In Mepham's conception, "social life is structured like language" arranged on a semantic field that is, in the main, beyond the scientific comprehension of those who participate. Mepham: "The natural self-understood meanings encountered in social life form a text which we need to decipher to discover its true meaning."[11] The comparison with Freud's theory of dreams comes to mind. Just as the dream speaks a language different from that of the conscious life and defies literalization, so people in everyday life speak an unconscious language that can only be translated by means of other, more scientific categories. The structure of Freud's thought entailed a "generative set" of concepts through which the dream work could be deciphered. These were grounded in the mechanism of the psychic structure and certain processes that followed from the contradictions among its elements. Freudian-Lacanian psychoanalysis constitutes itself much like the Althusserian definition of science. The unconscious speaks a language whose meanings are hidden to ordinary comprehension. It can only reveal itself through slips of the tongue, jokes, gesture, and dreamwork, which must be transcoded into the language of science in order to be understood. The "real" is a set of relations constituted as a structured discourse that is invisible to ordinary cognition but presents itself in a "phenomenal form." Social life perceives this phenomenal form and translates perception into the structured discourse of ideology, which is constituted as a symbolic order, maintaining the real as opaque, that is, concealing its generative set of relations.

In Freud's theory, science as ordered-discourse-deciphering is necessary, but not sufficient, for cracking the hidden code of the unconscious. The unconscious constructs mechanisms of defense (condensation, displacement, linear causality, clues that lead to blind alleys), just as the real masks itself in cognition. Thus, the Althusserian attack against the possibility of gaining scientific knowledge through observation since the "data of experience" yield only the real in its phenomenal form. Science must treat the observed perception of things with skepticism, treating these data as ideology, the critique of which will constitute the first step in the development of science. The transformation of the data of experience into the raw material upon which the knowledge-producing machine will labor is the way Freud hoped to make the manifest text of the dream part of the process by which the latent text is revealed.

Althusser's theory of the relation of science and ideology is that ideology is not produced by erroneous or even class-bound beliefs or value systems. If this were so, ideology would disappear with the end of class society. But the Althusserian variety of Marxism insists that ideology is structured discourse of lived experience whose variance from the "real" is not subject to historical change. The particular phenomenal form of real relations will surely change with the transformation of capitalism into socialism. But the gap between real relations and phenomenal forms is transhistorical, that is, ideology as "lived experience" does not disappear with the new social order because its source is not the distorted values and beliefs of bourgeois society; it is not false consciousness. Ideology transcends one mode of production because it is a structure of relations between lived experience and the real.

Once again, we are reminded of Freud's remonstrance against the "naive" Marxist belief that the collective ownership and control of the means of material production will abolish all conflict, certainly all human contradictions.[12] Just as Freud posited the invariance over time of the contradiction between the pleasure principle and the reality principle (or, in the earlier form, the id and the superego), so Althusser holds to the eternity of the distinction between ordinary ideological discourse and scientific discourse, where the former is considered a generative set of concepts inscribed in everyday language and interaction.

Science arises from the critique of ideology in every society. The theory machine will always be necessary to prevent the phenomenal form of real relations from rendering all reality opaque. Here, then, is the inevitable privileging of science, the necessity of its separation from ideology, the heart of the concept of its transhistoricity. While Kuhn does not, in principle, exclude social and historical determination, or at least influence, of the process and structure of scientific knowledge (indeed, his analogy between scientific and political revolutions is explicitly drawn), he has brought none of these relations inside the process of scientific development. The implication of this exclusion from the discourse of Kuhn's investigation of the history of science is close to Althusser's attempt to separate science from ideology. Both would agree, to be sure, that historical, ideological, and social considerations are part of the context of science, but the insistence on the autonomy of normative practices such as criteria for validity, constitution of the scientific object, etc., tends to neutralize the ideological influence/determination of science itself.

Kuhn is, of course, much more critical of the category of scientific truth than Althusser, whose insistence on the discontinuous, with his categories of relative autonomy and of the primacy of structure, render his complex argument philosophically naive, in the last instance. Although he does not go as far as Paul Feyerabend[13] in claiming that the dominance of a scientific paradigm in any historical period is arbitrary, that it has no historical necessity, Kuhn does argue that scientific "progress" remains in the eye of the beholder.[14] But what puts Althusser and Kuhn in the same theoretical camp is the notion that the relative autonomy of the scientific community from the "laity and everyday life" is the foundation of the insularity of science from ideology.

In this respect, Kuhn's ascription of insularity to the separateness of science in social terms from everyday life, its institutional autonomy, has the virtue of leaving the door open for an empirical investigation of whether this assertion holds for contemporary science, if it ever did for earlier periods. I shall examine in this and later chapters the thesis that such autonomy of science from ideology could ever be successfully argued, at least up to the present. Althusser, on the other hand, in his desire to show the scientificity of Marxism, an antiempiricist and antipositivist science to be sure, has been constrained to hermetically seal both Marxism and other 'scientific' practices from interaction with social, ideological, and historical relations that determine, in any measure, the content of scientific knowledge, except insofar as science emerges at the intersection of the epistemological break from ideology in terms of the constitution of the object of knowledge. But, for Althusser, the paradigms of science are debated, decisions arrived at, new theoretical norms agreed upon, entirely within the scientific community.

Consider the words of the philosopher of science Charles Sanders Peirce. Referring to the concept of truth, Peirce says:

> The opinion which is fated to be ultimately agreed to by all who investigate is what we mean by truth, and the object represented in this opinion is the real.[15]

Peirce was aware of the problem of the necessity of these opinions and the question whether it was possible to speak of infallible results of investigation. He viewed the object as independent of the processes of knowledge but remained fixed on the ideal of one necessary result of investigation by all those competent to conduct this work. For Peirce, as much as for Althusser, the mechanisms of science are the road to knowledge. It is their infallibility that must be relied on to yield truth. If these could be challenged, the relativity of scientific truth, its fallibility, and thus its ideological character would logically result.

Althusser has worked himself into a cul de sac on the road to asserting the scientificity of Marxism. Since he can only admit that Marxism may be an ideology from the point of view of the revolutionary movement but not in the rigorous terms in which scientific discourse is cast, and he rejects the reflection theory of knowledge and the correspondence theory of truth, he finds himself caught in a very un-Marxist idea, that is, the possibility that a sphere of social activity may be free of what he considers ideology to consist in—a structured asymmetric relation between humans and their objective world. In his conception, Althusser has posited a privileged mechanism that saves science from historical, political, and social determination. This mechanism is the category of structure or system of concepts that are ordered in a type of hierarchy where the succession of one concept by another has a specific form that is said to be scientific. The origins of these concepts are obscure, except insofar as we may trace their ideological roots, e.g., the development of modern chemistry from the phlogiston theory, or Galileo's radical critique of Ptolemic physics from within science. However, all true science adopts, according to Althusser, "an ordered discourse of proof" that enables it to grasp the real unmediated by social determinations.

In fact, Althusser finds the concept of mediation itself to be ideological since this is not a material concept. By its insertion into the process
of the production of knowledge, it undermines the certainty of knowing and reintroduces the question about the problematic relation of
science to its object. Thus, we find Althusser in the position, especially
curious for a Marxist theorist, of asserting that the mechanisms of scientific knowledge are ideologically neutral. This implies that technique, the experimental method, and technology as such are neutral as
well since, in Althusser's conception of the history of science, scientific
knowledge gives rise to technology. Consequently, Althusser equates
science and technology with the labor process as the sum of human
relations to nature, constituted as material practices that are entwined
with the structure of production. The forces of production, in Althusser's discourse, follow Marxist orthodoxy: they are (relatively) independent from and prior to the relations of production. When Althusser
uses the metaphor of production processes to understand knowledge,
it is not, in his own term, "innocent." The theoretical terminology
anticipates the result: science and technology are free of ideology
since the former are part of the base of society and the latter is part of
its superstructure. The superstructure, according to Althusser, is relatively autonomous from the economic base and may have some influence upon it, but the forces of production, of which science and technology are important elements, are the materializations of society's
grasp of the real world through a theoretical practice of which technique is a crucial part.

I have argued that the doctrine of the neutrality of technology is
untenable. We must now demonstrate that the ideological neutrality of
science is similarly untenable. Despite their antagonistic theoretical
frameworks, both Habermas and Althusser deny that technology and
science are ideology. But, where Habermas's argument rests on his
conception of the permanence of reification after the universalization
of the commodity by capitalism and has conflated reification with the
new given of rationality, Althusser has attacked the category of reification as such. For Althusser, science is a self-legitimating discourse
whose ordering of proof, apparatus of theory, and method of discovery
are unproblematic. His task is to show that Marxism is scientific
because it orders its concepts in a structural unity that is homologous
with other sciences—theoretical practices whose histories consist in
the overturning of the ideological character of prior paradigms. So
Marxism is now a social theory whose determinations are relatively

independent of their social, historical, and ideological contexts, since science cannot refer to its origins or its context, but is ultimately self-justifying.

The importance of Althusser's contribution to the Marxist theory of ideology is his insistence that ideology is situated within the forms of social life rather than within the realm of ideas alone. However, the freeing of the concept of ideology from the label of "mere illusions" was originally suggested by Georg Lukács, who found the basis of ideology, not in Weberian values and beliefs, or the earlier Marxist idea of "false consciousness," but in the ordinary apprehension of the forms of appearance of things.[16] But Lukács's argument stems from his theory of reification within capitalist society. To the degree that, historically, the commodity becomes dominant in the process of production, relations among persons (the "real relations" within the capitalist mode of production) are enshrouded in a fog of mystification. Their form of appearance is relation between things, an exchange of equivalents that seem to be grounded in the intrinsic properties of objects rather than in social relations. In this conception, the source of ideology, which is universal among all those who live within the capitalist mode of production, is the commodity form. The opacity of the material world is not a property of perception, but a "natural" cognitive effect of the transformation of use value into exchange value, the process of production into the process of exchange and the subsumption of the labor process under capital.

For Lukács, class relations *mediate* these fundamental sources of the production of ideology by giving ideology a specific character. But the values, beliefs, etc., of a certain class are necessarily variants of bourgeois beliefs, not because of the imposition of these ideas by concrete persons who may be their bearers but because the configuration of commodity production subsumes the concrete into the abstract, both at the level of labor, where labor time as a unit of measurement replaces the specific kind of labor (weaving, carpentry, cooking, waiting on tables, etc.), and at the level of the commodity, where use value is subsumed under exchange value. Thus, we measure ourselves in terms of how much wages the sale of our labor power will bring. We are "worth a definite quantity of money." Mepham points out that the transformation of the value of labor power (the amount of socially necessary labor time embodied in the commodities necessary for the reproduction of the worker and her/his family) into wages is a prime example of how the form of appearance of real social relations leads to ideology. The worker believes that her/his wages represent the number of hours for which labor power has been sold. Thus, the ideological category the "value of labor," as if the exchange of a certain quantity of labor power for wages was an exchange of equivalents. In turn, this mystification hides the source of capitalist profits that now appear to originate in the marketplace, the risk factor in investment, or in the morality of individual enterprise. The form of appearance of the commodity, in this case labor, hides the source of profit, the difference between the value of the commodity and the value of labor power. Marx uses the terminological transformation to point to his distinction between appearance and reality. The value of labor power becomes wages; surplus value becomes profits; production becomes a series of market exchanges. It is not that the perception of the social reality is false, but that the reality has two forms: its appearances and its real relations.

Lukács locates the appearance/reality problematic within a definite historical stage of development: the capitalist mode of production. Althusser, on the other hand, transforms Lukács's insistence on the historicity of the category of ideology into a structural principle. That is, he posits the distance of humans from the real and the *opacity* of the real relations as transhistorical, since these relations are rooted not in the commodity form but in the structural distinction between the language of appearances and the language of reality. Since these languages are ordered discourses that obey their own inner laws, and are quite separate from each other, the deciphering task is finally cognitive rather than a social and historical problem. The consequence of formulating the problem of ideology in these terms is to establish science as the only possible means by which ideology may be overcome. There are, for Althusser, no circumstances that may render social relations transparent to lived experience. We are *always* destined to live our lives ideologically, regardless of the social system.

Thus, the ideological is *naturalized* by Althusser's structural binaries. On the one hand, he has abolished the myth of the "integrated civilizations" which was, for Lukács, the basis for his critique of late capitalist society. In Althusser's hands, the critique of ideology no longer relies on the assumption that transparent social relations may be experienced under any circumstances. Our relation to the real will always be problematic because of the incommensurability of the symbolic order as discourse with the real; only science, by constructing a mechanism of knowledge ordered in the manner of the real by the structural rather than empirical homology, may grasp the real. On the other hand, the implicit assertion of Althusser's theory of science is that the real is the rational; that is, modern scientific knowledge is an ordered discourse that holds the secret of the real, and its concept will only be overthrown from within its practices, its theory machine. Habermas really ends with a variety of "end of ideology" critique that was typical of American and British sociology in the late 1950s.[17] At that time, the prospects for capitalism appeared limitless; the working class appeared safely integrated into the apparatus by the rewards of technological development and trade union successes; and the system seemed amenable to an infinite series of technical adjustments to keep it going at high levels of production and consumption. Since the source for the production of traditional ideology was the *reflection* in our minds of the real class relations, as these relations took on the character of a reified totality, that is, as it could no longer be argued that society was irreparably divided against itself, only an act of will could restore that which was lost by the capacity of technocratic consciousness to subsume ideology within the processes of technical problem solving—the utopian element in all ideological production.

Althusser's fundamentally rationalist framework privileges theory as a series of practices that, taken together, constitute the "real." Science is constituted by its nonessentialist categories which are developed by the critique of ideology. Materialism is combined with the Kantian preoccupation with epistemology to produce a theory of knowledge whose referent, but only in the last instance, is the real world. However, in Althusser's version, the real world, like economic relations, has no *practical* significance; it is a postulate that remains unexamined in an otherwise methodological inquiry that focuses on the problem of theory-formation. In this sense, the concrete empirical object is always supposed as theory dependent and has little status in scientific inquiry.

In contrast, Galvano Della Volpe seeks to restore to the concrete object the position of both starting point and end of the process of knowledge, at the same time insisting on a critique of categories as necessary meta-science.[18] That is, Della Volpe agrees with Althusser that Marxism's scientificity depends on more than the correspondence of its propositions to an objective world; the work of constituting categories precedes specific scientific hypotheses. However, his purpose is to clear Marxism of its metaphysical, irrationalist baggage, in a word, to free Marx from Hegel. The key move will be to install formal logic, especially the principle of noncontradiction as a positive science. Della Volpe argues that judgment always entails identity and noncontradiction. Thus Hegelian dialectics is effectively removed from Marxism.

Marxist structuralism is neither dialectical in the Hegelian sense nor, strictly speaking, materialist. Rather, it adopts the position of the Kantian tradition. While distancing himself from Kant's essentialism, Lucio

Colletti (who calls Della Volpe's *Logic as a Positive Science* "the most important work produced by European Marxism in the post-war era") still insists that "it is important to take epistemology as one's starting point in order to understand the genesis of the Marxist concept 'the social relations of production' in the very problems of classical philosophy."[19] According to Colletti, these problems are the relation of thought to its object, a relation which is, of course, always problematic for classical philosophy because of the nonidentity of thinking and being. Colletti shows that for Marx, the concept of work as simultaneously human beings' transformation of nature and themselves is the core of Marx's transformation of the Hegelian categories from idealism to materialism. But Colletti's main concern is not this, a point both unremarkable and unexceptionable in Marxist literature. More important is his implicit Kantian reading of the epistemological consequences of the labor process. Following Della Volpe's insistence on the continuity of the early and late Marx, Colletti reads the early Marx's *Economic and Philosophic Manuscript*, not as a derivative Feuerbachian text still mired in the Hegelian problematic (as Althusser claims), but as embryonic historical materialism, the key to Marx's theoretical revolution. But what is historical materialism? For Colletti, it is the doctrine according to which nature is "objectively sensuous," that is, ineluctably linked to "my own subjective sensitivity itself." The "object" nature does not take on significance, even if posited formally outside human consciousness, until it is appropriated through labor. And this appropriation is, at the same time, a cognitive act both of nature and of self. "There is no consciousness of the object without self-consciousness. What I see of the world is what my ideas predispose me to see. My relation to nature is conditioned by the level of social-historical development."[20]

There are many things to unpack from these sentences. First, there is Colletti's own framework of self-consciousness, of his focus on knowledge for the individual, which implies that the process of knowing is the subject of philosophical inquiry. With Engels, Colletti conceives the object always in the process of traversing the "in-itself" to the "for us" by means of the labor process. In this reprise, knowledge of nature is made possible not only because of the mediation of labor as equally historical and epistemological activity, but also because "man" is part of natural history, linked to nature by the totality of being, while admitting the heterogeneity of thought and being. Thus, Colletti with Della Volpe affirms that materialism always needs a concept of noncontradiction to establish the status of its epistemology. "Noncontradiction" becomes the "material determinacy" of thought, but only mediated by the social relations of production. For Colletti, consistent with both Marx and part of the Marxist tradition, these social relations are, first of all, relations with nature. Thus, "thought is not a self-contained entity of epistemology" but must necessarily complement the sciences of "man" as a "natural being," a being with imagination prior to labor.

Hence, labor is always purposive activity, with purposes that refer not only to the adaptation of humans to nature, but also to their mutual relations. Colletti's concept of social relations of production reveals the degree to which scientific knowledge is always mediated by social relations, although Colletti does not make this point explicit because, like Della Volpe, he stops with the explication of the logical structure of knowledge, a debt they both owe to Kant. Yet, in concert with Althusser, the efforts of Horkheimer and Adorno to show the ideological character of science are met with scorn. Science is nothing less than "higher achievements of human thought," which the authors of *Dialectic of the Enlightenment* subject to "nihilistic negation." The Frankfurt School has confused the romantic critique of the Enlightenment, a current ingrained in the phenomenology of Heidegger and Husserl, with the critique of capitalism. So, Colletti identifies himself, albeit implicitly, with the Marxist reverence of science as truth, a reverence that has marked philosophy as such since Bacon and, later, Kant. And it is quite reasonable, one would like to say "natural," for Colletti to dismiss the critique of science as an aspect of the domination of nature precisely because of his own Kantian framework for which the "respect" for nature as autonomous is a meaningless concept in light of his valorization of work as knowledge-producing activity. Since the theory of knowledge is the starting point of our comprehension of nature and social relations, Colletti falls into the hole of the radical separation of space and time, privileging, as do all structuralists, the former. Temporality becomes a purely intellectual category and history subordinate to epistemology. The fundamental relation between the priority of the problem of knowledge and ahistorical ideology becomes the basis of the structuralist attack against the Marxism of Lukács, the Frankfurt School, and Jean-Paul Sartre. What united these otherwise heterogeneous modes of thought is their conception of dialectical reason as nonidentical with scientific practice. Sartre makes this distinction most explicit: "Bachelard has shown clearly how modern physics *is* in itself a new rationalism: the only presupposition of the *praxis* of the natural sciences is an assertion of *unity* conceived as the perpetual unification of an increasingly real diversity. But the unity depends on human activity rather than on the diversity of phenomena. Moreover, it is neither a knowledge, nor a postulate, nor a Kantian a priori. It is action asserting itself within the undertaking"[21] in which the ends of the activity take precedence over means. For Sartre, the key distinction between scientific rationality and dialectical reason consists in the latter's situating itself in the world, rather than making the radical separation of subject and object which is a basic presupposition of modern scientific ideology. Sartre also attacks the substitution of analytic for historical reason which he accuses many of his contemporaries, notably Claude Lévi-Strauss, for having done. Here, Jonathan Rée defines analytical reason as "the form of reason appropriate to the external relations which are the object of the natural sciences,"[22] or, to put it another way, the rationalistic core of these sciences. Yet Colletti, who wishes to comprehend the object subjectively, argues for this concept of knowledge by defining activity as chiefly *appropriation*, in contrast to which Sartre maintains his earlier position of *Being and Nothingness* by locating human activity within the world. Colletti understands this as human self-production but through appropriation, thus reproducing the assumptions of scientific rationality. As the mediation between us and nature, the core concept of labor is made, like much of the Marxist tradition, into a neutral activity; and human purposes, which must always be ideological, are purged of their antagonisms.

In effect, Della Volpe and Colletti seek to provide a philosophical basis for Marxist science by bringing the propositions of Marxism into conformity with scientific method. Della Volpe argues for a conception of "matter" that is quite distinct from thought. Moreover, this matter is intelligible, unlike the presupposition in Althusserian thought that, following a more orthodox Kantianism, holds the reverse. But, as Martin Jay has remarked, Althusser's antipositivism treads dangerously close to idealism, despite his materialist intentions. Della Volpe veers toward positivism when he declares that experimental science would verify the basic propositions of Marxism or it could not claim the mantle of science. At this juncture in the argument, Della Volpe invokes Galilean science as the model to which Marx adheres, at least morally.[23]

The internal debate among anti-Hegelian Marxists fails to resolve the antinomy of idealism and positivism precisely because its starting point remains ensconced in the problem of knowledge. Since Althusser's rationalism is ultimately unacceptable to those who would insist on the scientificity of Marxism in the usual sense of the term, it is only a short step to an unabashed positivism. And that is precisely what the recent spate of philosophical and social theoretical attempts to resolve the crisis in Marxism have done. Gerry Cohen, John Roemer, Jon Elster, and Eric Olin Wright[24] have insisted that Marxist categories and propositions concerning the social world be subjected to the same analytic scrutiny and empirical falsifiability to which any assertion of normal natural science must suffer. Finding no empirical basis for a given category or proposition (the labor theory of value or surplus value, for example), Jon Elster takes up the basic axioms of "Marxian economics" as if they were those of positive science and concludes that while many of Marx's own discoveries remain valid, others do not. Especially important in Elster's *Making Sense of Marx* is his own acknowledgment that the validity of any set of propositions depends on their methodological assumptions, which must be justified by whether they can be demonstrated as "theorems that would otherwise be unsubstantiated postulates." In other words, the criterion of operationality governs the validity of the assumption of methodological individualism upon which neoclassical economics is based. Although Elster is careful to avoid the equation of "rational" economic "man" with this methodological theorem, arguing that individual action and belief rather than human nature is the starting point, Marx is faulted for his frequent, although never consistent, assumption of methodological "collectivism," which Elster equates with Hegelian essentialism.

As Elster explains, the labor theory of value makes the collectivist assumption of homogeneous labor. Elster argues that if it can be shown that labor is irreducibly heterogeneous, value theory in Marxist economics is inevitably undermined since the entire foundation of his hypothetic-deductive system depends on this assumption. Elster always assumes that the question is the scientificity of Marx's work, especially whether he is a social scientist in the contemporary, positive sense. Are the various elements of historical materialism and the political economy of capitalism demonstrable as rigorously empirical propositions? As Elster surveys Marx's work from his own analytic perspective, three closely related criticisms emerge: Marx employs metaphysical assumptions upon which empirical assertions are based; he is guilty of teleological thinking, in which intentionality is held as a presupposition of social action; and Marx stands condemned of functionalism, which is, of course, counter to the nominalist a priori of methodological individualism.

Of course, the key metaphysical assumption is dialectics, a Hegelian hangover which is the basis of methodological collectivism. On the other hand, Elster tries to save Marx from condemnation as just another metaphysical philosopher by examining his theories to show that when he adopts principles such as those inherent in the mechanical, positivist perspective, he produces much that is of scientific value.

Since this body of work draws its fundamental inspiration as well as core categories from concepts that are, to say the least, as contestable as those of Marx and Hegel, one would want Elster, Cohen, et al., to provide justification that would take account of the criticism leveled by, say, the Frankfurt School, against positivism. Instead, the criterion of operational, behavioral nonteleological science is invoked sui generis as if no debate exists within natural and social sciences concerning these issues. Elster shows that Marx is not methodologically consistent, but Gerry Cohen finds no such inconsistency. Instead, he chooses to justify, or, in his words, "defend" Marx's theory of history using the presuppositions and logical categories of analytic philosophy. Consequently, he tries to assert a technological determinist interpretation of historical materialism by showing that the forces of production are not part of the economic infrastructure but, on the contrary, determine them. Cohen has a substantive as well as a methodological theory of science. Science can in no way be an ideology since it is not part of the superstructure but is, for Marx, located in the productive forces insofar as the theory of history is concerned. (Cohen acknowledges that science is not totally subsumed by productive forces, but that portion of scientific discovery which contributes to the growth of the productive forces is.) It is instructive to examine Cohen's argument in some detail because it reveals precisely what the issues are in the effort to read Marx's theory of historical materialism as positive science.

One of Cohen's key objectives is to purge Marx-interpretation of its vague, ambiguous elements. He wants to employ the "standards of clarity and rigor which distinguishes twentieth century analytic philosophy" to this task, not only because of the mistakes of his followers but also because of the confusions in Marx's own language and sometimes his formulations of theory. In his pursuit of rigor, none of the distinctions Cohen wishes to make is more apposite than that between material relations of production and social relations of production. Quite apart from Cohen's claim that this distinction is crucial for understanding Marx's own technological determinism, an analysis of what he means by it will instruct us best concerning his defense of historical materialism. For, although Cohen performs much textual reading, particularly Marx's Preface, to show that he wanted to separate the forces from the relations of production, it is his own categorical separations that give force to the argument. Quoting Marx, he argues that machinery is no more an economic category than the bullock that drags the plough. Machinery is merely a productive force. . . . The modern workshop which depends on the application of machinery is a social production relation, an economic category.[25]

Although Cohen acknowledges that the forces of production change over time, depending increasingly on scientific discovery and the application of the results to technology, and that productive relations may retard or advance this process, Cohen's point is that to confuse material relations of production with social relations of production violates the rules of logic. More particularly, the problem is the relation of form to content. Machinery is not capital, which is a social relation, until it enters exchange. Until then, it is only a means of material production, a use value. For Cohen (as for many in the orthodox tradition), machines are tools that help humans negotiate their relationship to nature. But since modern machinery is indirectly the product of scientific development and, according to Cohen, despite the fact that it shares the quality of being a mental activity with ideology ("science is not ideology"), machinery can be understood in its material substance as a thing, except under certain conditions. This means, we may infer, that whereas capitalism is responsible for setting social conditions that spur or retard material production and its preconditions such as science and technology, machinery is not a social relation.

Similarly, the relations that workers enter as they saw wood, for example, are material relations. The mode of cooperation between two workers has nothing to do with social relations of production until their product enters the marketplace. Unless we grant these distinctions, the weight of Cohen's argument for a technological determinist interpretation of historical materialism is seriously impaired. "History is the growth of human productive power, and forms of society rise and fall accordingly as they enable or impede that growth."[26]

Is machinery a thing, or are work relations merely a material relation that becomes social relations only under specific circumstances? The core of Cohen's methodological argument rests on the assertion of the ideological neutrality of science and technology, that machinery, for example, would be the same under different and even competing social systems that employ different economic structures. Presumably, the different systems using the same technology would or would not maximize its most effective impact, but the technology would not, in the last instance, be dependent on these different social systems with respect to its character as a material productive force.

If the socially neutral covering of science and technology is removed, the entire picture changes. We might find that machinery is not merely "machinery" in Cohen's sense, that its embeddedness in social relations of production extends to its core design as well as to its function. This would not obviate the statement that a worker operating a drill press makes a hole, but it must be pointed out that the drill press is a design that presupposes a division of labor marked by increasing specialization. This division of labor is not "technical" in the sense of being neutral with respect to class and other social relations of production. The single valence machine is the product of a long process in the class struggle and is *embodied social interest*. The object "drill press" is constituted by social relations of domination, and its material configuration is not outside these relations. Of course, in order to undertake this kind of analysis, one would have to set aside the analytic "rigor" that supposes an object can be identical only with itself and takes on a new form only when it enters a different context. For the object itself possesses a twofold character: material production contains its social character within its (non)identical material form.

This formulation entails that an economic and social system that includes its overt and covert ideological premises in all of its mental activity, including science, will be found embedded in its material relations of production, such as machinery, forms of cooperation in the labor process, and in the organization of work. In the case cited above, we no longer assume that the drill press, the computer, and laser technology are devoid of social/discursive premises. Thus, when the Soviet state chooses to replicate the Ford production process in its giant Kama truck plant, this technology transfer has social consequences because the production of trucks in Ford entails a division of labor, forms of specialization, and management control that are intimately linked to social domination.[27]

That the assembly line is a socially constituted technology should not be news to anyone who has perused the recent Marxist and non-Marxist literature on the labor process. But there is no evidence in *Karl Marx's Theory of History* either that Cohen has considered this literature or that he has taken it into account in his reading of Marx himself. As we have seen in chapters 2 and 3, Marx's own study of the labor process led him to understand the embeddedness of social relations in the material process of production. That he elided the question of science is, of course, consistent with a long tradition that is now in the process of being reexamined from many perspectives, including that of neo-Marxism.

One of the crucial concepts in the recent understandings of science and technology is the close link between professional engineering and the struggle waged by capital in the industrializing era to wrest control of the workplace from craftpersons. This is an instance of the power/knowledge fusion spoken about by sociologists of science such as Bruno Latour, historians such as Michel Foucault, and the philosopher Alfred Sohn-Rethel. That knowledge is bound up with power, and in the late nineteenth and twentieth centuries increasingly seeks this tie as a condition of its own growth, does not obviate the proposition that it is constituted through discourse/knowledge. These concatenations take place in the sphere of "pure" science as much as in technology insofar as concept and object are constructed and not merely "there" to be discovered.

For these propositions to hold, it is not necessary to show that Pasteur's discoveries of a serum for anthrax became de rigeur in agricultural practice because he successfully "marketed" his product within the scientific community, agricultural business, and the state (although Latour shows that this was the case). More to the point is the fact that by the eighteenth century, a nexus of cause and cure for disease was established, and that medical science incessantly sought explanations for the spread of disease that biochemical substances could thwart. Thus, National Institutes for Health in the United States allocate large sums to scientists for the purpose of investigating how multiples of individuals may be cured of cancer by means of some medical procedure that arrests the growth of cancer cells in the individual body.

What counts, therefore, is that medical science has defined the object of inquiry as the individual who "has" the disease. Much research is devoted to discovery of why some people are susceptible and others relatively immune. Thus, the body becomes a material object upon which scientific efforts to oppose disease are focused. Is this choice "innocent" of the tradition of both medical science as well as philosophy? The question is whether the metaphysics of Descartes, Hobbes, Concordet, and others bears at all on science in defining its object and its conceptual apparatus of inquiry.

This question is particularly apposite to biological and social sciences, but of course has relevance to physics as well. Even if scientists do not seek a fusion with large-scale pharmaceutical firms, regarding their work as pure discovery, their theory and practice are themselves formed by these discourses. For these reasons, the ontological distinctions made between material, social, and discursive relations are not tenable. Cohen's defense gives us only a rigor that ossifies the categories, and a clarity that obscures the relationships, particularly between thought and object.

There are other specific problematic aspects of Cohen's defense. One is the assertion that human nature is inherently rational; another is that "the historical situation of men is one of scarcity," which places work at the center of their life activity.[28] A third is that society does not replace one set of productive forces with inferior ones (this is stated as a "fact"). To elaborate:

a. The idea of rationality is identical with what I have called instrumental rationality, in this case, the ability of humans to identify and set about to meet their needs and wants, to wrest from nature the required raw materials from which to fashion products of consumption.

b. Cohen's work contains no idea of the historicity of the category of scarcity.[29] There is no intimation that scarcity is not, in contemporary societies, a function of material scarcity in the sense of being separate from social relations, although scarcity manifests itself for Third World societies and individuals everywhere as a material phenomenon.

c. What is superior or inferior in Cohen's defense of the idea of "progress" in the development of the productive forces is almost exclusively linked to productivity and the scientific and technical conditions for it. There are, for example, no ecological mediations to progress in productivity, and no sense of the environmental impact of technologies that have revolutionized the workplace.

Most astounding is the statement that "when knowledge provides the opportunity of expanding productive power, they [humans] will take it, for not to do so would be irrational."[30] So, it turns out that what is real is rational; a society is to be evaluated by the degree to which it succeeds in expanding productive power. Even if a ruling class promotes this development, Cohen admits that this coincides with the general human interest.

Cohen has given us nineteenth-century evolutionary Marxism without the reservations that the ghastly consequences of development have generated, even among enthusiasts of science and technology. We would have expected that analytic rigor would have helped introduce these mediations into consideration of the primacy of the productive forces. That some of our key scientific and technological advances are bound up with the accretion of forces of destruction (not only military weapons but also chemicals, fuels, the mass automobile) seems to have escaped Cohen's understanding of rationality criteria. What Cohen has described accurately is the degree to which a specific conception of rationality may lead to a general celebration of the "scientific technological" revolution as the motor of history.

In sum, there are two points to be made about the new analytic Marxism: first, its relentless effort to purge the dialectic from Marx's discourse by demonstrating, even when not asserting explicitly, Della Volpe's insistence on the logical principle of noncontradiction as the central a priori of social theory. Second, in Cohen's work, the remarkable reassertion of unilinear causality and the idea of progress based, not so much on evolutionary theory, but on the formal logic of recent analytic philosophy for which two opposing concepts cannot occupy the same temporal space. This implies that concepts such as multiple determination, "overdetermination," etc., are simply metaphysical propositions. For Cohen, technological change is conceived in terms of autonomy from social determination—an objectivist, internalist account upon which rises forms of society. Absent is more than the Hegelian dialectic: also missing is any reference, except in derision, to social relations as the "material" framework within which scientific and technological change occurs. As we shall see in the next chapter, the primacy of the scientific technological revolution over social relations and the preeminence of history as the embodiment of "progressive reason" are also the twin pillars of Soviet discourse on science and technology. This is invoked not to show the convergence of the structuralist and analytic tendencies (in different ways) with Soviet Marxism, but to assert the unity of an orthodoxy based on scientific and technological determinism regardless of particular political tendency.[31]

As we discovered in chapter 5, even if the Frankfurt School failed to complete its critique of modern science and technology by showing *concretely* how they were constituted as hegemonies and refused to specify what a new science might be, their singular contribution was to have shown the ideological underside of the scientific worldview. In contrast, Jürgen Habermas, as perhaps the major legatee of the Critical Theory tradition, abandoned the project by declaring that science and technology corresponded to the *general*, i.e. human, interest, notwithstanding its ideological features. His category "rational-purposive" action to describe science merely removed this sphere from the searchlight of ideology critique. Henceforth, theory became an inquiry into the condition for undistorted communication, a language game whose space was confined to the surplus. Now, in a major respect, this shift from the sphere of production to "interaction" followed the suggestion of Marcuse at the conclusion of *One Dimensional Man* that there was no longer a *practical* critique of technology since it had succeeded in completely dominating the contemporary social world. Marcuse, in effect, calls for a politics of marginality, a new focus on the *remainder* where emancipation is possible precisely because of its trivialization by the forces of domination. But where Marcuse speaks of the Third World and art as fields of political contestation, Habermas tries to integrate traditional sociological and psychological theory, particularly functionalism, with the philosophy of language to discover the "human" interest per se. In this respect, his work moves away from the "Critical" tradition. Theory is no longer deconstructive, a means to critique the forms of social domination, but becomes an algorithm for establishing the harmonious community in which social distinctions no longer regulate "higher" activity.

What unites this program with that of Marxist structuralism is not common purposes, but a common *will to scientificity* that is shared with the analytic school. The common category that spans these otherwise disparate discourses is the principle of noncontradiction, that is, the return to Kant and scientific philosophy. In these theories, science and technology are not regarded as forms of ideology because of their inescapable rationality, their universality, their correspondence with human interests taken in their totality. For analytic Marxism and the structuralists, the task is to integrate Marxism with science, to ruthlessly expunge its metaphysical elements carried over from Hegel. While Habermas seems to have abandoned Marxism, at least for the most part, his dedication to universal and rationalistic principle is no less fervent. In fact, his program for undistorted communication varies from the scientific ideal insofar as it retains traces of hermeneutic interactionism, but his tendency is toward positivism.

As Marxism becomes dissociated from any possible critical, revolutionary social movements of the working class or others, its main referent as a school of social thought is increasingly the universities and the intellectuals who inhabit them. Now the university is among the premier sites of scientific and technological inquiry. Far from being a pole of critical discourses opposed to prevailing political and economic forces, major universities are today places of knowledge production, much of which is destined for the industrial workshop, health institutions, and military installations. Not only natural sciences but social sciences as well have become crucial elements of national poli-

cies of all industrial and industrializing nations. The reasons, cited earlier, are that these discourses are understood as the central preconditions for both economic growth (capital accumulation) and political
stability. In short, science and technology are not merely knowledges
that contend in the "marketplace of ideas": they are inextricably
bound with power. To employ a metaphor invented by Althusser to
show the materiality of ideologies, science and technologies are *apparatuses of power*. Following from this, even Marxism is obliged,
whether or not willingly, to accommodate to that power. In its most
supine manifestations, Marxism becomes official knowledge, as in
Eastern Europe and China; in Western countries, its relationship to
mainstream science varies from enthusiastic integration, as with the
analytic school and Italian structuralism, to reluctant, almost shame-
faced collaboration, as in post-Frankfurt School social theory. Thus, in
the 1940s, one of the paragons of Critical Theory, Adorno, directed a
major study of the authoritarian personality and employed some of the
characteristic techniques of the sociology of his time—the structured
interview, "scales" of behavioral variation, and so on. Adorno, anxious
to find both income and some academic status in the United States,
understood that one of its main requirements was to do "normal"
social science. I am not claiming that the value of this study is thereby
diminished. The construction of the F scale, the measuring instrument
to determine degrees to which individuals and groups correspond to
characterological authoritarianism, may have been a useful innovation,
and the interview material is often fascinating. Taken as a whole, one
may learn a great deal about America and Americans from this study.
The point is that the work is entirely uncharacteristic of Adorno's position, stated most sharply in his essays in the *Positivist Dispute in
German Sociology*, that the presupposition of such empirical methods
is that they are closely associated with the commercial and administrative interests from which they have emanated and have, at best, limited
value.[32] The empirical methods most typically used in social research
can only skim the surfaces and, contrary to their claim to objectivity,
are nearly always bound by their own subjectively-wrought aims.

Adorno argues for metatheoretical presuppositions that are close to
those of dialectical, speculative reason. The empirical is not defaced
but is relegated to a contingent place in the pantheon of social inquiry.
Faced with the overwhelming fact that the context for knowledge is
inescapably linked to interest, the intellectual possesses few exits.
Against his/her will, the intellectual becomes complicitous—this is the
price one pays for space in the precincts of knowledge production.

To the extent that the avant-garde has passed into history, hastened by the closing of critical, marginal spaces, both physically and intellectually, artists and intellectuals typically turn to teaching to support their work. Their communities are dispersed by high rents brought on by the transformation of traditional cultural cities such as New York, London, and Paris into international financial and administrative centers. Consequently, alternative presses, art forms, and journals occupy an ever-shrinking space in cultural life. Thus, the university becomes a refuge, but it exacts a price that is calculable in terms of both the decline of the audience for critical discourse in recent years, in comparison to the nineteenth and first half of the twentieth centuries, and the pressure, now open, now covert, on intellectuals within the university to produce work that conforms to the ethical and formal precepts of modern scientific knowledge.

Within the university, the humanities, particularly literary criticism, have constituted an alternative cultural sphere precisely because of their marginal position with respect to the main aspects of technological and social policy (which are, for practical purposes, merged). But with the organization of national and international foundations, ministries, and professional organizations to support the arts and humanities, the chance of bureaucratization of these discourses grows, even as cultural apparatuses protest that their mission is to support, not direct, the arts and humanities.

Only the most myopic would claim that Marxism or any other nineteenth-century theoretical paradigm remains unaltered in the wake of criticism, both internal and external, and parallel developments in competing and complementary paradigms. Since social sciences are not marked by evolutionary development in which one paradigm displaces another, but characteristically retains their many systems, Marxism stands alongside others such as functionalism and empiricism. In different historical periods, Marxism is insurgent and in many Western countries becomes the semi-official social knowledge even where Marxist-oriented political parties do not enjoy state power.

Such was the case in France in the two decades immediately following the last World War, and in Austria and Germany in the 1920s and early 1930s. Today, in Great Britain, academic Marxism enjoys a degree of ascendancy it has never before experienced, despite the long-term decline of the Labour party as a political and social force. But in nearly all instances, Marxism does not flourish (or wane) in some mythic pristine form. Its existence as a major paradigm depends, in the main, on the degree to which its axiomatic structure and its methodological framework (both technical and epistemological) conform with that of prevailing, normal science. Marxism adapts itself to other paradigms and adopts them as a condition of its own legitimacy within the academy. Far from constituting an alternative to "bourgeois" theory, it becomes a variety of this mode of theorizing.

I do not want to imply that, for these reasons, Marxism offers nothing to the accumulated treasure of social knowledge. Its insistence that capitalism is structured by class relations, that what we mean by the "economic" is entwined with class struggle, its powerful theories of ideology, its pathbreaking work on the nature of the state—particularly the capitalist state and its truly powerful historiography which, at times during the last twenty years has dominated the American as well as world discourse on history—are contributions that should not be demeaned, even in the course of a critique. But one cannot ignore the overwhelming evidence for a convergence thesis, that is, my claim that Marxism resembles more a normal social science, especially in its discourse on the epistemology and methodology of science and technology. For it is in this discursive space that Marxism must face the positivism of its own axiomatic structure, particularly its tendency to posit science and technology (forces of production) as knowledges and material relations that stand outside the matrix of social relations that can somehow be exempted from "ideology," even when self-critical. That contemporary Marxism exhibits this tendency is prepared, as we have seen, by Marx himself and especially Engels. What has disappeared in the current conjuncture of theory with its institutionalization is Marxism's subversive side, its insistence on historical agency—that "history is made by men" and not by reassembled structures. For it is precisely the indeterminacy of social actors formed as social movements that ruins the technological theater in which causality is produced by purpose and organization. Social actors who play outside the rules of the political game, an event that is posited by Marxism as a consequence of the contradictions of accumulation and political struggle, have the capacity to belie forecasts by their own refusal.

But in scientific Marxism, the "actor" disappears or is made a dependent variable of the accumulation process. Just as Marx was constrained to acknowledge that although "men make their own history, they do not do so as they please," modern Marxism takes the cue and takes determination away from agents. This absence is due, in no small measure, to the project—flagrant since the late Marx—of removing ethics from social inquiry. Social science may study the role of moral precepts as a "mediation" of the determination by strict material causes of historical events but may not impute to them a crucial moment of independent determination. Ethics are always an efflux.

Recent work on social movements that wishes to restore to social theory its critical edge has insisted on the relative autonomy of the agents or the historical actors from the social situation within which they operate. Of course, such writers as Alain Touraine, Henri Lefebvre, and Fredy Perlman,[33] who examine the uprisings in France in May 1968, do not ignore or minimize the social structural contraints of "new" political activity. The issue is where the emphasis lies. For scientific Marxism, these constraints are what counts in the acquisition of social knowledge for they provide a measurable object in contrast to the movements that resist such rationality. Again, the issue is not one of scientific accuracy, but the standpoint from which inquiry is conducted. Does the "observer" intend to discover irregularities, want to find whether and how the reductions that inhere in structure are destroyed or circumvented? Or, in terms of Parsonian sociology, are we trying to find continuities in social life, a biological or even physical analogy to explain why and how movements are reintegrated by the social order?

These are important questions for social theory which bear on the distinction between critical or transformative scientific work and normal science which always takes the point of view of the established order, however insightful its findings. Of course, it is unrealistic to expect Marxism to place itself outside the prevailing order, given the conditions of its own existence as an important discourse of mainstream social science. Yet, it must be noted that most, if not all, radical theoretical and empirical investigations are conducted by those whose Marxist roots are revealed in the choice of the object as well as the referent of class relations, constituting a point of departure from which the claims of autonomy for social movements not rooted in these relations derive.

Post-Marxism cannot avoid class, ideology, and the state as crucial categories of social inquiry. What distinguishes its works is the way in which these are concatenated as relations and relations of relations. As we shall see in Part III, even those who ostensibly owe little or nothing to the Marxist tradition must, when confronted with the question of the social relations of science, situate their own work in relation to the categories of ideology, power, and state. So, it is not so much the specific propositions of Marxism that account for its enduring influence as its categorical structure, which, like Kant's categories of judgment, appear to be the only possible framework from which "science" is done.

The contribution of Antonio Gramsci to the discussion of ideology begins to correct the limitations of recent Marxist theories. At the outset, it is important to recognize the similarity between the thinking of Gramsci and the structural conceptions of ideology advanced by Lukács and Althusser. Gramsci's polemic is directed against those who maintain that ideology is a mere "reflex" of the economic infrastructure, a distorted image in the minds of persons and groups of underlying processes that has no material effect. Gramsci also attacks the pejorative use of the concept of ideology by insisting that it is "necessary to a given structure" and is thus an element of power to the extent that "ideologies mobilize human masses," but also because they "create a terrain on which men move, acquire consciousness of their position, struggle, etc." For Gramsci, ideology is the form of which material forces is the content (of social structure), although this distinction between form and content has purely "didactic" value since each is, in his account, necessary and is inconceivable without the other.[34]

So Gramsci refuses the concept of ideology as error or false consciousness as merely superstructural (which effectively connotes its status as an epiphenomenon). At the same time, he situates the ideological in the processes of social life, in the sinews of politics and revolutionary action. But he also situates science within the same context:

If it is true that man cannot be conceived of except as
historically determined man—i.e., man who has developed,
and who lives in certain conditions, in a particular social
complex or totality of social relations, is it then possible to take
sociology as meaning simply the study of these conditions and
the laws which regulate their development? Since the will and
initiative of men themselves cannot be left out of account, this
notion must be false. The problem of what "science" itself is
has to be posed. Is not science itself "political activity" and
political thought, as much as it transforms men and makes
them different from what they were before? If everything is
"politics," then it is necessary—in order to avoid lapsing into a
wearisome and tautological catalogue of platitudes—to
distinguish by means of new concepts between on the one
hand the politics which correspond to the science which is
traditionally called "philosophy" and on the other between the
politics which is called political science in the strict sense. If
science is the "discovery" of formerly unknown reality, is this
reality not conceived of in a certain sense as transcendent? And
does the concept of science as "creation" not then mean that it
too is "politics"? Everything depends on seeing whether the
creation involved is "arbitrary," or whether it is rational—i.e.,

"useful" to men in that it enlarges their concept of life, and raised to a higher level develops life itself.[35]

Much of Gramsci's conception of science was developed in a polemic against Nicholai Bukharin's "popular manual" of Marxist "science" published in the Soviet Union in the 1920s. Here, Gramsci opposes Bukharin's repetition of the now orthodox view that science and ideology can be strictly opposed by showing that neither can the methodologies of the natural sciences be mechanically applied to the social sphere, nor can science itself be abstracted from the totality of social relations that produce it. For Gramsci, there is a distinction between ideology as the false conceptions of a few individuals, and ideology and science as different sides of the material forces of historical change. According to Gramsci, science is a form of politics, i.e., an ideology which "discovers" a formerly unknown reality, not as a discourse separate from the social and historical context that gives rise to it but as a function of that context.

Consistent with one list of Marxist theory of ideology, for Gramsci, every class that contends for political and social power generates its ideologies that compete for "hegemony" within civil society and the state. The dominant class establishes its reign over intellectuals, who contend for moral and intellectual leadership in society because no class may rule without the "spontaneous consent" given by the great masses to the general direction of social life given by the dominant social group. The social group does not gain ascendancy arbitrarily but dominates because of its position in production. While the proletariat cannot gain power or achieve hegemony over the producers of ideologies (the intellectuals) until the crisis of the existing order has loosened the hegemony of the bourgeoisie, it proposes its science against bourgeois science, whose adequacy may be measured by the degree to which it can "uncover" reality.

In the sense in which ideology with a small "i" is deeply political, its close relation with science, which is also political, is evident. Since, for Gramsci, science is a "praxis," that is, a set of material practices infused with the political ideologies of social classes, which seek political, social, and economic hegemony over other social classes, the idea of the neutrality of science is simply not in accordance with the conditions for the possibility of science as it actually exists.

Gramsci implies that the "truth" of science arises as an outcome of the material conditions, the complex of social relations that give rise to it and within which it functions. When Gramsci states that science "transforms" people and constitutes a form of "creation" itself, the

"will and initiative" of scientific praxis, that is, the subjective teleological element, cannot be abstracted from either the process of knowledge, its production, or its results.[36] And, since sciences as much as
ideologies mobilize masses, we are talking about a conception of science that does not require a special machine to legitimate itself, does
not need to distinguish itself from ideology, except insofar as the neutrality of science is a type of ideology when, grasped by ruling groups,
it becomes an aspect of their effort to gain the "spontaneous consent"
of the masses.

According to Gramsci, the proletariat gives rise to an ideology that
becomes scientific because it does not require a self-legitimating
dogma to mask the coercive basis of the power of its apparatus over
civil society. In this respect, Gramsci locates the possibility of a Marxist
science that can "transcend" the opacity of social reality because its
class praxis is "interested" in emancipatory discoveries. But this creative process of discovery may not be construed as nonideological.
Even if it arises as a critique of the dominant ideology, it is a type of
political discourse, one that has a teleological element and is limited by
its historical and social circumstances.

The theory of hegemony, according to which systems of ideas are
produced by dominant social groups or groups seeking power whose
object is to create a new terrain "on which men move" to mobilize
masses to struggle or not, I call the *general* theory of ideology. Not *only*
is ideology produced by the metonymic extrapolation from the forms
of appearance of "real relations," which in Althusser's theory become
materialized in social institutions that are self-reproductive (the famous concept of ideological state apparatuses—religion, education,
trade unions, the family). It is also produced by the process by which
the ruling class or the oppositional class in capitalist society gains
hegemony over a group of intellectuals who generate a language, a
cognitive apparatus of investigation and understanding which expresses its specific relation to the world and tends to reproduce it as
"natural." The degree to which its language and apparatus successfully
penetrate the material world of society and "nature" is an aspect of its
collective relation to social life.

The binary structure of Althusser's thought—the "real" and the
imaginary, appearance and real relations as *formal* opositions, and science and ideology—is overcome in this conception by the common
root of "science and "ideology." Both terms refer to the material
world. They are not chiefly distinguished by their ordered discourse
(language), which, in any case, may be the same although their vocab-
ularies may differ. For example, mathematical and other theoretical

concepts may be, in fact, as political as a strike slogan, but expressed entirely in the discourse of "normal" science. So it is not the language or even the mechanism of knowledge that distinguishes science from ideology, for the mechanical metaphor for knowledge is, at the bottom, a bourgeois metaphor, one that arose from a system of ideas that had definite historical origins, and that succeeded in creating a terrain upon which social perception and social struggle were obliged to enter. That is, the "forms of appearance" in which the universe appeared to be a machine not only were created by the commodity but were entwined with the praxis of the rising bourgeoisie, its theory as much as its institutional matrix. The praxis of the bourgeoisie in the transition from feudalism to capitalism was linked to its ability to make the world a "universal marketplace," a metaphor for the subsumption of nature and humans under the laws of commodity production. Thus, scientific theory became an instrument of human domination over nature, and techniques the instrumentalization of science. The debate within the history of technology concerning the relations of science, technology, and craft-wrought invention has focused on the question of succession—i.e., which came first? Or, more exactly, did modern industrial technology arise out of sciences or the practice of craftpersons in production? Evidence has been adduced to support both theses,[37] but I believe that the argument may be subsumed by the priority of the domination of nature and humans by capital. Both technology and science were able to remain relatively autonomous while, at the same time, infused with the dominant ideology regarding nature within the bourgeois epoch, according to which (1) the "book of nature" was a text deciphered only by mathematics, and (2) nature could be pictured as a machine.

CHAPTER 8
SOVIET SCIENCE:
The Scientific and Technological Revolution

From its inception, the Soviet Union has harbored an enormous conflict that refuses to disappear. Even in the time of Lenin (1917-24), a period of incessant debate and plurality of ideologies and policies, not only between the Bolsheviks and their critics, but also within their own ranks, the social structure was rent. Lenin and his colleagues insisted that the transition from a "backward, semi-feudal" capitalism, enclosed in a decrepit monarchy unable to adequately promote the development of the productive forces, required a unique form of state power. Broadly interpreting Marx's "Critique of the Gotha Program" (1875), Lenin argued that the task of bringing Russia and its peripheral nations into the twentieth century required a "dictatorship" of the proletariat against the most retrograde forces of society. For Lenin and the "old Bolsheviks" around him, this form of state rule did not imply the suppression of differences as long as the framework of debate remained within the socialist parameters set by the party and government. On the contrary, even harsh critics of Stalin and his successors, like the biologist Zhores Medvedev, regard the early Soviet period as generally favorable to the development of science. Medvedev labels the years 1922-28 a "golden age," primarily because the regime had adopted the "new economic policy," a program that recognized that a mix of state and private initiatives was necessary to promote development.[1] From this followed freedom for scientists and technicians. The regime encouraged international contacts, inviting some distin-

201

guished scientists from abroad to join Soviet working groups, and permitted many Soviet scientists to study abroad in what were regarded as more "advanced" scientific communities in Europe and the United States.

Lenin always made a clear distinction between the need for intellectual freedom for science and art and the equally powerful requirement that, in general, political and ideological freedoms must be restricted. Even in this regard, policy and even ideological debates both within and beyond Communist party circles were permitted to rage as long as openly counterrevolutionary political activity was not involved. The freedoms enjoyed by scientists did not, of course, relieve them of "outside" party criticism or, more to the point, the intervention of philosophers and other theorists into scientific issues. The reason for this flow of debate must be explained, especially to those in societies where the scientific community has seemed traditionally inchoate. First, in Russia, the continuity between science and philosophy was never broken, even as experimental science of the Western type became normative in scientific practice. Second, and equally important, the universal claims of Marxism include its character as a science; some strands of orthodoxy claim that dialectical materialism is a metascience that has the task of unifying the disparate scientific disciplines, which, because of their fragmentation, are unable to *generalize* the results of investigation. Third, the concept of "partyness" which Lenin had introduced to signify that philosophy could not be held immune from ideological criteria was broadly applied to science, especially after his death. But, for the old Bolsheviks, science and art are to be exempted, not only because political "interference" might result in holding back the development of the economy, but because the cultural level of the entire society—its degree of sophistication, education, capacity for innovation in all respects—depends on intellectual freedom. Historian David Joravsky chronicled the tortured relationship between Soviet Marxism and natural science in the 1920s. His account demonstrates the unease that the concept of partyness engendered among scientists, in the early as well as the late 1920s when the growing power of Stalin limited scientific freedom without entirely restricting it. As early as 1922, Kalinin, the President of the Executive Committee of the Supreme Soviet and a major Bolshevik figure, could not assure scientists of unrestricted freedom.

"The use of repressions cannot be renounced," he told the Twelfth Party Congress, "not only in relation to S-R's [Social-Revolutionaries] and Mensheviks, but also in relation to the intriguing upper strata of pseudo non-party, bourgeois-democratic intelligentsia who, for counter-revolutionary purposes, abuse the fundamental interests of entire bodies and for whom the true interests of science, technology, pedagogy, cooperation, etc., are only a hollow phrase, a political screen."[2]

Joravsky comments: "Nevertheless, the evidence indicates pretty clearly that until 1929 [the end of Medvedev's "Golden Age"], repressive measures were not the rule for the government's policy toward the natural scientists and their institutions."[3] This from a vehemently anti-Soviet scholar. Yet, the scientists were justified in their concern by the profound ambivalence between Lenin's acceptance of the idea of scientific freedom within the regime's priority of cultural and economic development, and the idea of partyness which is rooted in the univeralist claims of Marxism as interpreted by the Bolsheviks. That Stalin carried this concept to its logical conclusion is not aberrant. The ontologizing of dialectics, as we have seen, cannot be laid exclusively upon his doorstep; these tendencies are present in Plekhanov, Lenin, as well as Marx and Engels, despite efforts by many Marxist scholars to exempt Marx and Lenin from this stigma attached to Stalin's rule.

The question of the position of the intelligentsia has remained perplexing to Marxist theory as well as Soviet and Eastern European practice. The relation of science to the Soviet regime cannot be ascribed merely to political and ideological repression, as many scholars are prone to understand it. That on occasion scientists suffered from the direct intervention of the Soviet state and the party in the past half-century or more is beyond dispute. But to argue, as Medvedev does, that what occurred after 1928 were simply instances of Stalinist and neo-Stalinist terror not only distorts the picture but leaves much to be explained. While the famous Lysenko controversy of the 1930s and 1940s illustrates the worst features of the link between science and the state, it is also true that the Soviet scientific establishment grew rapidly during this period, its achievements not entirely encircled by the Stalinist compass. According to Medvedev, a stern critic of the Stalinist "terror," one of the greatest crimes against science of the regime was to annoint certain theoretical positions with the mantle of "truth," suppressing competing paradigms. Yet, by his own description, not all of the favored figures and their theories, even in this period, were fakes. Pavlovian psychology and physiology enjoyed supremacy in these fields. As limited as conditioned reflex theory is for explaining the psychological functions, its value is recognized by the international scientific community; indeed, Pavlov received the Nobel Prize. Similarly, physics and chemistry, especially in applied areas, enjoyed considerable range even during the Stalin era, although, for a short period, debates concerning relativity and quantum mechanics were deeply infused with ideological discourses in a manner resembling the claims of Lysenko to have applied dialectical materialism to biology. In the 1930s, relativity and quantum theories were condemned semi-officially as instances of idealist and metaphysical ideology masquerading as science, just as genetics was labeled in the same way, but proponents of these interpretations were not always repressed.

At issue here is more than an episode of Stalinist terror. Foremost is the relation of the state to science and technology, a partnership which, as we have noted, is not confined to authoritarian regimes but has become global in its application. "Official" science exists in the West as well as the East. Science legitimates itself by linking its discoveries with power, a connection which *determines* (not merely influences) what counts as reliable knowledge and, even more basic, who performs scientific research. In both the United States and the Soviet Union, national institutes in various branches of science award resources to universities and independent research groups in accordance with the priorities of state science policy as established by executive bodies. In both countries, committees of scientists make judgments about the value of the work, but each committee is, implicitly or explicitly, bound by broad policy directives.

So, the question of scientific "freedom" is more complex than is represented by the position that, in the West, scientists are free to expound dissident views. This fact is fundamental, but does not exhaust the issue. For if resources are not made available in Western countries to those whose dissidence consists in *theoretical* opposition to the dominant paradigms, the validity of scientific claims cannot be established by normal procedures since these entail collective work in institutionally based facilities.

The Lysenko episode illustrates the dangers of eliminating criticism as the signpost of legitimate scientific inquiry, not whether ideological considerations are included in the process of scientific discovery. For ideology is ineluctable in all intellectual labor. That ideological considerations are invoked in the judgment of the results of scientific investigation does not in itself constitute a measure of repression. Those who would characterize Western science as "free" must also explain the ubiquity of the military in United States and British physical science, the configurations in medical science that focus on treating disease by inventing drugs that are designed for individual consumption (and thereby starve approaches to disease that focus on prevention by addressing issues bearing on epidemiology), and the progressively close relationships that have evolved between science and technology in both countries.

Having said all this, I want to argue that the subsumption of science and technology under the state is a crucial feature of the entire history of Soviet society, and that what is implied by this subordination is the position of the scientific and technical intelligentsia in Soviet society. This intelligentsia is privileged in all developing countries as well as advanced capitalist societies where economic supremacy in world markets is perceived (correctly) as dependent on developments in science and technology. In both the United States and the Soviet Union, superpower conflict remains a major restriction on such development because the imperative war preparation determines the course of scientific discovery in many leading discliplines. In the United States and, to a limited degree, in Britain, France, and Germany as well, some scientists perceive such restrictions as a limitation of intellectual freedom. More specifically, the mobilization of institutions, funds, and scientists themselves to serve the war and corporate machines is regarded by a minority within the scientific communities as a perversion of their work, as a kind of regimentation. Recent declarations by U.S. physicists, chemists, and biologists refusing to participate in President Reagan's strategic defense initiative (Star Wars) illustrate the presence of dissension within the American scientific community regarding the relation of knowledge to state power.

Similarly, some of the most celebrated dissenters in the Soviet Union have expressed themselves in terms of the use of their discoveries for purposes of nuclear destruction. The most famous of these, of course, is the physicist Andrei Sakharov, a leading figure in Soviet nuclear science, widely known for his key role in the development of the hydrogen bomb. It is important to observe that Sakharov was not harassed for his opposition to nuclear policy until 1980, when he was subjected to internal exile in the city of Gorki for opposing the Soviet invasion of Afghanistan and not ostensibly for his scientific views. Yet, what is clear about Sakharov's dissent is that he objects to the uses to which scientific research are put; in effect, he opposes the subsumption of science under state defense policy, arguing vehemently for alternative uses of science, particularly in areas of human welfare.

Beyond the obvious moral issues, the position of the intelligentsia within Soviet society is at stake in scientific/political dissent. Soviet science and technology are so intimately bound with the development of the Soviet economy, military capability, and the political stability of the regime that the position of the intelligentsia as a social class rather than a category may be said to constitute a major political problematic of the entire society. This problematic underlies ideological and political debates within science as well as the relation between the scientific community and the state. What is at issue here is the question whether the scientific and technical intelligentsia will assert their political or intellectual autonomy from the party and their freedom to act independently within the state, especially regarding science policy. But science policy is not merely a question of resource allocation between competing paradigms; as was the case during the Lysenko affair, it entails considerations affecting the entire economic and military programs of the Soviet Union.

In order to explicate these issues more fully, I want to undertake a brief digression concerning the intelligentsia as a class. Recent sociological theories of class have departed sharply from the conventional Marxist position according to which economic identification is both the primary determinant of class formation and a predictor of political, ideological, and cultural relations. Nor does recent theory hold, with Weber, that class is a temporary and mobile phenomenon having little relation to political and ideological power. Instead, classes are formed by economic, political, and ideological relations taken as a totality, and the dominant set of relations among these may vary according to the historical context. Intellectuals as a social category predate the rise of capitalism. The social role of those possessing knowledge varies according to the social formation. The owners of knowledge may differ from possession, so that the use of technical labor for production and administration does not differ in substance from the employment of manual labor until the late nineteenth century. More significantly, intellectual discourse moves from the margins to the center as scientific and technical knowledge becomes a productive force.

This historical transformation of the significance of intellectual labor is a question of heated dispute among social theorists. One tendency links the centrality of knowledge to the labor process and its consequent marginalization of manual labor with the emergence of the intelligentsia as a class. In state socialist societies, intellectuals recruited both from technical and managerial categories and from "traditional" liberal circles increasingly play a crucial role in the party and state apparatuses. George Konrad and Ivan Szelenyi have theorized this development by asserting that these intellectuals constitute themselves as contenders for "class power."[4] Surely, at the dawn of the Bolshevik Revolution, the key figures in party and state were the intellectuals. Professional revolutionaries, they quickly realized that the intricate problems of production and state administration could not be solved by politics alone. The new regime required a layer of technically trained experts to manage a large, unskilled labor force. In the absence of a revolutionary category of experts, the party was obliged, especially during the period of the New Economic policy, to accommodate to the old bourgeois technical intelligentsia; partyness coexisted uneasily with technocratic thinking. For many, in and out of the Bolshevik ranks, Marxism was not a universal theory but a theory of political and economic development, an explanation of human history on the basis of social laws. Science and art belonged to different spheres. This expressed the *dominant* interpretation of Marxism in the post-Engels period of socialist growth prior to World War I. Those like Trotsky and Lenin, for whom science was a vindication of dialectical as well as historical materialism (even if spontaneously), saw no contradiction for socialism in adopting Western science and technology to the task of "socialist construction," although Lenin had some doubts about "capitalist" methods of organization of the labor process. Therefore, when the so-called workers' opposition led by Alexandra Kollontai raised objections to the introduction of a Taylorist system of factory production and its accompanying authoritarian system of management, Lenin and Trotsky responded by calling attention to the practical tasks of securing the existence of the revolution, tasks which required the most rationalized systems of production and management. The demand for "workers' control," a program for placing the manual working class in a position of primacy in the revolution, was regarded by the majority of leading Bolsheviks as "infantile" in the wake of the pressing problems of the regime.

But these pragmatic considerations were supported by the idea of "partyness" and the elevation of "dialectical" as opposed to historical materialism as the universal side of Marxist theory—a controversial philosophical tenet from the point of view of the history of Marxism but also disputed among the scientific, literary, and technical intellectuals, many of whom gradually enunciated an ideology supporting the concept of the autonomy of science and art. In the 1930s, these tendencies were suppressed in favor of universalistic arguments that stressed the primacy of manual as opposed to intellectual labor, the party as opposed to technical experts, ideology as opposed to the autonomy of science and the arts. Thus, the "new class" arises not on the basis of its centrality to the productive and administrative processes of society, but in relation to the *primacy of politics* over all social relations.

In this conjuncture, "civil society"—an arena of political, cultural, and theoretical discourse—disappears, and, as Ferenc Feher, Agnes Heller, and George Markus argue, the party asserts its "dictatorship over needs" at every level.[5] Under these conditions, dissent is increasingly identical to exclusion from society. The intelligentsia becomes integrated as an important, but subordinate, variable in production and administration. In the sciences and art, it manages to gain more room, but only at the highest level and in some sectors. But there is no consideration of class formation, and no culture of independent, critical discourse among the intelligentsia.

As the division of intellectual and manual labor becomes more pronounced in industrial societies, including the Soviet Union, two strata emerge as "productive forces" alongside manual labor: the managers, and the scientific and technical workers. Together, these categories constitute the *intelligentsia*, whose sociopolitical position is characterized, during the industrializing era of capitalist and Soviet development, by its subordination. This fraction of the intelligentsia emerges in the nineteenth century not as an independent social category or class, but as an arm of capital. Even as it becomes more central to the development of not only the productive forces but also the state, in proportion as the bureaucracy emerges as the key to the latter, it resembles more a privileged caste than a class because its power is contingent on the expansion of production, on the one hand, and the functions of the state, on the other. Many writers have argued that the position of the intelligentsia has radically changed with the incorporation of knowledge as a crucial productive force in the twentieth century. On the one hand, the old traditional intellectual gradually disappears as a social category. The independent scientist, literary critic, philosopher, journalist—the main occupations to which this intellectual stratum attached itself—are replaced by intellectual laborers who perform specialized functions in large organizations. In the main, these categories are incorporated into universities, the state bureaucracy, and, in the West, the laboratories and headquarters' staffs of large corporations.

On the other hand, as Alvin Gouldner has argued, a new culture of critical discourse ("CCD") becomes more typical among intellectual labor than was true of the old system, where a narrow group of intellectuals recruited from among the aristocrats and upper middle classes constituted a privileged circle within every country. Gouldner's point is that there has been a relative democratization of CCD so that we may speak of a *class* of intellectuals, regardless of its economic position within the productive and ideological apparatuses of late capitalism and state socialism. CCD constitutes a class ideology that is "more situation free, more context or field 'independent'."[6] Gouldner adopts the categories of language theory research, which posits discourse as, among other things, a social as well as cultural marker. According to this view, context-bound relations, such as those that limit scientific and technological research to state/corporate priorities, do not exhaust class determination. On the contrary, the development of intellectuals as a broad social category signifies a new historical situation. According to Gouldner, intellectuals are self-authorized and have the power to de-authorize others. The theory of CCD offers a new type of explanation for the growing tension between intellectuals and established economic and political power. If knowledge becomes a new site of power that is valorized by political and economic power as crucial for its own reproduction, the stage is set for confrontations, the outcomes of which are indeterminate. In short, Gouldner argues that CCD is a power form incommensurate with that exercised in previous social formations.

The Soviet case differs from that of Western countries insofar as the contexts for the power struggle change: the state/party axis replaces the corporation as the major institution of economic power. This axis constitutes an oligarchy rather than a bureaucracy, the difference being that an oligarchy functions without benefit or deterrence of a set of rules that govern its power in contrast to a bureaucracy which, however undemocratic, observes "objective criteria" in the exercise of its authority from which there is a basis of appeal. The burden of historical and social research demonstrates the pervasive politicization of the Soviet state and its bureaucracy. This politicization makes inaccurate characterizations of the Soviet state as "state capitalism" and "bureaucratic collectivist." For if the bureaucracy has any precise significance, it is that qualifications, divisions of labor, "careers" whose trajectories are subject to charting in accordance with definite rules, determine in fundamental respects how the state is organized. Such rules vitiate the intervention of political bodies and ostensibly ensure (for better or worse) a more or less autonomous bureaucratic apparatus. Of course, this characterization does not vitiate the existence of what Kendall Bailes has called a "technostructure" in the Soviet Union.[7] At the levels of local management, engineering, and applied sciences, such a structure emerges during the 1929–55 period, but its autonomy is mediated by political bodies. Careers are never secure from political decisions made at the national as well as local levels by politically appointed officials. Surely, the Soviet leadership has more or less deliberately bridled the autonomous development of the technostructure upon which production and administration increasingly depend. My contention is that such subordination, possible during the industrialization period in which expansion was achieved chiefly through such traditional devices as speedup, rationalization, and lengthening of the working day (all of which depend on the extensive utilization of manual labor), becomes anomalous in the period of the intensive regime—the stage of production where scientifically motivated technology replaces manual labor.[8]

Politicization creates an uncertain ambiance for the intelligentsia. It obliges them to make increasing efforts to influence both state politics and the party, if only to preserve a measure of autonomy. As the "productive" intelligentsia becomes more central to the entire social structure, as its work bears on issues of political stability, economic development, scientific and technological progress, as its ranks become more numerous in both absolute and relative terms (relative to the size of the industrial working class), we can foresee the transition from intellectuals as a social category to a class.

Referring to Western Europe, Gramsci argued that the intellectuals were a *category*. They were not a class, in his view, because their effectivity was inescapably linked to key classes of capitalist society: the proletariat and the bourgeoisie. Or, in societies undergoing the transition from feudalism to capitalism, they might be "organic" to old classes, particularly the fading but by no means impotent aristocracy, peasantry, and small landholders. The designation of intellectuals as a category rather than a class signifies for Gramsci that they are unable to act politically on their own, incapable of generating what Marx called "ideological flags" that were self-representing. Instead, their historical mission is to exercise "moral and intellectual leadership" on behalf of contending classes that hegemonized them and made them, in Gramsci's terms, representatives of the class. For Gramsci, these classes are formed, finally, by their relation to ownership and control of the means of production.[9]

This conceptual apparatus for understanding the location of intellectuals presupposes social formations in which science and technology do not yet appear as leading productive forces or, more exactly, are perceived to be subsumed by capital in a manner resembling the industrial phase of its development. Technology, in Gramsci's discourse, is identified closely with Taylorism and Fordism, that is, takes on primarily an organizational character. In turn-of-the-century Italy, industrial production is still dominated by the clash of the crafts with management; natural science is entwined with art, a cultural phenomenon that is still somewhat removed from economic practice. This is the period where traditional intellectuals, that is, those trained in the humanities and sciences, attach themselves to political parties and scientific ideologies such as liberalism and Marxism as a choice and sign of their political will.

For Gramsci, there is no question of the working class or the bourgeoisie bidding for political or economic power without the benefit of ideological struggle. The intellectuals, bearers of ideas, especially moral and scientific concepts, constitute a sufficient condition for the successful struggle for power. Intellectual hegemony is a material force. Therefore, competing classes also contend for the hearts and minds of the intellectuals. But, as Andre Gorz has argued, the traditional intellectual passed from history some time in the interwar period, yielding to an intelligentsia whose primary possession is technical knowledge, itself determined by the division of labor.[10] Recently, especially in France which was somewhat behind in the transformation of its educational system to conform to this development, we have witnessed a policy of the state acting on behalf of the new setup to reform the system by restricting the liberal arts degree to a rather small elite, expanding technical education, and promoting specialized learning. In 1986, students and some university faculty opposed these reforms with some success. In most other countries, the weakness of the traditional intellectuals and their ideology of the liberal arts resulted in similar "reforms" years ago. The United States, which has a decentralized school sytem, shows considerable unevenness from state to state, yet, in the larger picture, the decline of "traditional" education is evident; technical training has overtaken most state-run universities, even as private elite schools struggle to retain their trappings of tradition. (Columbia University is a notable exception; it appears to have effectively renounced its claim to tradition in favor of a policy of almost complete professionalization.)

Today, the "intellectual" in Gramscian terms, or even as understood by Karl Mannheim, who posits the marginality of intellectuals, is eclipsed by the emergence of the technical intelligentsia. The social weight of this "category" has increased substantially since the 1920s and 1930s when Thorsten Veblen, Emile Lederer, and others theorized the emergence of the technical intelligentsia as a major force in "advanced" capitalist societies. Recall that Veblen, in arguing that potentially the engineers and not the dequalified industrial working class were historical agents owing to the centrality of knowledges as productive forces, nevertheless noted that the likelihood of a professional class contending for political and economic power was severely diminished by its privileged material circumstances.[11]

The Soviet party and state oligarchy recognized that scientific and technical strata had to be treated with special care if development goals were to be realized. Moreover, Soviet society is marked by its respect for material and artistic culture, leading to considerable resources having always been made available for education, "pure" scientific research, and the arts. At the same time, the Soviet Union

suffers an absence of "civil society" in which critical discourse circu-
lates. That is, the state/party "dictatorship over needs" extends not
only to the economy but also to social life. In the absence of a public
sphere of (relatively) free scientific and ideological discourse, the intelligentsia becomes restive, not primarily because of its comparative
economic disadvantage to the West (this is not always the case, except
for superstars), but because "partyness" in areas that have traditionally
been considered autonomous such as the arts is believed onerous to
the development of knowledge. The fundamental demand of the intelligentsia—"freedom"—has never been entirely abrogated by the party.
As the Lysenko affair illustrates, it was not until Lysenko and his associates had entered the front rank of Soviet biology for twenty years that
their doctrines were christened official science. One of Lysenko's main
opponents within the scientific community, N. Vavilov, disappeared in
1940, but other dissenters survived until 1948, when to refuse to
acknowledge the truth of Lysenko's doctrine amounted to treason. In
physics, figures like V. Fok and A. D. Alexandrov openly defended
quantum mechanics and the theory of relativity throughout the 1930s
and 1940s when these views were in disfavor, especially within the
party ideological leadership; however, as we shall see, Fok and Alexandrov, among other Soviet proponents of the new physics, interpreted
the results in terms of dialectical materialism, a move that tried to preserve rather than overturn Newtonian concepts of absolute space and
absolute time. In effect, Fok and others understood relativity as a special case of classical physics, whose propositions limit its broad ap-
plicablility.[12]

What has become different since Khruschev's secret report to the
Twentieth Congress of the Soviet Communist party on Stalin's "crimes"
is the gradual development of a politics of intervention by some scientists in state policy. In 1969, Andrei Sakharov issued a manifesto for scientific freedom, by which he meant both freedom to criticize state policies bearing on the sciences and freedom to debate scientific issues
without the threat of state intervention. This document appeared early
in the Brezhnev era, which scientists perceived as a partial reversal of
Khruschev's post-Stalin reforms.

Even more dramatic is the emergence of a distinctly technocratic
ideology in the Soviet Union, one that has semi-official recognition but
emanates from the technical and managerial intelligentsia. This ideology has two aspects: the doctrine of the Scientific-Technological Revolution (STR) as the social, political, and economic imperative of Soviet
society; and the Scientific Management of Society. This ideology is, of
course, shared by some Western social theorists and scientists. It takes

many forms, the most important of which has been elaborated by Daniel Bell, Alain Touraine, and others as the concept of *postindustrial society*, one in which the traditional classes—proletariat and bourgeoisie—no longer occupy the commanding heights of social power. Touraine argues for the emergence of new historical agents, the scientific/ technical categories that represent the new forces of communications, information, and so on. Bell leans toward a managerial interpretation of these forces. For him, every social problem is now subject to technical, rather than political, solutions. Those who possess expertise, particularly those linked to organization, are the real social agents. Despite these differences, the totality of which is represented by the two sides of Soviet technocratic ideology, the old human agents are equally held to be surpassed. In the Soviet case, the argument does not directly address class issues. For to enter this arena would, at the present time, still be politically hazardous. Nevertheless, a reading of the Soviet literature in this respect reveals the configuration of ideological formation.

Let us recall Marx's discovery, presented most forcefully in his notes for *Capital*, the *Grundisse*, that science becomes the decisive productive force displacing human (manual) labor owing to the transformation from the "extensive" to the "intensive" regime in the labor process. This change is a socioeconomic development which signifies both the growing strength of the labor movement to limit working hours, with its concomitant successful struggle to improve wages, and the sharpening competition for commodities and capital on a world scale that matures in the late nineteenth century. What in Marx was a brilliant prognostication becomes a widespread *fact* of industrial production in the twentieth century. The Soviet discourse on these developments remains, however, relatively muted in the long industrialization phase (1921–50) which focused on the semi-militarization of labor as the primary productive force in the interwar period. Although Soviet ideology recognized the importance of scientific and technical education in this period, the critical task was defined by the Soviets as the organization of production—the managerial tasks associated with mobilizing a vast unskilled labor force recruited from the peasantry. Until World War II, industrial production was symbolized by the figure of Stakhanov, the miner who showed how to achieve incredible levels of manually driven production, that is, to speed himself up. For decades, workers' productivity was seen as a task of *motivation* by an effective management because the level of technological development in many industrial sectors was admittedly far behind Western competitors. During the 1930s, Soviet imports of Western technology declined in comparison to the 1920s. This reversal of modernization was produced by both the growing isolation of the Soviets in world economic and political life and the ideology that generated the rise of Lysenko — the need for a Soviet development pattern that was autarkic, that is, no longer linked to dependence on Western science and technology. That these policies were tied to Stalin's consolidation of his power at the expense of the old Bolsheviks (the cadre of the revolution) is not accidental. The slogans of "proletarian" science, combined with extensive exploitation of labor associated with the Stakhanovite movement, constituted a program imposed as much by necessity as by political will.

Bailes has shown that among the most regressive consequences of the Stalin period was the resistance, common among local managers, to the introduction of new technology. However, by the 1950s, the romance of manual labor faded from Soviet industrial discourse. The concept of the scientific/technological revolution replaced the focus on images of manual labor as the Soviets resumed imports of machinery and transport equipment after 1955. In 1955, the value of these imports from Western countries stood at 104 million U.S. dollars. Within a decade, this figure jumped by 350 percent, and, by 1975, it reached 4 billion dollars.[13] This sum does not include technical assistance contracts, "turnkey" deals (where a foreign company builds a plant or other facility and turns it over to the "host" country or company), and other types of technology transfer such as foreign visitors and Soviet trips abroad, particularly to research and manufacturing facilities. In 1955, the Soviet Union initiated cultural and technical exchange agreements with the United States and other Western industrial countries that have continued to this day, although expansion has been subject to the vicissitudes of international political relations.

Thus, after the war, the Soviets once again became inveterate borrowers in science and technology, except in selected industries connected to military preparations. During the 1950s and early 1960s, at the height of the Cold War, the issue of technology transfer became central to intelligence and other Western concerns. Dedicated to destabilizing the Soviet regime, Western governments sought, unsuccessfully, to restrict the export of industrial techniques to the Soviet Union. Well into the 1970s, major firms in the United States and Western Europe entered agreements with the Soviet government to build plants, sell technology, and even manage production for limited periods of start-up.

By the late 1960s, Soviet and Eastern European intellectuals were enunciating versions of a new doctrine of the scientific/technological revolution (STR), which now represented the crucial historic conjuncture upon which the economic, political, and social future of socialism depended. Although the literature emanating from Eastern Europe on STR stressed that its emergence was due to socioeconomic developments, and that the degree and types of its application depended in the words of one theorist on "different social systems," the overriding theme was to stress the autonomy and the universality of STR as a determining influence. What is STR?

Fedoseyev, a vice president of the Academy of Sciences and former director of the Marxist-Leninist Institute of the Communist Party, describes STR in 1974:

> The scientific-technological revolution is basically the radical qualitative reorganization of the productive forces *as a result of the transformation of science into a key factor in the development of social production.* Increasingly eliminating manual labor by utilizing the forces of nature in technology, and replacing man's direct participation in the production process by the functioning of his materialized knowledge, the scientific and technological revolution radically changes the entire structure and components of the productive forces, the conditions, nature and the content of labor. While embodying the growing integration of science, technology and production, the scientific and technological revolution at the same time influences all aspects of life in present-day society, including industrial management, education, everyday life, culture, the psychology of people, the relationship between nature and society. . . .
> The Marxist methodology in analyzing this phenomenon is characterized by a comprehensive, integrated and systems approach. [My emphasis.][14]

This is an authoritative statement by a leading party intellectual speaking at an international conference which includes many "Western" experts in the social relations of technology, including such luminaries as Ralf Dahrendorf and Alain Touraine. Dahrendorf remarks that in addition to being a productive force, "science is an element of the relations of production" because "science and technology are the epitome of a society that has made instrumental rationality its guiding value."[15]

In none of the Soviet debates about STR are science and technology identified as an element of production relations. True to orthodox Marxist interpretations, the forces of production are mediated by these relations but are considered both logically and existentially separate. Thus, as we have seen, the developing technological determinism of this brand of Marxism is reproduced in Soviet discussions of STR. Reading Fedoseyev, one is struck by the objective, "subjectless" discourse. "STR is . . . " and thereby itself becomes a subject to which social relations must respond and adjust. This tone of inexorability is barely qualified by Fedoseyev's statement that STR influences the fate of labor differentially, depending on the social system within which it operates. In socialist societies STR, according to this perspective, will not disturb existing social relationships but will open new opportunities for individuals to use their "creative abilities." STR requires a higher educational and cultural level in order to bring "the masses into the administration of state affairs." On the other hand, STR will aggravate capitalist contradictions, throwing millions of workers onto the street. But for Fedoseyev, there is no question of challenging, much less reversing, the trend toward the closer integration of science, production, and social life. The major question is how to harness science and technology for social benefits. For him, the promise of this "revolution" is self-evident; only bourgeois apologists utter words of alarm.

Fedoseyev's introductory article to a major collection of essays entitled *Lenin and Modern Natural Science*, published four years later, makes more explicit the official view of the significance of STR.[16] Although the article is replete with familiar statements defending the "materialist" (actually realist) theory of knowledge concerning reflection and truth, for our purposes here the key point is that STR has become so politically crucial that dialectical materialism faces the task of articulating its methods, if not its doctrine, in the face of "new scientific data." Marxism must generalize the results of natural science but must not interpret these results in a way that would in the least alter the propositions of materialism. In other words, science is to be considered "data" requiring integration into a priori materialist doctrine. The theme of the integration of science into socialist relations parallels Soviet discourse on technology transfer. Surely, both science and technology present new challenges to Marxist theory and Soviet practice, but these challenges cannot disturb the fundamental social paradigm because material and social forces are logically separate.

Thus, in vain we seek evidence that technology transfer from Western countries where production is organized on a hierarchical principle presents problems for socialist democracy. The typical Soviet factory, although culturally distinct from similar Western plants, is organized on similar principles. The manager and staff direct the work forces, mediated only by spontaneously imposed counterrules initiated by workers on the shop floor. These include resistance to production norms, frequent break times, and local reluctance to transform the labor process in accordance with technological norms and other measures. Yet, the social organization of production is normatively Western in the Soviet Union, and machinery, conceived as "material relation" to nature, is considered socially neutral with respect to relations of power.

Nor can Fedoseyev acknowledge that science is capable of generating paradigms of physical reality that diverge from those of dialectical materialism. Philosophically, the "integration of science" becomes one of the key tasks of Marxist theory. What appears to contradict the precepts of dialectical materialism is the fault of erroneous interpretation to which scientists, when they engage in philosophical reflection, are as prone as philosophers. The polemic resembles that of Lenin in relation to relativity theory in his *Materialism and Empirico-Criticism*. This is, in large measure, a polemic against those in the contemporary Boleshevik party who demand that Marxism conform to the latest development of science, particularly to the concept that "matter" disappears as a category in relativity physics and the corollary that *relations* rather than *things* constitute the object of physical knowledge. Of course, Lenin was unwilling to attack science for its nonconformity with dialectical materialism; rather, he interpreted relativity theory as a confirmation of Marxist philosophy. Quoting Engels, Lenin admitted that "with each epoch-making discovery in the sphere of natural science, materialism has to change its form."[17] But this does not signify the need to reexamine the fundamental tenets. What is required, according to Lenin, is an adjustment: now the correspondence theory of truth is amended to hold that our knowledge of the external world is approximate, probabilistic, and relative—relative, that is, to the development of the productive forces. In fact, the new physics is unconsciously dialectical and materialist, and its problems—its "crisis"— result from empiricist, positivist interpretations by idealistic philosophers like Ernst Mach, Pierre Duhem, and Richard Avenarius.

Fedoseyev adopts this strategy with respect to STR, and, as we shall discover below, one way is to "adopt" systems theory—to call it materialist and scientific. Another is to assimilate new scientific discoveries within a dialectical materialist frame. These strategies have become the key to the Soviet discourse on science throughout the country's history. When neither of these can be accomplished, as in the case of the debate on genetics, the position of dialectical materialism as officially interpreted prevails.

The ideology of STR has become contested terrain in the debate concerning the social and political perspectives of the intelligentsia. To a large degree the concept, although corresponding to Marx's assessment of the transition from industrial to what has been called advanced capitalism, takes on a specifically new meaning in the context of the Soviet debate. Fedoseyev is preemptively accepting STR as a phenomenon mediated by social relations and asserting that its effects are determined by the context within which it functions. At the same time, he acknowledges that advances in science and technology themselves hold the key to the future progress of socialist societies. The simultaneous effort to reassert the subordination of STR to socialist relations of production and the relative independence of its effects constitutes a compromise with the mainstream of the Soviet and East European discourse on STR.

Against what ideas does the Soviet leadership invoke STR? Against the contention widely shared by the scientific and technical intelligentsia that the fundamental problematic of contemporary economic and social development is the ability of different social systems to cope with the scientific-technological revolutions of our time. In its pristine form, ideologists in Eastern Europe who have advanced this thesis insist that "socialist production relations born of revolutions in the realm of power and property cannot be expected alone to provide a solution to the problems inherent in industrial civilization or to eliminate conflicts it engenders." What is needed, according to this view, is a "transformation of the entire foundation of civilization" by the relentless application of science to technology, their integration with other aspects of social life, and, most especially, "to invite scientists, technologists and other specialists, economists and working groups to put forward their own suggestions [independent of the party] on how to advance the civilization base of human life, how best to gear the country to the scientific and technological revolution in the world to a process of shaping a socialist mode of life." Further, "to initiate a long term *open discussion* in which the public could assess the alternatives, giving all citizens the opportunity to voice their proposals for the future" (my emphasis).[18]

These proposals were advanced in a remarkable book, *Civilization at the Crossroads*, published by members of the Czech Academy of Sciences in 1967 and edited by Radovan Richta. The document is closely associated with the short-lived Dubcek government, which was dissolved a year later by Soviet tanks, and is possibly the clearest and most comprehensive statement of critical discourse on STR produced in Eastern Europe. Arguing that STR signifies a new penetration "into the foundations of human life" that is at once a "historical upheaval" and makes necessary an entirely new approach to social relations, notably a turn in "really existing" socialist countries toward democracy, the book represents the most technologically determinist of all recent works with a Marxist perspective on social transformation:

> The structure of productive forces—and consequently every specific type of production and technology—always possesses an impicit social attribution and, in turn, demands an appropriate structure of social life. The relations of production are no more nor less than the mobile form of the productive forces, which are always productive forces of a specific type of human life and a source of specific relationships among people.[19]

For Richta, the entwining of forces and relations of production does not imply their mutual determination, even if they are inseparable. That the relations of production are seen as a "mobile form" of the productive forces signifies the priority of the latter. And, knowledge is the heart of the new productive forces generated by STR, and the "entire foundation of civilization" is transformed by it. The focus on the new scientific and technological intelligentsia as historical agents of socialism pervades the discussion in this book of STR. Therefore, "democracy," understood as the demand for autonomy for the members of this category and their effectivity over decision making at all levels of social life, constitutes a new political program. Democracy does not appear as an ethical demand but as a necessary feature of the STR.

This emphasis on scientifically wrought politics is squarely in the Leninist tradition which seeks to remove affairs of the state and political life from the realm of ideology, that is, as a contestable and contingent set of relations. The Scientific Management of Society in this discourse corresponds to Marxism's scientific will, its attempt to sidetrack ethical considerations which are considered vulnerable because they are "subjective." There are, of course, many elements of agreement between the Soviet and the dissident discourse on STR. What distinguishes them is the insistence of the dissidents, in themes repeated in *Civilization at the Crossroads* and other Eastern European technocratic statements and by Sakharov, on democracy and "open discussion" of STR and its consequences.

Sakharov's "Peace, Co-existence and Intellectual Freedom" (1969) repeats this slogan, making it a central theme of the opposition. *Civilization at the Crossroads* calls for "public" participation in debates about STR as a means of beginning the process of democratization. This contrasts sharply with the Soviet meaning of the term "democracy," especially with respect to STR.

Soviet theoretical interventions into the discussion of the STR acknowledge its unprecedented character, especially the consequences of the displacement of manual labor from the center of the production process and its replacement by knowledge. For the most part, the Soviet discourse on STR focuses on its implications for production, especially the need to upgrade the qualifications of labor through a new and more sophisticated approach to education. In their contribution to a symposium organized by the Soviet Academy of Sciences on the social effects of STR, several writers alluded to the possibility for broader participation by scientific and technical labor in decision-making processes in industry.[20] But these comments are nearly all combined with identifying the "development of democracy" with the Scientific Management of Society, the second element of technocratic ideology. On the one hand, writers repeatedly call for greater participation by scientists and technicians in sectoral decision making so as to ensure greater efficiency in production. There are also descriptions of greater administrative involvement by these professionals on a sectoral basis.

Far more pervasive, however, is the possibility of more accurate and scientifically based state planning, using economic forecasting to a greater degree, facilitating the integration of science into both production and administration, and placing more emphasis on scientific management. Although Soviet literature on the Scientific Management of Society includes these rationalization measures, some of which resemble the Taylorist principles developed at the turn of the twentieth century, STR has introduced new concepts into the Soviet discussion of management.[21] The concept of "scientific control" includes more rational methods of controlling labor resources, materials allocation, and planning objectives. But the Scientific Management of Society ultimately focuses on the "control over science," principally "between theoretical, long-term research and applied research in many branches." The work of control is still firmly lodged in the Communist party, according to most writers. However, V. G. Afanysev, while paying obeisance to this principle of party control, joins other writers in asking for greater managerial autonomy, particularly through training and efficient administration of the state and industrial enterprises. Most important, Afanysev introduces a theme that permeates the discussion of Scientific Management of Society—the need for the global application of systems theory, or organizational integration of basic and applied research into planning processes. This is an implicit call for more "scientifically" based control as compared to its *silent* antinomy, political control. If my thesis that Soviet society is more or less entirely politicized is right, two crucial features of advanced industrial societies are missing: an active and influential scientific intelligentsia with effectivity in economic and social policy, and an efficient bureaucracy in the traditional, Weberian meaning of the term.

Severyn Bialer, in his otherwise excellent study of recent Soviet politics, offers contradictory arguments regarding the power of the Soviet intelligentsia. On the one hand, he states that "with some exceptions, [the Soviet intelligentsia] is only a statistical category" since its "old intelligentsia" artists, writers, scientists, etc., remain unorganized precisely because of the concessions to their autonomy made during the Brezhnev era. The bulk of the intelligentsia "is highly materialistic and primarily interested in pursuing careers."[22]

On the other hand, as a result of broad international and Soviet support for Sakharov after he was denounced for his Manifesto, "the party has shown increasing anxiety about the political and social attitudes of scientists and has moved to tighten its supervision of personnel policy in the scientific establishment and to denigrate the expertise of scientists in social and political matters."[23] Since Bialer denies that the scientific elite has "significant social and political influence," he has difficulty explaining party "anxiety" about their views. Although the Sakharov affair can be explained by his international prominence, this is not enough to account for the series of measures to restrict scientific autonomy which Bialer describes.

Bialer's study was published on the eve of Gorbachev's dramatic release of Sakharov from internal exile in Gorki and his return to Moscow in December 1986. In subsequent interviews with the American press and on television, Sakharov was still staunchly advocating intellectual freedom in the Soviet Union, although he was cautious about radical democracy. Sakharov's position is close to that of the Czech group insofar as he emphasizes freeing intellectuals from party restrictions without making a sweeping statement in favor of grass roots democracy, pluralism, or a multi-party political system. Bialer's view of the limited perspectives of the Soviet intelligentsia is confirmed by Sakharov's position that carefully addresses such issues as arms control, the need for peace (even criticizing the Soviet government for its position in 1986 that an arms agreement could not be negotiated until the United States abandoned its Strategic Defense Initiative). These are important political questions, but they could be interpreted as appropriate issues of concern for the scientific intelligentsia under a system that remains profoundly undemocratic.

Technocratic ideology does not address broad issues of democratization, East or West. On the contrary. STR is a doctrine that argues discursively for the authority of knowledge and knowledge bearers over social life. If the foundations of civilization will be transformed by the relentless introduction of scientificity into social life, and if this entails the domination of intellectual over manual labor (a theme repeated over and over by official and dissident proponents alike), the real question for economic and cultural development is whether the party will share power with the intelligentsia.

This is the dilemma of the Soviet leadership today. Constituted as a thoroughly politicized society, it has placed power over knowledge in social relations, although this hegemony may not be in the long-term interest of the leadership itself. Yet, neither the Khruschev reforms of the late 1950s and early 1960s, nor the Brezhnev reforms of the 1970s, have succeeded in producing dramatic changes in power relations. Bialer can still judge the Soviet intelligentsia a dependent variable in the social equation, even as he recognizes the "anxiety" of the party.

Seen in this context, Gorbachev's rise to power may constitute an opening to the intelligentsia, notwithstanding his conventional political rather than professional background.[24] Like his precedessors, he has expressed concern over the state of Soviet relations of production. Speeches denouncing alcoholism and corruption in connection with lagging productivity rates are clearly criticisms not only of the working class but of the party and managerial oligarchies as well. His bold proposals for nuclear disarmament in the summit meetings with Ronald Reagan in the fall of 1986 echoed those of the intelligentsia. Moreover, by freeing Sakharov, he has shown considerable sensitivity to the Soviet and international scientific communities that have regarded the government's treatment of this famed scientist as a sign of its intention, announced by Gorbachev, of renovating Soviet society.

It is still too early to know whether the new government intends to share power with the technical intelligentsia, especially over production and its concomitant, education, as well as in the field of foreign affairs. Surely, compared to the "creative" intelligentsia (artists and writers), this is a more important stratum in terms of the priority placed on STR. One is chastened against undue optimism by the failures of equally powerful attempts in the past. The oligarchy of the Communist party is not likely to relinquish even a measure of its absolute power without a severe struggle. Indeed, the resistance of the oligarchy may be the condition for the transformation of the intelligentsia from a category into a class, especially if by this resistance the party perpetuates the stagnation that currently afflicts the Soviet economy. This is why arms reduction has taken a front rank in Soviet diplomacy, even as it hesitates by offering obstacles to settlement. The huge military establishment inhibits rapid economic growth and holds a virtual monopoly over scientific and technological resources.

Despite these political contradictions, what is clear is that Soviet ideology is deeply enmeshed with STR and that implicit in the logic of its propositions is the need for wide intellectual freedom and power sharing by those possessing knowledge relevant to production. Thus far, the party and state oligarchy have recognized only that a measure of intellectual freedom is desirable for the technical intelligentsia, and have sought its expertise for decision making in specific sectors of the economy. Further, this intelligentsia has been permitted to draw liberally on Western science and technology within limits imposed by the party and the relatively tight economic situation. But the broad participation of the intelligentsia remains an unfulfilled promise, unfulfilled because of "anxiety" that what is at stake is a transfer of power. Although it would be a mistake to calculate that the party's absolute control over all elements of political and economic life will collapse of its own weight, there is no question that it is on the defensive.

From the perspective of the history of Soviet discourse about science and technology, I discern at least three distinct periods. The first, which embraced philosophers as well as scientists, corresponds to the beginning of the New Economic Policy and ends somewhere in the late 1920s before Stalin consolidated his power, ended the policy, and declared the era of ideological war against capitalism. During this period, scientists, especially those engaged in fundamental research, enjoyed a large measure of intellectual freedom, but, as I have argued, were not exempt from the problems associated with partyness. As Joravsky shows, the discourse was dominated by the struggle between those who espoused the mechanical view (scientists themselves and a substantial section of philosophers) and those led by Avram Deborin for whom Hegel's dialectics were the condition for any Marxist discourse on science.[25]

The mechanical view, which included among its supporters the majority of natural scientists, declared philosophy superfluous in the wake of the findings of natural science. Philosophy consisted in their statements that the course of scientific discovery confirmed materialist epistemology and, in so declaring, opposed those philosophers and physicists who interpreted the emergence of relativity and quantum physics as evidence that the materialist hypothesis was extraneous to a science for which measurement and observation were, by their nature, relative to any given framework and to the instruments employed. The "idealism" that some contended was inherent in the new physics—

particularly its denial of absolute space and absolute time, and its modality of probabilistic results—did not become characteristic positions of Soviet Marxism until the introduction of partyness in science after 1929.

The second period is characterized by the attempt to oblige science to conform to the precepts of dialectical materialism as a matter of policy in the 1930s. During this period, a new norm was established for theoretical and philosophical discourses on science: they were required to integrate science with dialectical materialist philosophy. The ideological a priori associated with science's politicization did not entirely destroy the independent judgment of leading scientists, nor the development of some original contributions by Soviet scientists to physical, chemical and anthropological theory. Indeed, Fok, Pavlov, Ioffe, Oparin, and other internationally reputed scientists continued to work with the approbation of the authorities, although Bolshevization seems to have wrecked havoc most completely in biology. Nevertheless, many of those who continued to work in the mainstream of their disciplines, such as Fok and Ioffe, felt obliged to attempt to rationalize Soviet Marxist philosophy with developments in contemporary science.

This tendency has outlived the Stalin era, so it cannot be argued that such philosophical discourse is merely a product of the imposition of an ideological line on the scientific community. Indeed, the history of Russian science before and after the Russian Revolution shows a definite philosophical bent that parallels that in Western Europe. That most Soviet scientists in the 1920s declared themselves epistemological realists reflects not party pressure, but the assumptions of the new sciences that physics and biology refer to "real" objects. The status of the "real" is not to be confused here with materialism. What is at issue is whether the laws of physics refer, as Heisenberg argued, to the relation between science and nature or whether they refer to self-contained natural phenomena, the knowledge of which is mediated by the frame of reference but in no way, as writers such as Arthur Eddington argued, vitiates the need for the postulate of matter itself. Heisenberg's position, that quantum mechanics demonstrates the indeterminacy of knowledge of phenomena apart from the mediation of measurement, was certainly controversial to Einstein, for whom physical theory referred to nature even if knowledge could be specified only in relation to its frame of reference. So, it may be argued that debates within the Soviet scientific and philosophical communities on questions of modern physics are not substantially isolated from the rest of the international communities examining these questions. What is different, of course, is the Marxist assumptions of those engaged in the controversies. Since the early 1930s, Lenin's *Materialism and Empirico-Criticism* has been virtually the universal reference point for Soviet discourse about science, because his is the only *classical* treatment within Marxism of the results of the new physics. As we have seen, Lenin is true to Marxism's tendency, since Engels's extensive writings in the 1870s and 1880s, to accept the results of science and insist that these confirm dialectic and materialist interpretations. In contrast to this tradition, the 1930s were marked by an attempt to establish "proletarian" science in opposition to the main tenets of "bourgeois" science, a path advocated in culture by Lenin's arch philosophical adversary, A. Bogdanov, in the 1920s but initially repudiated by the Bolsheviks.

The concept of "class" science is scandalous, from a Marxist viewpoint, precisely because it challenges the universalistic claims of the Enlightenment to have found a method ensuring that the propositions of experimental science are true because verifiable by experience and calculation. Interestingly, the most eloquent work demonstrating that science arises from economic and technical conditions is not provided by an attack on the new physics as "bourgeois," but by B. Hessen's paper, "The Social and Economic Roots of Newton's Principia," delivered at a 1931 international conference on the history of science. This paper argues that war and war industry, navigation and mining were the problems connected with bourgeois commerce and conquest which constituted the foundations of Newton's discovery. Hessen contends that economic and technical problems had to be solved in a way different from traditional solutions offered by medieval scholasticism, the dominant paradigm of knowledge until the seventeenth century. "Systematic experimental knowledge," including a renovation of mathematics, owes its appearance to the emergence of the new practical problems associated with the development of industrial capitalism, according to Hessen.

Of course, Hessen does not equate the considerable growth of applied mechanics and "instruction on simply machinery" that Newton himself refers to in the introduction to his *Principia*. Hessen's argument is that these developments, although mentioned and dismissed by Newton, are the "earthly core" of the work. From this issues evidence that Newton's pedagogy to his students consisted in the instruction to study technical problems of navigation, metallurgy, and other practical sources of knowledge.

After undertaking a detailed examination of Newton's law of gravity and other aspects of his physics, Hessen concludes that his breakthrough and limitation consisted precisely in having developed a mechanical view of nature, reflecting the practical problems of the development of the productive forces of his day. This was a progressive development, according to Hessen, because it reflected the "great struggle of the European bourgeoisie against feudalism." At the same time, Newton is shown to have refused an interpretation of his own work that verified its materialist underpinning. Instead, like Descartes, Newton gives a religious connotation to his discovery. The key element is his refusal to consider time and space intrinsic properties of matter. By separating space, time, and matter, Newton revealed himself "a child of his (bourgeois) class." He was mechanistic and relatively naturalistic in comparison to the static, Aristotelian conceptions of the feudal scholastics, but, at the same time, Newton left room for God by refusing an evolutionary theory of the origin of the universe. "When I wrote the third book of the *Principia*," writes Newton to Bentley, "I paid special attention to those principles which could prove to intellectual people the existence of Divine Power." Hessen criticizes Newton for abstracting bodies from motion. If matter does not possess the "immanently inherent attribute" of motion, "an outside impulse is always necessary to bring it into movement or to alter or end this movement."[26]

Hessen's essay is perhaps the most detailed exposition of a theory of science that takes seriously the contemporary Soviet notion that science is determined by the social relations within which it functions. From Engels, Hessen derives the idea that the laws of physics are abstractions from the techniques of production and respond to problems associated with various industrial practices. At the same time, the nature of physical theory expresses, according to Hessen, the worldview of the class that has come to dominate production relations, in this case the bourgeois class. Thus, both the technical and ideological components make up the content of physical theory. Newton liberates science from certain theological imperatives associated with the feudal order because his "class" requires practical solutions to economically and technically based problems, at the same time retaining the need for religion and preventing science from adopting the standpoint of a thoroughgoing materialism.

Hessen is a Deborinite in the debates between those for whom science requires no philosophical interpretation beyond pointing out that most working scientists are "spontaneous" materialists, and those like Abram Deborin who insist that the results and methods of science suffer from their lack of dialectical synthesis. Although Deborin absented himself from the debates concerning science for most of the 1920s, his students and followers were obliged to enter this controversy after

1929. Hessen represents a profoundly antimechanist tendency that, at the same time, attempts to argue from historical and social premises that science is neither universal in its origins nor neutral in its ideological character. Further, he joins the debate on the crisis in physics. Against those who, from mechanist premises, oppose quantum mechanics and relativity theory as evidence that bourgeois science cannot resolve its contradictions and suffers from lingering "Machism" (an epithet that condemns the accused to bourgeois idealism, in Lenin's terms), Hessen joins Fok, Alexandrov, A. F. Ioffe, and other physicists in declaring the alleged contradictions and "dualisms" of modern physics a vindication of the dialectical nature of matter. If bourgeois science cannot synthesize its own discoveries because it has fallen victim to specialization and fragmentation as well as to philosophical idealism, the task of dialectical materialism is to assert, from the point of view of monism, the essential unitary nature of matter in motion. Seen in this light, the Lysenko controversy, which matures during the period of Stalin's ideological and political rule, takes on quite a different significance than Medvedev's ascription of "terror." I want to separate the experimental side of Lysenko's work from the political and philosophical context within which it was performed. That Lysenko rose to a position at the commanding heights of Soviet biology, despite the questionable nature of his theoretical position, attests to the importance of the links between knowledge and power as criteria for truth in the Soviet context. The form of the ideological debate, however, is continuous with a much older and ultimately more serious issue: can scientific knowledge liberate itself from social relations, including their ideological aspects?

Writing from the perspective of a victim, Medvedev sees little or no merit in the attempt to relate science to its social and economic roots. For him, as well as many other scientists and technical experts in the Soviet Union and other "really existing" socialist countries, the question is intellectual freedom from state intervention. Partyness has no place in controlling science and technology; it inhibits explorations of new ideas and, from the perspective of Soviet priority on economic growth, prevents technological development necessary for increasing labor productivity. Thus, the freedom of science is justified on more than moral grounds; its proponents argue that Stalin's ideological rigors thwarted socialist economic development. Surely, fundamental research in biological and physical sciences was deeply affected by the impositions of the regime. Moreover, scientists who refused to bow to pressure, or whose accommodation was incomplete, paid the price of their lives or at least their careers.

As in the cultural sphere, the few scientists relatively exempt from the demand for submission to the dominant ideology were usually international figures. Yet, the efforts to link Marxism with scientific and technological problems reflected as much a set of valid intellectual issues as they did questions of raw power. For there is no doubt that Western inquiry in physics and biology has always been intimately connected to philosophy and that, in this century, prominent scientists in many countries have been attracted to historical materialism as a way of explaining their work or even of guiding their scientific activity. For this reason, it is necessary to make the distinction between the general argument that heredity cannot be severed from environmental conditions, or, to be more precise, the conditions of social life, and the ways in which this argument has been translated into research.

This is precisely what Richard Lewontin and Richard Levins, American Marxist biologists, have done with respect to the Lysenko controversy.[27] They marshal considerable evidence to show that the Weissmann "refutation" of Lamarckian theory was experimentally flawed. The issue is not whether severed tails of rats can be inherited by succeeding generations but the totality of environmental effects. Weissmann's celebrated experiment allowed geneticists to "sidestep" the issue of the impact of environment on heredity, but the theory, according to them, has never really been confronted.

Moreover, Medvedev admits that the *practical* experiments in hybridization conducted by Lysenko and his associates benefited wheat production in colder climates, however limited the *theoretical* interests of these efforts were. The monstrosity of the affair consisted, therefore, not in the value of Lysenko's work but in its installation as official doctrine against which dissent became politically dangerous. Lewontin and Levins offer an explanation for the Lysenko affair that goes far beyond the simple ascription of the issues to "terror." The basis of their argument is that the situation in Soviet agriculture is chronically serious, apart from the distortions generated by Lysenko and the official reception of his work. The permanent crisis is due, primarily, to an adverse climate which shortens the season during which grains can be planted, in comparison to European and United States agriculture. That the government should support innovations that seemed *practically* effective is no surprise. Moreover, genetics had not yet proved itself able to make sufficient practical strides to constitute a viable alternative to Lysenko's practical successes.

Beyond this, Lewontin and Levins adduce evidence to show that in the era of Lysenko's ascendancy, 1930-64, the rates of increase in wheat yields in the United States and the Soviet Union were approximately equal—2 1/2 times. That the level of Soviet wheat yields was substantially lower is explained, not by the failure either of the collectivization or of Lysenko's disastrous theories, but by endemic climatic conditions that transcend the regime. Lewontin and Levins want to preserve the utility of dialectical materialism heuristically and to refute the purely Cold War arguments that proceed from the proposition that a free society must respect the ideological neutrality of science in order to advance the general interest. In this regard, I think that Jerome Ravetz is probably right in commenting that ideological neutrality is a "luxury" which is realized in few historical periods.[28] Indeed, his list of instances of the link between knowledge and power in different historical contexts illustrates that the Lysenko affair was no extraordinary event. Witness the evidence provided by another instance, Arthur Koestler's account of the lesser-known Kammerer affair.[29] Paul Kammerer, a Viennese biologist, conducted experiments with midwife toads to test his own Lamarckian hypothesis. Since he was an experienced animal breeder and a respected scientist, Kammerer's work in establishing the inheritance of acquired characteristics attracted widespread attention and won support within the scientific community both in his own country and abroad, but also elicited the antagonism of the genetics establishment, particularly a major Mendelian of the turn of the century, William Bateson. Kammerer's experiments consisted in severing the tubelike siphons of a seq squirt called *Ciona intestinalis*, finding that the next generation grew longer siphons. These experiments were conducted between 1905 and 1910, and were followed by other experiments that showed to his satisfaction that characteristics acquired under "radically different environmental conditions" were passed on to succeeding generations. These characteristics included mating habits, changes of color, and so on. He did not claim that the very next generation inherited all of the changes that had been acquired by varying conditions such as soil. But he was satisfied that his work had shown that the relation of heredity to environment was significant and that mutations, produced accidentally by such interventions as X rays, were not the only mechanism of variability of species.

In subsequent years, his work was subjected to merciless attack by geneticists in England and on the continent. Despite these trials, he maintained his position and was never definitively refuted. Nevertheless, he became something of a pariah, his reputation destroyed among the mainstream biology community and his financial and personal life ruined. In 1926, he was invited to establish a laboratory by the Soviet government, which was seeking innovations that could improve both its fundamental and applied research capability. On the eve of his voyage, he committed suicide. Until Koestler's narrative, historians of science, following scientists themselves, chose to sidestep the incident and the issues entailed by it. In contrast to the Lysenko affair, where the knowledge/power axis resulted in what has become a scandal, the Kammerer affair affirmed the orthodoxy of classical genetics and warned dissenters that there is more than intellectual disgrace awaiting their normative departures. Bateson, Bauer, and other pillars of the genetics faith apparently judged Kammerer's work prior to its experimental falsification. The incident gives a tragic and ironic twist to the prevailing ethic of pure science. Koestler, a relentless critic of Marxism and the Soviet regime, notes that Kammerer was treated as a "hero and martyr" by the followers of Lysenko and inevitably the "fraud" associated with Lysenko's work rubbed off on Kammerer.[30] Although the struggle for hegemony over biological science was "bloodless" in Kammerer's case, it serves as a fair illustration of the argument that the link between scientific theory and ideology/power is not confined to so-called authoritarian regimes but is part of the everyday life of the sciences.

It is easy to dismiss dialectical materialism's claim to play a positive part in the process of scientific discovery by referring to the Lysenko affair as a counterfactual example. This is comforting to those in the West and in the Soviet Union who would "free" science from state intervention and "political" ideologies. I have argued that the use of such arguments by the scientific and technical intelligentsia is part of a severe and perhaps successful political struggle in the Soviet Union. The ascendancy of Gorbachev and his attempt to vitiate the absolute power of the party and state oligarchies over culture—in the workplace, the laboratory, or the concert hall—may develop under the banner of freedom, just as the "revolution from above" takes place in China under the banner of democracy (and its chief agents are students and other intellectuals).

But the doctrine that science and its laws are free of metaphysical or doctrinal presuppositions is no more valid for Galileo, Mendel, or Einstein than they are for Lysenko and Kammerer. Facts and the methods by which they are adduced presuppose a theoretical structure or paradigm that constitutes their *axial* structure. As I shall argue later in this book, the requirement that "observation" underlies facts and that "facts" are the inductive basis of theory is itself an ideology. Moreover, this is precisely the ideology of Soviet science, which, in this respect, is thoroughly Western. Falsification of experiments, excessive claims derived from modest practical innovations, and other "frauds" cannot detract from the seriousness of the Soviet claim, following only one strand of Marxist ideology, that scientific theory has not only metaphysical foundations (an argument also made by the English philosopher E. A. Burtt in 1924) but also social and political preconditions.[31] Thus, the Soviet attempt to reconstruct science on the basis of an alternate metatheoretical system to that of empiricism or positivism. These ideas became official doctrine under Stalin, an occasion that became for detractors of the Soviet regime during his years of power a sufficient argument for dismissing these substantive claims.

The Lysenko affair showed what Ravetz calls "tragedy" in the history of science. Tragedy consists in the influence of ideology on a science that wishes to be free of it, at the same time being unavoidably intertwined with value-laden choices. The virtue of Soviet discourse on science and technology is that it has always acknowedged the relevance of "interest" to both theory and applied sciences. For this reason, dialectical materialism has remained, even in our day of relative scientific freedom in the Soviet Union, a metascience within which scientific discovery is assimilated.

This Soviet penchant for assimilation of "bourgeois" science into Marxism runs like a red thread in the writings of both scientists and philosophers. Two streams are particularly important. On the one hand are those who, from time to time, discover that even the most hallowed discoveries of Western science betray signs of bourgeois metaphysics. Such was a frequent but by no means a dominant tendency during the Stalin era. For, in the cases of theoretical physics and chemistry, debates raged in the Soviet scientific community without relations of hegemony being established by the state. These contrast with the situation in the biological and psychological sciences, where the reverse occurred. Yet, in physics, the supporters of relativity and quantum theories invoked versions of Marxism not only to justify their support for doctrines that were condemned by others as "bourgeois," exemplars of the "crisis" in physics but also to inform their own contributions to the field.

Thus, V. Fok and A. D. Alexandrov, one an internationally recognized physicist and the other a major mathematician, were militant supporters of the new physics throughout the Stalin era and beyond. Alexandrov, writing in the 1970s, unconditionally supports the special theory of relativity but argues that there is no "general" theory that has the same level of validity. Whereas the special theory establishes the relativity of space and time to the frame of reference adopted, Alexandrov argues that the general theory is not really a "law" of the same magnitude but constitutes merely a rule for writing equations. Alexandrov insists that while no physical law may be enunciated in absolute terms but only expresses a series of relations, the sum of these relative laws constitutes "absolute" truth or its approximation. The postulates of absolute space and absolute time have been overturned but not for "matter as a whole."[32] Fok insists that Lenin's account of dialectical materialism suggests that the new physics is the limiting case of classical physics, which is the more general theory. This limiting case is the description of microobjects: "at the same time, it cannot be simply discarded since an objective description requires as a basis (which may be used directly or indirectly) something approximately independent of the way observation is performed, and this is just the 'absolute' mode of description used in classical physics."[33] Fok and Alexandrov, together with the philosopher of science M. E. Omelyanovsky, hold to the position that "dialectics is an adequate form of thinking for modern physics and science as a whole that completely corresponds to the character and constantly varying content of modern science."[34] Heisenberg's views—that the new physics is a confirmation of relativism and, more strikingly, that cognition is an active intervention and not a passive observation—are explained by the categories of dialectics, rather than suggesting the need for a whole new philosophical interpretation, as N. R. Hanson asserts. Dialectics shows that the apparent relativism (the ineluctability of measurement to the object) is explained in terms of a larger macrotheory which converts chaos to order, relative to absolute truth. Thus, Omelyanovsky adds the criterion of *adequacy* to scientific law, where the term connotes a correspondence of theory to an object existing independently of human consciousness or intervention. Modern physics, in this view, makes observation indirectly and suggests the possibility of interpreting physical law as indeterminate of its object, moves toward concepts such as statistical probability. These interpretations are, according to Omelyanovsky, due primarily to the inadequate conceptual apparatus with which scientists and some Western philosophers approach the results of the new physics.

It would be a mistake to dismiss the correspondence theory of truth as expressed by Soviet philosophy of science as merely a political a priori invoked for the purpose of legitimating the ability of Soviet physics to remain part of the international community and, at the same time, to maintain support of the Soviet Union's own government. Fok and Alexandrov make substantive arguments for their antirelativism and are Einsteinian in their perspectives against those, like Bohr and Heisenberg in the 1920s, for whom quantum mechanics was evidence for the view that physical law referred to a relation between method and measurement and the object. For Omelyanovsky, who has published a full philosophical statement on the relevance of dialectics to the new physics, science consists in the adequacy of its theoretical system to the object; that is, he is unwilling to jettison the idea of truth from scientific discourse. This is clearly a widely shared perspective among Western scientists and philosophers. The Soviet distinction between micro- and macroobjects and the "laws" corresponding to these levels of matter can be seen within the framework of the effort to validate the concept of the "unified field," to which Lenin's idea of successive approximations refers. Fok's reference to the system "as a whole" is consistent with the refusal of Soviet discourse on physics to accept the puzzle-solving approach to scientific discovery in which totalities play no role; such perspectives prefigure the relativistic result.

Consequently, what may seem quaint to postmodern theorists is entirely comprehensible in terms of dialectical materialism, namely, its insistence that matter in motion is the referent of physics. These Soviet discussions do not correspond to the earlier effort to found a proletarian science but are relevant to the agreement among those who discourse on the new physics that philosophical isuses are immanent in the content of quantum and relativity theories. Those for whom relations and processes replace "matter" as the object of scientific knowledge in physics, chemistry, and biology have adopted an epistemological standpoint whose status cannot be separated from the presuppositions from which it springs. Nor can the laws of physics, no matter how widely held, be invoked as proof of adequacy since this implies correspondence from which Bohr's principle of complementarity departs. Having said this, it remains obvious that materialism is placed on the defensive by quantum mechanics, if not by relativity. Dialectical materialism as an interpretive discourse is pressed to find legitimacy for its outlook in the "spontaneous" dialectical and materialist practices of scientists themselves. Thus, in the long odyssey of the Soviet discourse on science, the mechanists have set the tone, despite the apparent ideological victory of "partyness" after 1929. As with Marx and Engels, contemporary Soviet Marxism looks to science for confirmation and retains only its interpretive apparatus. In this respect, Fok and Alexandrov are exceptions; they have *assimilated* rather than merely interpreted the new physics and effected a partial synthesis with materialist dialectics. While accepting the fundamental approaches of quantum and relativity theories, they have tried to work with the categories of these subdisciplines in terms of the concept of simultaneity. Thus, rather than understanding complementarity as demonstrating the validity of wave and particle theory at two different moments of measurement, true within specific referents, Fok has attempted to argue that the dialectical principle of contradiction, in which opposing forces may occupy the same ontological (in contrast to logical) space, provides a more adequate explanation of the alleged paradox of quanta.

In chapter 9, we shall see how deeply antifoundationalism has penetrated Western science and philosophy. For our investigation, such considerations seem to have no place in Soviet Marxism. For this reason, Marxism appears controversial in the philosophy of science even though, when compared with the Nietzschean concerns of much of present-day European philosophy, it appears conservative. At the same time, Soviet discourse on science is philosophically complex, virtually unique in the recognition of its own partisanship as well as in its insistence on objectivity in the classical sense.

The current, third period of Soviet science is marked by its twofold character: on one hand, the universalist historical claim of the scientific-technological revolution to primacy over social relations; on the other hand, the *convergence* of Soviet science with Western science on the practical level. Theoretically, dialectical materialism has, indeed, changed its form. Gone are the efforts, except in specific instances such as physics, to found new science on the basis of dialectical materialism as a metascience and a "guide" to action. Instead, Marxist philosophy becomes a mode of philosophy *of* science; its interpretative apparatus is posterior to scientific discovery. As a *discursive formation*, there is no specifically Soviet science. This follows the abandonment by the Soviet state of the program of technological independence which, although never fully implemented, was the stated policy of the Stalin period. For if science has become a crucial component of the development of the productive forces, and if these, in turn, are regarded as historically autonomous from production relations, the argument for an epistemologically distinct science loses its force. In fact, it may be argued that the principal areas where distinct "Soviet" sciences were developed, or at least sciences that made original contributions to the international scientific communities, were those in which the practical applications of Western theoretical paradigms had not (yet) been demonstrated, such as biology; or, in the human sciences, such as anthropology, sociology, and economics, where Marxism has never yielded to "Western" science in part because Marx himself enjoys virtual mainstream status as a founder of these disciplines. Although the situation is not the same in psychology insofar as this discipline is considered a biological or social science whose major discoveries occur in the late nineteenth and twentieth centuries, Marxism refuses to yield.

In this, the third phase of Soviet science, the dominant tendency is to retain the distinction between Marxist contributions to the human sciences and "bourgeois" theories which are regarded as ideological. However, since the 1960s social scientists in Eastern Europe, including the Soviet Union, have "discovered" the work of Robert Merton, especially his insistence that grand theory must give way to theories of the "middle range" because of their testability. For example, although holding precariously to a Marxist framework, a leading Polish sociologist Piorr Sztompka has argued that Western, non-Marxist sociology has a definite place in social science because its focus on conditions of system stability is largely commensurable with Marxism, which, however, concentrates, properly, on conditions for system-crisis and transformation. In a single stroke, he makes room for functionalist social theory within a socialist context by arguing that such issues of general theory as the interdependence of social institutions within a social system, the existence of normative structures within a system, problems of consensus and conflict at the system-level of analysis, etc., are valid. Moreover, he argues that these concerns can be historically situated and can be integrated into Marxist theory.

It is not my concern here to discuss the merits of these assertions. For my purposes, it is important to note the parallels between the effort to assimilate Western natural science and social science into Soviet and Eastern European scientific work in the post-Khruschev era. To be sure, the same a priori obtains in this area as it does in physics. The key move must be to show no substantial conflict between these theories and Marxism, and to demonstrate the utility to current problems of socialist "construction."

In the Soviet Union, the crucial mode of assimilation is not Western, particularly American sociology but systems theory, which has much wider applications in technology, specifically planning and other mechanisms of control in both the biological and policy sciences. It is noteworthy that concepts such as teleology, hitherto excluded from scientific discourse, reenter through systems theory, following the biological discovery that *function* is an integrative mechanism that connects organs with one another. As we shall see in chapter 11, on the philosophy of biology, the view of nature is radically changed by the concept of the interdependence of system functions from that characteristic of both dialectical materialism and post-Kantian philosophies of science. Similarly, when social theory adopts the idea of the coherence of a social system through ideas of shared norms and values among its elements, which in turn introduce the notion of function, "system" is used interchangeably by its proponents with the ideas of totality more appropriate to Marxist discourse or wholeness which has become a leading ecological concept. In terms of the conflict generated by the reductionist tendencies in both physics and molecular biology, systems theory has become a major ideological intervention in Soviet science. In their major statement *Systems Theory: Philosophical and Methodological Approaches*, Blauberg, Sadovsky, and Yudin make explicit the link between ecology, system, and dialectical materialist ideas and more or less wholeheartedly support organismic interpretations that stem from Ludwig Bertanaffly and other leading systems theorists. While emphatically denying the claims of system theory to have founded a new general theory embracing all levels and all fields, the authors are quite enthusiastic about the pertinence of this metatheory for the advance of science. Their interpretation is framed in terms of the resemblance of systems theory with dialectics, especially the concept that the object of knowledge must be understood as a system, not an isolated "thing."

The spread of Soviet adaptations in the human sciences has not, to my knowledge, extended to such areas as psychoanalysis, principally, I believe, because the tendency of the work in the third period is to a more explicitly functionalist and positivist definition of science. In this context, neither psychoanalysis nor hermeneutically oriented neo-Marxisms are welcome to what is almost frankly admitted to be a technocraticization of social theory.

In sum, the movement of Soviet Marxism as well as the intellectuals is away from dogmatism in both natural and social sciences. But it is also continuous with important strands in Western Marxist orthodoxy as well as with some of the major tenets of Leninism as an interpretation of science. Thus Westernization extends to functionalist, analytic tendencies in both the Marxist and non-Marxist theories of science. In the next chapter, I want to examine the nature of the non-Marxist theory of science. I shall demonstrate that just as Marxism has several, often conflicting strands, so analytic philosophy is internally divided, and opposing pictures of scientific development offer other alternatives.

PART III

CHAPTER 9

THE BREAKUP
OF CERTAINTY:

History and Philosophy of Modern Physics

In this chapter, I wish to examine Anglo-American theories of science, in their philosophical and historical modes. Contrary to some interpretations, I claim that while they hold tenaciously to a logical empiricist and positivist *ideology* (in which science is defined as the concatenation of empirical and a priori mathematical knowledge whose determining moment is independent experience), the developments in modern physics, generally privileged by philosophers and historians as models of all knowledge, have forced a serious crisis among both philosophers and physicists. The crisis appears to have shaken both the rationalist and the empiricist sides of the equation.

Since Kant, all philosophy, except ethics, has been about science. Having eschewed the construction of metaphysical systems on the ground that the study of reality is a systematic enterprise consisting in a formal procedure that entails both observation and mathematics, philosophers have generally followed Locke's self-description as underlaborer of the sciences. Henceforth, philosophy's task is limited to the clarification of various aspects of scientific investigation that might be grouped under the rubric of metascience. These include issues connected with processes taken as central to science—i.e., the nature of evidence, and problems connected to the formal apparatus of scientific measurement. On the other hand, there is vigorous debate concerning the implications of the material results of science—

implications for the general theory of knowledge, the "nature" of reality, etc., and the ontological status of formal systems such as mathematics.

In the twentieth century, under the sign of evidence, primary attention is directed to the questions of what is observation, what is inference, and how valid are experimental and quantitative methods for adducing evidence and for verifying scientific statements. These questions reflect a fundamental shift in physical theory. The new physics has raised the problem of the reliability of scientific law as representation. W. V. O. Quine, a self-described lay physicist, has posited that reality itself is *assumed* by any act of knowing. Ludwig Wittgenstein has argued that since representations refer not to an object but to paradigmatic instances of language as scientific laws and particular statements, the proper study of metascience is language in use. Thus, even if semantics and linguistic philosophy appear to be concerned with "abstract" issues, their referent is indirectly the sciences. These are disciplines of edification.

Philosophy rediscovered its task at the turn of the twentieth century with the refinement of positivism by the Vienna circle, Wittgenstein, Rudolph Carnap, and Popper, among others, because the paradigm shift in physics produced a high degree of indecision about the character of scientific law itself. That is, although physicists are always philosophers, the more theoretical the more philosophical, the "new physics" obliged many to reexamine the underlying premises, the epistemological and ontological roots of the classical paradigm. Consequently, in this century we encounter treatises on philosophical aspects of science from "professional" physicists and biologists precisely because their practice seems to overturn many of the concepts with which they began. Einstein, Bohr, and Heisenberg are not merely leading figures about whose discoveries philosophers speculate—they engage in speculation themselves. Typically, it is not their work that is challenged by those outside the scientific community proper, but what they make of it, the philosophical significance of the work, particularly statements about the referent of scientific law.

The main difference between philosopher-scientists and scientific philosophers is this: almost invariably, professional philosophy concerns itself with problems of method following the pattern set by Descartes. Although it would be excessive to claim that philosopher-scientists rarely discuss method, these discussions are nearly always placed in the wider context of epistemological and ontological discourses. This is particularly true of Europeans such as Einstein, Bohr, Heisenberg, and von Weizsacker, for whom relativity theory and quantum mechanics presented new problems concerning what we mean when we speak of a "law" of nature. As von Weizsacker complains when reporting on a meeting between Bohr and a group of "positivist" philosophers:

Bohr was deeply disappointed by their friendly acceptance of all he said about quantum theory, and he told us: "if a man does not feel dizzy when he first learns of the quantum of action, he has not understood a word." They accepted quantum theory as an expression of experience, and it was their *Weltanschauung* to accept experience; Bohr's problem, however, was precisely how such a thing as quantum of action can possibly be an experience.[1]

von Weizsacker goes on to cite the well-known Heisenberg "thought experiments," which assumed experiment cannot observe all that is presupposed in a field, and that thought can produce knowledge on the basis of warranted inferences. The positivists expect that anything that can be theorized is based on corresponding "sense-data." von Weizsacker's point is that theoretical physics may posit phenomena for which the data not only are unavailable but cannot be observed or measured. He distinguishes Bohr's view from positivism as follows: "The fact that classical physics breaks down on the quantum level means we cannot describe atoms as 'little things.' "[2] This does not seem very far from Ernst Mach's view that we should not invent "things" behind phenomena, but rather than dissolving things entirely, as Mach proposes, he believes that things are "in" phenomena. A second difference with positivism is that Bohr and other modern physicists refuse to describe events in terms of properties of objects independent of the situation of the observer.

von Weizsacker has somewhat overstated the case. Influenced by the uncertainty that is intrinsic to quantum theory, contemporary philosophers of science, while framing their insights within the language of method, have increasingly addressed themselves to problems arising from relativity and quantum mechanics. Among the critical issues in Anglo-American philosophy of science has become the question of observation, i.e., what is "seen" in the experimental situation. Further, "facts" no longer enjoy an unproblematic existence, even in empiricist philosophy. The reason is quite clear. Physics has provided the opening for speculation; permission has been given. Even more fundamental is the growing recognition among philosophers that the new physics has raised epistemological problems that cry out for attention, if not solution. Therefore, while logical questions, such as the meaning of inference or the traditional issues of language philosophy such as "must we mean what we say?" are still topics for dispute, over the last twenty years increasing attention had been paid to "metaphysical" questions, framed largely in the discourse of language analysis.

Thus, W. V. O. Quine, perhaps America's leading philsosopher of science in the analytic school, has distanced himself from many of the major propositions of logical empiricism. He argues that the distinction between analytic and synthetic statements, the cornerstones of Kant's epistemology, does not hold. Further, he attacks the hypothesis of reductionism that underlies the "verification" theory of truth. The first proposition claims that there exist analytic statements that are *logically* true, i.e., that do not require verification because they do not depend upon experience. A statement like "no unmarried man is married" is true prior to any possible experience of this. On the other hand, the statement "no bachelor is married" becomes analytic only by substituting "unmarried man" for "bachelor." Synonymy depends on definitions that are arrived at conventionally, i.e., that are constructed in use or by lexicographers who perform etymological/historical research which amounts to discovering their usage:

> Just what it means to affirm synonymy, just what the
> interconnections be which are necessary and sufficient in order
> that two linguistic forms be properly describable as
> synonymous is far from clear; but whatever interconnections
> may be, ordinarily they are grounded in these usage.
> Definitions reporting selected instances of synonymy come
> then as reports of usage.[3]

Thus, substituting one word or linguistic form for another demonstrates that, except for a rather narrow class of statements, all others are actually synthetic, that is, we must refer to historical, empirical, practical uses in order to justify substitution. But experience, as we shall see, does not stand in a causal relation to what are taken as factual statements.

Quine ostensibly stays in the logical and empiricist lexical matrix but introduces a relativist position on the basis of a deconstruction of the two dogmas that he opposes. Quine's first refutation, that of analytic statements, rests on a pragmatic argument: linguistic form refers to experience outside logic. Difference is ineluctably structured into the language's syntax, and identities can only be posited by convention. Accordingly, all statements are synthetic but not in the way Kant and his followers have it. Quine insists that a statement is not, except under what he calls "boundary conditions," a synthesis of logic and experience, whether derived from the senses or otherwise. For all intents and purposes, scientific statements are formed as a "man-made fabric."

> The totality of our so-called knowledge or beliefs, from the most casual matters of geography and history to the profoundest laws of atomic physics or even of pure mathematics, is a man-made fabric which impinges on experience only along the edges. . . . But the total field is so underdetermined by its boundary conditions, experience, that there is much latitude of choice as to what statement to reevaluate in the light of any single contrary experience. No particular experiences are linked to any particular statements in the interior of the field, except indirectly through considerations of equilibrium affecting the field as a whole.[4]

The reference of any scientific statement is only putatively "experience" of the physical objects, which, in any case, are merely "convenient intermediaries," "cultural posits," articles of faith that the external world exists and is "superior epistemologically" to the cultural posit of Homeric gods.

In another place,[5] Quine argues that the "truths" of mathematics have no empirical referent because they are synonyms for categories of logic, which, in traditional Kantian terms, are a priori analytic. Quine's claim that he remains an empiricist despite his thesis that points to the de facto self-referentiality of scientific investigation has contributed to major controversies in the logical empiricist camp. Beyond his refusal to accept that most classes of statements considered "analytic" are true prior to any possible "confirmation," Quine further tries to demonstrate that the verification theory of confirmation contains the errors of the supposed analytic/synthetic distinction. The final judge of scientific statements, in the Popperian or indeed the entire empiricist tradition, is "experience" understood as the correspondence of individual sense experience with a scientific statement. Following Duhem, Quine argues that "our statements about the external world face the tribune of sense experience not individually but only as a corporate body."[6] But since, for Quine, experience constitutes only the "boundary edges" of science and "there is so much latitude of choice," it is difficult to sustain the belief that he is really an empiricist at all. On the contrary, the notion of "choice" introduces a major cultural arbitrary into the process of scientific discovery, especially in physics. "Underdetermination" by boundary conditions points to the proposition that correspondence, even of the "corporate body" of scientific statements with a physical world that is known by means of experience and which is capable of refutation, is improbable. Quine's framework is not so much empiricist as it is conventional. What Kuhn calls "anomalies" in normal science is similar to Quine's "a conflict (of a statement) with experience at the periphery." But, even if experience is a real boundary limiting the arbitrariness of reference, the idea of underdetermination indicates the function of "experience" is to force reevaluation, but cannot determine the nature of adjustments. Accordingly, since an individual scientific statement presupposes a theoretical system upon which it depends, algorithms of refutation can apply only to the system as a whole. But it cannot be a question of adequacy if by that term we mean agreement of statements with experience. Instead, one might interpret Quine's position to be this: at the boundary, conditions of scientific activity, events we shall designate by the name "experience," occur that conflict with the corpus of scientific statements upon which research depends. By logical necessity, this conflict demands resolution by means of reevaluation and subsequent adjustment in the corpus. Lacking this possibility, scientists are obliged to transform the basis of investigation, the theor(ies) themselves, in order to resolve the conflict. But the nature of the resolution is underdetermined by the event(s). Hence, for the most part, science is self-referential inquiry. Either it refers to its own logical presuppositions, themselves man-made, or to extralogical influences that may be designated as "cultural," "political/ideological," or whatever.

Quine: "Science is a continuation of common sense and it continues the common sense expedient to swell ontology to simplify theory."[7] As a self-confessed pragmatist, Quine relies on the practical value of the cultural posits of experience, the external world, and repudiates Carnap, upon whose work much of his own relativism is based, because "their [Rudolph Carnap and C. I. Lewis] pragmatism leaves off at the imagined boundary between the analytic and the synthetic."[8] In other words, we choose those suppositions that satisfy the requirement that (1) theory be simple, and, more to the point, (2) refer to something beyond itself. These are heuristic hypotheses incapable of confirmation. Richard Rorty rightly points out that Quine's refutation of the distinction between analytic and synthetic reason undermines the entire project of Anglo-American philosophy. Bertrand Russell tried to preserve philosophy, through its claim to the turf of analyticity, as a discipline concerned with "pure" knowledge, as opposed to science, which, inevitably, must be considered synthetic and provisional.[9] For, if there are no statements that can be removed from convention (which is the only "certain" way Quine employs the concept of experience—i.e., social experience as opposed to "observation"), the way is opened to collapse philosophy into other discourses, in the first place science, both what is called natural and what is designated social.

One may discern a crisis in analytic philosophy as well as physics on many of these points. The crisis consists not so much in the reflection on duality between wave and particle explanations for the motion of light or in the relativity of measurement to its frame of reference, but in the nature of scientific law itself. For if the observer and the apparatus of investigation constitute an intervention in the field, and scientific theory is thereby "man-made fabric" underdetermined by the events we call experience, or by the results of methods of falsification except for the theory as a whole (the paradigm), then the results of "normal" science, whether history or physics, constitute a discourse incapable of verification. The belief in the validity of results turns out to be a cultural posit which, despite the protests of Einstein and the later Bohr, refers to the relation of humans to the object of inquiry, not to a unified field independent of our activity.

Rorty, following Nietzsche, wants to abolish foundationalism in philosophy, the doctrine according to which a priori grounds exist for determining truth. Instead, philosophy is to become a "conversation" of a special kind, whose principal activity is hermeneutic investigation of texts to determine their significance in relation to a series of questions posed by the reader. This redefinition is a restatement of Dewey's conception of the role of philosophy. Recall that it was Dewey's reading of Heisenberg's indeterminacy principle that led to his radical conclusions about the production of the "real."[10] In this sense, Dewey points to the later conclusion of Norwood Hanson that there is a philosophy generated by modern physics that sidesteps nearly all the traditional epistemological questions. For Dewey, philosophy may not speculate on natural law except on the basis of the actual result of scientific inquiry. One of his conclusions points to the end of the sharp separation, characteristic of modern science, between organism and environment, between the animate and inanimate. The refusal of such distinction leads, as we shall see in chapter 11, to a radically different picture from either that proposed by mechanistic thinking or some major tendencies of "life" philosophy, which, in the name of securing a space for human sciences, insists that its vision and methods are distinct from those of natural science. Dewey's naturalism is, however, continuous in many respects with movements of thought in the first decades of the twentieth century that rejected the mechanical world picture and its reductionist conceptions, including linear causality, the denial of intentionality in nature and history, and its idea of nature as self-contained entity.

Stephen Toulmin points out that this is precisely the point of what is, perhaps, John Dewey's major contribution to the philosophy of science, his 1929 Gifford Lectures published as *The Quest for Certainty*. The work closely follows on Heisenberg's indeterminacy principle (1927). As Toulmin rightly points out, Dewey's interpretation of the new physics goes beyond Heisenberg by insisting that human action has been shown to "participate" in the formation of the object. It is not merely that our knowledge of microparticles can no longer fulfill the requirement of precision established by Newtonian mechanics. The object itself, according to Dewey, is "eventual," itself the outcome of inquiry.[11]

This formulation considerably revises Dewey's theory of experience. We do not merely "observe" phenomena, but phenomena are constituted, in part, by the consequence of intervention. The subject/object split becomes as untenable as the materialist hypothesis that our knowledge refers to an external world that is self-contained. What we mean by experience can no longer be identical with sensationalism but is intimately linked to what Toulmin calls "praxis," more accurately, recorded as "practice." It is clear that the epistemological project of empiricism is abolished; the act of knowing is reflexively mediated production. Experimental method is identical to action, whose aim is control through the intentional construction of a future. But intention, or telos in the nonmetaphysical sense, connotes not only that "experience" is constituted out of interaction and no longer has a subjective significance, but also that "truth" is a relation between the intention of control, of construction of a future and the (indeterminate) object that results from inquiry. Dewey's argument that science, like morality and social policy, is ultimately *practical* removes a huge gulf that inured it from the judgments freely suffered by other discourses.[12] For if knowledge *is* interaction, neither observation nor mathematics is free of conventional uses. They refer not to the *independent* object, but to the relation between the object as constituted by inquiry and the purposes for which inquiry is undertaken. "Experience" loses its originary and pristine existence. It no longer can be invoked as the arbiter of theory since it is implicated in theorizing. "Interactions produce changes in what is experienced," and if we are to follow Dewey's theory of interaction, intention underlies what the known brings to scientific activity. If the consequences of this interaction are indeterminate from the point of view of intention, the "construction of a future," which is the real object of scientific inquiry, is always problematic as well as *the* problematic of science. With experience, the meaning of "discovery" is also changed. For the referent of scientific inquiry is not really "nature." Nature becomes, semantically, the *scene* of interaction, as well as one of its terms. On the one hand, "man is continuous with nature." On the other, nature takes on meaning only through our interaction, both "blind and intelligent." Although Dewey seems to retain the category of nature as something "there" to be discovered, its history is coincident with the consequences of intelligent action that make impossible a naturalistic definition.

I believe this helps clarify how Quine employs the term "experience" as a boundary condition for knowledge. While he appears to accord to experience a subordinate but independent role, in fact the science shows that experience is not only theory-dependent but practice-dependent. Quine's concept that the components of statements refer to terms whose truth value is derived from "usage" corresponds to Dewey's operational theory of truth as *validity*. To speak of truth is to submit to an essentialist conception. Rather, Dewey wants to consider scientific work provisionally valid on the basis, not of its "internal elaboration and consistency"; nor are "scientific conceptions . . . a revelation of prior and independent reality. They are a system of hypotheses, worked out under conditions of definite test."[13] Taken out of context, these statements can be easily confused with philosophical idealism. But Dewey's relational and practical theory of validity refutes this charge. For Dewey, "tests" are not constructed apart from the intention of directing changes in reality. Thus, the common theme that scientific theory "discloses" an immanent object is transformed into a set of practical operations relative to the concept of control over reality. If an hypothesis is valid, the consequences of its interaction with the object of knowledge is its ability to construct a future on the basis of specific operations on that object. The "operators" of his theory of science are intention-intervention-experience-provisional validity, and nature is understood as the context of these relations. It is difficult to see how any proposition constructed under this regime can be refuted/falsified apart from the referents, which, to say the least, cannot remain unexamined.

Quine's concept of "underdetermination" of propositions by events/experience is homologous with Dewey's practical teleology. Dewey specifies that intention is linked to the development of science, a consequence of which is to view "defects" in perception and knowledge not as natural facts to be accepted as immutable but as practical defects to be overcome. Experience is no longer empirical but has become experimental, active, and purposeful. The key to understanding the significance of any scientific "law," therefore, is its larger context in the relation of knowledge to human need. Dewey's reprise of the history of science since the Copernican Revolution is framed in the transformation from knowing as a condition of acceptance of the natural world to experimentalism which seeks power over that world, to shape it in light of practical tasks. Just as we "make" the world through experiment/intervention, so experience is our own product, which, however it appears to emanate from without, is at the base a social relation, social in the sense that interaction is the condition of its construction. We may infer that "reality" is no longer a broad expanse, the secrets of which can be unlocked by reason, but the outcome of specific acts of inquiry whose presuppositions are skepticism about the inherent powers of reason. Quine's latitude of choice of theories becomes for Dewey the practical conditions that motivate scientific action.

Although Dewey is a celebrant of scientific progress and a determined opponent of metaphysical claims to positive knowledge, there can be no doubt that his conception of science implicates it within the social relations that form the context of inquiry. These social relations demand that control over the object, rather than passive contemplation, governs experiment and theorizing; critical for a discourse on scientific theory, he refuses narrow inferences concerning indeterminacy but understands this to confirm the proposition that the distinction between sciences of nature and sciences of "man" is untenable. This, of course, does not imply for Dewey that human sciences may be reduced to natural science by means of mathematics. Instead, as I read Dewey, naturalism refers only to the fact that as organisms, we are part of nature. But, since we are practical nature, all sciences are human sciences since the final referent is practice and its intentionality.

Dewey's blunt challenge to Cartesian premises becomes an invitation to an inquiry concerning the social relations of science, or, to be more precise, to view science as social relations. This invitation, construed as a critical discourse, underlies the work of Part III of the present volume.

The separation of science from the *theory* of science is still controversial, especially for those who insist that science needs no metatheoretical reflection. If a discourse on science is at all valid, according to this view, scientists themselves should undertake it. This standpoint takes its validity from the now normative statement that philosophy, history, and sociology can contribute nothing to physical or biological sciences because knowledge is gained exclusively from observation and the logical investigations of mathematics.

Thomas Kuhn suggests that the academic field of history of science arose as "a byproduct of pedagogy." Traditionally, scientists told the story of scientific discovery as a means "to elucidate the concepts of their specialty, to establish its tradition and to attract students."[14] As a separate field, historical studies of science are a product of the twentieth century and, in the United States, do not appear as a profession before 1950. What Kuhn suggests, but does not elaborate, is the idea of history as an adjunct to science's effort to consolidate its position as a discourse that can be distinguished by more than mere difference of its object of knowledge from humanities and the social sciences. Unlike the other disciplines, (physical) science claims a method of knowing that is superior not only because it is based on "observation of nature"—a privileged territory in technological society—but also because of its declared methodological precision by means of mathematics, a precision that is denied the human sciences or evolutionary biology. History becomes the story of the evolution of science from its magical stage to the present logical-empiricist, when all elements of superstition are relentlessly purged and speculation must confront rigorous procedures of refutation.

The history of science has been written as a story of intellectual influences. In its most "pure form," these influences are entirely encapsulated within scientific discourses. Having liberated itself from the restraints of religious superstition in the seventeenth century, science becomes self-contained in images proposed by historians. Within the panoply of scientific disciplines, physics occupies a unique place not only because it addresses the most elementary features of natural phenomena, but also because it is the one science that fully incorporates the "ideal" tool of pure abstract analysis—mathematics. Norwood Hanson, writing from a critical empiricist perspective similar to Duhem and Quine, argues that physics contains its own philosophy which is thoroughly incommensurable with traditional metaphysics. While the nature of discovery is sufficiently ambiguous to invite philosophical analysis, one may not appeal either to the Anglo-American empiricist tradition or to German metaphysics for solutions. Accordingly, Hanson's chef d'oeuvre *Patterns of Discourse* addresses problems of the philosophy of physics from the "lens of particle theory," that is, from within modern physics itself.[15]

Thus, for Hanson and other philosophers of physics, there is no question of so-called externalist accounts for physical laws. Explorations of *method* constitute the starting point and the horizon of philosophy that addresses natural science. Hanson argues that both observation and hypothetical/deductive thinking are necessary for theory formation; theories are observation based, but observations are, in turn, theory laden. Hanson was not the first to argue that observation is not free of presuppositions; after all, Dewey, Wittgenstein, and Duhem made the same argument. But his description of the history of physics establishes a unique space for philosophy. Philosophy teases out "ways of seeing," or what Hanson calls paradigms of scientific discovery, the "how" rather than the "why" of science. This is, of course, what distinguishes scientific philosophy since the seventeenth century from traditional metaphysics. The "why" is renounced as a legitimate object of inquiry since the answer to this question leads to quasireligious modes of explanation—teleology in the bad, nonscientific sense.

Observation is "theory laden," and theory entails language that frames what is seen. Science generates its own unique *names* for the observed, and, insofar as thought is inconceivable without language (Wittgenstein), language is a condition of knowledge. In addition, in the course of the paradigm shift from classical to modern physics, investigators have discovered that the apparatus of observation not only becomes an indispensable element of discovery but enters into logical, i.e., mathematical, calculation. Measurement may not be separated from observation.

Therefore, as Heisenberg reminds us, physical laws refer not to "nature," but to our relation to it. Facts, then, are theory, language, and technique laden, making *relations*, not things, the true object of inquiry in contemporary science. As Hanson argues, behind these categories lie people. But it is precisely this addition to the "pattern" that makes the entire process of discovery indeterminate, not merely in relation to the measurement of place and velocity, but, as Dewey asserts, in relation to the intention of inquiry.

Hanson describes the paradigm shift that occured between the assumptions of classical and modern particle physics. According to his view, all the categories of discovery are relative to the frame of reference. But, for Hanson, the referent is really physics itself, or, more precisely, its theoretical, linguistic, and mechanical apparati. Although the history of science attempts to trace the intellectual process by which this shift occurs, giving it definite temporality, it may be true that from the point of view of the scientific community, a series of events led to the discoveries of Einstein, Bohr, and Heisenberg, events such as the various nineteenth-century failed efforts to find ether, Mach's argument in his *Science of Mechanics* against the concepts of absolute time and absolute space, and so on. However, these events were themselves conditioned by a broader cultural and ideological context. Just as the fundamental discovery of relativity and quantum physics is that significance (e.g., observation and experiment) depends on the *context* within which these activities take place, so the methods and physics itself may be determined by the context of discovery. The nature of this context cannot be specified in advance, but the relation of history to context must become the metatheoretical principle for understanding not only how discoveries are made, but the significance of these discoveries. If this proposition has validity, it should to be possible to find parallel developments in other fields in proximately the same space/time frame.

This does not imply a theory in which physical paradigm shifts are determined "externally" by what Kuhn calls "nonintellectual" factors such as the economy, technique, and politics. At issue is the question of what is the source of anomalies that trigger paradigm shifts in science and culture? From the point of view of the scientific communities, the appearance of these "observations" or "facts" that cannot be integrated within the prevailing theoretical arsenal constitutes "puzzles" that must be solved. But, as Gerald Holton has argued, creating paradigms depends in turn on recurrent themata that are transhistorical. Continuity and discontinuity, for example, are themes that recur in all periods of the history of science; relative and absolute time are not antinomies that appear uniquely at the turn of the twentieth century, nor are they confined to problems in physics.[16] Moreover, the concept that the subject is part of the object of inquiry has a long history. Similarly, the "mechanical world picture" of a self-contained world turns out to be a particular metaphysics generated under historical conditions informed by a complex of economic, political, and ideological developments. Philosophy overturns this picture long before science, but Hegel and Marx are metaphysicists, not physicists. And, as is well known, Greek science contains both the dialectical and mechanical worldviews.

As I have argued earlier, the dialectical world picture has been construed in two conflicting dimensions. Its objectivist form is the idea that nature *in itself* is contradictory, that opposing forces constituted the natural order in dynamic tension. But Lukács, following Hegel, argues that the dialectic signifies the interaction of subject and object, a relation that is mutually determining and, through praxis, is constituted as an identity. It is obvious that this metascientific principle is partially consistent with the new physics, quite apart from the existence of direct influences on specific physicists (Einstein read and admired

Mach; Bohr studied with Hoffding). Yet, there can be no doubt that the vehement arguments of Nietzsche against determinism prefigure a causal theme of modern quantum physics by a half-century.

Paul Forman's study of the intellectual climate surrounding the appearance of quantum mechanics in Weimar of the 1920s deserves detailed treatment here because it is the boldest attempt, from a thoroughly rationalist viewpoint, to link the process of discovery with a culturalist hypothesis. Forman's thesis is that the development of quantum mechanics, with its statistical, acausal, indeterminate suppositions, cannot be explained on an internalist account, that is, one that focuses exclusively on the anomalies appearing in the course of investigation:

> While it is undoubtedly true that the internal developments in atomic physics were important in precipitating this widespread sense of crisis among German speaking central European physicists and that these internal developments were necessary to give the crisis a sharp focus, nonetheless it now seems evident to me that these internal developments were not in themselves sufficient conditions. The possibility of the crisis in the old quantum theory was, I think, dependent upon the physicists' own craving for crisis arising from participation in and adaptation to the Weimar intellectual milieu.[17]

In turn, this milieu was shaped by the horrendous outcome of World War I, particularly Germany's utter defeat, a defeat that temporarily reversed the profoundly nationalist sentiments of the intellectual communities, especially the scientists whose prestige within Weimar culture had reached a pinnacle in the years immediately preceding the war. Along with many other social categories, scientists were mobilized spiritually as well as professionally on behalf of the war effort. Its failure produced a deep disillusionment among the "educated public" with the certainties not only of the old politics but of the old science as well. While Forman shows that these developments were in a small measure prepared by relativity theory (although Einstein remained after the war a firm advocate of causality as the indicator of scientific law) and by the widely known lectures and articles of Hermann Weyl, among others, he attributes the sharp change in philosophical temperaments of virtually all leading members of the central European "hard" scientific community to the antiscientific mood of the educated public. This mood expressed itself most eloquently in Oswald Spengler's widely read *Decline of the West*, published in 1918 and in its thirty-second edition by 1922. Forman adduces evidence to show that Spengler's views greatly influenced the movement called lebenphilosophie (life philosophy) and had such great effects on physicists that he is prepared to claim that Spengler's arguments in this book became the embodiment of the metapropositions of the new physics, even though he finds no "prima facie" evidence for the link. Nevertheless, insofar as the Weimar scientists were persuaded of the "collapse of the belief in the certainty of the experimental principle,"[18] that nature's phenomena were ultimately products of chance occurrences, and, most significant of all, were influenced by Spengler's concept that physics was "historically, psychologically conditioned by the character of the epoch"[19] and that it "represented that character more or less completely," the idea that measurement was a human intervention that could no longer be radically severed from its object became entirely plausible.

Forman argues that physicists responded to their public's demand for an end to a science of certainty. Although some, like the mathematician David Hilbert, resisted the call for a "new science" of acausality, they were clearly overruled by a demand for "an end to the causal monism of positivistic methods in wissenschaft."[20] Accordingly, Ernst Troelsch exhorted: "the crisis of learning must be followed by a revolution which liquidates this barren and intolerable mechanism in favor of a new wissenschaft of values, intuition, feeling of the living, the organic."[21] Forman's impressive evidence that the crisis in physics was, first and foremost, a crisis in confidence in the old assumptions generated by historical circumstances is, on the whole, convincing. Yet, it remains, at base, incomplete because its conception of context remains rooted in a single event—the war. To be sure, it may be argued that Freud's amendation to his theory of the psyche—the postulate of the death instinct—owes as much to the devastations suffered by central Europe in the war as it does to his growing infirmity, his perception fueled by psychoanalytic practice that human aggression was a fact *sui generis* and not dependent on repression of the pleasure principle. The rediscovery of evil as a biological trait summarizes the end of hope, or rather the end of innocence of a generation that had grown accustomed to the optimism of enlightenment thought, particularly the idea of the inevitablility of progress.

Among the leading suppositions of this idea is the universe as a self-regulating system whose "mysteries" are ultimately knowable by reason and guided by practice. Marx and Marxism were aware that our knowledge was always limited by human practice, especially the level of development of the productive forces. But, for Marxism as well as for classical liberal progressivism, *in principle* there is no limitation to knowledge (since the object is inexhaustible) and, therefore, to our

capacity to rationally plan our social order. Indeed, the two great revolutions of the bourgeois epoch, the French and the Russian, expressed the confidence of reason in its capacity for truth. And science,
the systematic form of reason that combined observation, experiment,
and rational calculation, became the key to human emancipation from
ignorance and the pestilence and disorder that accompanied it.

I argue that this process of disillusionment was already present
in the Paris bohemias of the midnineteenth century, in the dark ruminations of Poe and Melville in the United States, in Dostoyevski's
underground man. These manifestations of what Lukács labels "irrationalism," which he identifies with literary modernism and phenomenology, anticipated Weimar culture by at least three-quarters of a
century.[22] Moreover, as Holton has demonstrated, the critique of
mechanism, a necessary prelude to the development of modern physics, appeared in Germany in the last two decades of the nineteenth
century. The writings of Mach and Oswald, which had a deep influence
on Einstein, did not reveal a loss of faith in reason but replaced objectivism by phenomenology, a tendency developed by the leading
figures of Geisteswissenshaften such as Dilthey, who posited the significance of the cultural context for understanding intellectual expressions.[23]

In short, to understand the context for scientific ideas entails a
broader time frame and field than Forman permits himself to undertake, even if we remain on the level of purely "cultural" or intellectual
influences. Even though the war and its aftermath are often invoked to
help explain the "irrationalist" mood of Weimar's intellectual communities, we would have to ask, what explains the equally problematic
character of fin-de-siècle Vienna, the environment of Freud as well as
one of the most powerful social democratic parties in Europe? What
emerges from Carl Shorske's acute account of this milieu is that the
most profound political question of the time was the breakup of liberal
hegemony. Shorske sees social democracy as a continuation of the
underlying rationality, both moral and scientific, that defines nineteenth-century liberalism. However, the political break occurs in the
rise of various nationalisms in Austria and Germany, mostly right-wing
movements, the "crisis of the liberal ego" that gave rise to the experimental painting of Gustav Klimmt, and finally Wittgenstein's critique of
language and, more broadly, of the entire philosophical enterprise.
Just as Siegfried Kracauer finds in the film of the Wiemar years a pre-
figuration, or, more accurately, a collective premonition, of the rise of
fascism, we may see in the probes of the moral and scientific premises
of European liberalism undertaken by Austrian artists, philosophers,

historians, and especially of the musical innovations of Mahler and Schoenberg, the breakup of the old worldview.[24]

It is not that economic, political, or ideological (read: cultural) structures and the phenomenal forms of their development *determine* any specific discourse. Rather, discourses of science, art, politics, etc., are understood as a multiplicity of relations that are overdetermined as well as underdetermined. Overdetermined not only by the replacement of unilinear causality by multiplicity of relations, but also by the indeterminacy of the historical moment in which the effectivity of any given discourse may be dominant (temporarily) over others. Underdetermination refers to the concepts of chance and necessity. An illustration is provided by Quine's remark that the inferences drawn by investigators as a result of the appearance of a contrary observation to that expected under a given theory are never necessary because more than one plausible choice is available. The question posed by Forman's account of the rise of quantum mechanics is precisely whether this was the only choice. Obviously not. Wave and particle theories coexist well into the 1930s, and the interpretations of the Copenhagen school—in particular complementarity and indeterminacy—remain controversial even among central European physicists, especially Einstein. To Bohr's more or less complete acceptance of the idea originally advanced by Mach that the distinction between phenomena and an "underlying" reality must be abandoned since our knowledge of the external world is always obtained "under specific circumstances" of observation and experiment and warrants no further assumptions, Einstein poses the thesis that objective reality "is incompatible with the assumption that quantum mechanics is complete."[25] As we have seen, Soviet physicists agree with Einstein, at the same time agreeing that quantum mechanics holds as a valid description of the situation of measurement and that the dualism of wave and particle theories can be resolved by means of dialectical analysis. Nevertheless, Bohr's interpretation has prevailed: most physicists dispense with the assumption of objective reality that lies beyond the investigative situation.

This dispute poses the question of the survival of Bohr's perspective beyond the immediate cultural situation that influenced it. I shall reserve a fuller discussion of this point for the next chapter. In the present context, it is enough to note that a powerful tendency in contemporary Western European philosophy, emanating from what has been called poststructuralism, follows the antiessentialism of the Copenhagen school by holding that phenomenology and Marxism are both continuous with the metaphysical fallacies of traditional German thought, despite their contrary intentions. Consequently, poststructuralism abandons positing "things" in favor of discourses which have no fixed referent of "external" reality. Like Quine, the analytic a priori is jettisoned in favor of the cultural arbitrary.

This example is invoked to show that the milieu surrounding modern physics is part of an *epistēmē* that transcends particular cultural contexts. Among its central features is antifoundationalism, the refusal of a logical or epistemological ground from which theory emanates. Therefore, however antideterminist and even antiscientific the Weimar culture became after the war, this would not be an adequate explanation for quantum mechanics, even if supplemented by the anomalies that appeared in the laboratory.

The tradition of skepticism stemming from Nietzsche's antifoundationalism had already penetrated German thought, even if unacknowledged by those most deeply affected. The most important figure in phenomenological philosophy, Edmund Husserl, although owing a great deal to Kantian epistemology, owes as much to Nietzsche's call for an end to the metaphysical *ground* of "reason" upon which knowledge had stood. Moreover, Husserl refuses the search for origins until his last work, because to adopt this stance implies first principles to which knowledge must subordinate itself. Philosophy must suspend, as much as possible, a priori principles, for to presuppose anything but a method of knowing is to foredoom the task of finding a way forward to the things themselves.

The relativism of twentieth-century physics, particularly its methodological probabilism, is influenced by antifoundationalism of lebenphilosophie and of Nietzsche, who demands that science free itself of the moral judgments inherited from Christianity. At the same time, he insists that every fact entails an interpretation, that it does not stand by itself. For our purposes, there is no need to seek "prima facie" evidence that these ideas directly link with the conscious views of physicists at the turn of the century. More to the point, Nietzsche's philosophy signals a "crisis" in the metaphysical foundations of science that had dominated the international scientific community since Galileo. Even as Nietzsche propounds his doctrine of the decline and fall of truth, the American Charles Sanders Peirce happily announces the inherent uncertainty of scientific results and defines truth as that proposition about which members of the "legitimate" scientific community agree. At the same time, he asserts that the a prior "truths" of mathematics should be subject to criticism, not only the results of scientific laws adduced, in part, on the basis of these procedures. This assertion remains controversial to this day. Popper has argued, following Peirce, that any scientific proposition must be framed in a form that permits falsifiability to be demarcated from sheer speculation. On the other hand, few positivists or materialists have followed Peirce's skeptical view of the "certainty" of the results of scientific discovery, especially its "rational," i.e., nonempirical, side. Instead, Peirce makes a case that is not dissimilar from that of J. S. Mill that mathematics has an empirical, or at least a cultural, referent, that its "self-evident" truths are conditioned on the system of numbers of which they are a part. So, Peirce, despite his apparent empiricism (the unproblematic character of observation, the final appeal to empirically induced "facts"), prefigures later developments in physics. He is relativistic both with respect to the certitude with which scientific law may be declared "true" and with respect to nature. Nature does not behave in a deterministic manner. Chance is as much a part of the universe as necessity, and exact results are dependent on the "configuration" within which they are calculated. Peirce stresses "irregularities" in nature by referring to "a certain amount of absolute spontaneity in nature, despite all laws" and declares "our metaphysical pidgeon-holes should not be so limited as to exclude this hypothesis provided any general phenomena should appear which might be explained by such spontaneity."[26]

Peirce argues from the acknowledged perspective of a realist. This means that while the apparatus of inquiry may be a point of demarcation in the history of science (astronomy before or after the telescope, chemistry and biology before and after the microscope), he appears to insist that the instruments merely improve the accuracy of observation, changing the reliability of results but not the results themselves. Moreover, from this follows his argument that the postulate of continuity, a relation of past, present, and future, is tenable and even necessary for proper scientific work. This emanates from his unconditional acceptance of the fundamental evolutionist perspective. Yet, lurking beneath these doctrinal realist articles of faith is his flat assertion that all knowledge is uncertain with respect to its validity, and that all "proofs" are mediated by the rules of observation to which they conform.

Peirce's theory of science by no means influences German physics in the 1920s. I have invoked it to show that "scientifically" informed societies at the turn of the century were already imbued with a skeptical intellectual climate that did not depend on the shattering effects of World War I, or, more exactly, the disillusionment suffered by the defeated. The themes of determinism and indeterminacy, continuity and discontinuity, certainty and uncertainty were played out in different contexts in different ways. Forman tries to demonstrate that certain conditions obtained during the 1920s in Weimar culture that directly affected the development of quantum mechanics. But organicism was much more pervasive than its particular form in Germany. What may be termed the romantic movement in philosophy and the sciences is an event possessed of a multiplicity of forms and expressions. The revolt against the tendency of nineteenth-century science, following some dubious assertions of Darwin's followers, to *reduce* human relations to biologically determined characteristics (the ineluctability of competitiveness, the survival of the fittest, and other ideas of social Darwinism), or the equally reductionist views according to which all manner of phenomena could be transcoded into physicalist terms, brought forth a strong reaction from many quarters: (1) in Germany, the leading axiomatic proposition of the historicist school led by Dilthey was the radical separation of the human sciences from the natural sciences and its corollary, that human relations entail intersubjective understanding; (2) the Freudian theory of the psyche as physiologically rooted but as a relatively autonomous process from the conditions imposed by this fact; (3) the vitalism of Bergson, which makes the distinction between the organic which has feelings and the inorganic which does not and which is, therefore, subjugated with impunity. Bergson takes a direction characteristic of turn-of-the-century organicism: to let the *positive* physical sciences have their domain, but to deny to physicalism authority over biological or the human and moral sciences.[27] That is, antipositivism takes the form of a turf fight. This contrasts with Marxism's early attempt to establish itself, until Lukács, as a science conforming to the authority of natural science.

The refusal of physicalism is a sign of demarcation in Marxist and non-Marxist philosophy of science. Recall that Lukács, deeply influenced through his relationship with Simmel by neo-Kantian historicism, denied that natural science was a praxis, since praxis connoted reflexivity, or, more exactly, the notion of intersubjectivity in which the "object" was nothing more than a name associated with a particular frame of reference, but in a different context became a subject. Quantum mechanics, even more than relativity theory, acknowledges that the referent is provisional, that the "field" includes the apparatus and the investigator, who, in Lukács's terms, "mediate" and "constitute" reality.

Quantum mechanics remains tied to the methodological dual matrix of traditional physics—observation and rational (i.e., mathematical) calculation. Holton calls these the x and the y of scientific knowledge. To these he adds z, the themes or metascientific categories that inform, if not determine, the choice of object as well as how scientists see it.[28] Holton's amendment is valuable because it introduces metascience, or, more precisely, metaphysics, as an intrinsic component of science, not at all "external" to its processes. To these must be added the *apparatus* which mediates what is seen in the following ways: most important, what counts as observation is really a meter reading, a highly mediated activity; second, as Peirce and others following him have recognized, the apparatus stands as a boundary condition of investigation, especially in the twentieth century when the limits of the mechanism of observation constitute the limit of the (observed) world. So epistemological inquiry must take into account the effects of the indirect nature of observation, for direct empirical observation literally does not exist in contemporary physics. Second, physical knowledge is intertwined with the mathematical means at the disposal of science. If these refer to the processes of scientific inquiry, even if indirectly, mathematics is not merely a tool of observation or experiment but is a component of them, taken as an a priori axiomatic discourse. A theory of science must ask not only under what theoretical authority the world is reduced to its quantitative specifications (the point of proponents of organicism and other modalities of life-world philosophy), but also what are the consequences of reliance on mathematics for knowledge? For mathematics is a language, and language is not merely a means but a *mode of life.* Language then becomes a limit of the scientific world, as Wittgenstein argues.[29] Likewise, the apparatuses of observation, machines, are metalanguages; mechanism as ideology entails an instrumental view of nature that is inserted in the processes of investigation.

Thus, the romantic objection to objectivist natural science, which also refers to the objectivization and instrumentalization of nature, finds its validity not only on the discursive plane but in the practices of science. Mechanism is the other side of logical empiricism, which, with few exceptions, reifies observation and mathematics, and rigorously occludes thematic or critical metatheoretical reflection on scientific practices. Instead, it regards them as data, literally the given of reflection.

Popper's work represents the attempt from a "soft" positivist perspective to take account of these objections. Published in 1934, but reflecting the 1920s decade, his *Logic of Scientific Discovery* vigorously argues that science incorporates into itself the critical intention of the dialectical/hermeneutic concept of natural science. As he argued in his reflection, published twenty-five years later, his theory attempts, above all, to establish what is unique about science. He is concerned, therefore, with the problem of demarcation. Science is distinguished from nonscience by the falsifiability of its hypotheses, whether established inductively (he admits that this is not characteristic), or deductively through reason. Neither reason nor observation taken by itself is capable of producing objective knowledge; the language of discovery must be written in a form that permits refutation. That is, a hypothesis must subject itself to experimental "proof," which often amounts to logical falsification. This ingenious theory permits Popper to admit that the themes and even the problems of science may be derived external to the process of discovery, that ideologies may inform the goals (but not the process) of discovery. By subjecting these apparently "extrinsic" influences to an algorithm of refutation, they are neutralized and scientific truth is preserved. But, Popper's attempted fail-safe formula entails taking the axioms of mechanically mediated observation and mathematical reduction for granted; his critical edge is, in Hanson's terms, theory laden. And, the theory is roughly comparable to positivist premises.

Many historians, sociologists, and philosophers of science regard the Hanson-Kuhn theory of science as "liberating" to disciplines that have been shackled by the internalist explanations of paradigm shifts. Hanson's is, indeed, an internalist explanation, or, to be more precise, his theory of science is a discourse analysis of scientific procedure. His "close reading" of what is entailed by the terms "observation" and "fact" reveals their dependent character. Following his argument, "critical" science must inquire into the axiomatic structure of science, not only the falsifiability of a particular hypothesis. For any hypothesis or specific theory presupposes a repertoire of antecedent theories which together constitute a unique way of seeing (paradigm). If we understand "paradigm" in Hanson's and Wittgenstein's sense, it is the underlying *mode of life* of science, a mode of seeing, a language, and procedures by which the validity of a proposition is established.[30] The contribution of the much discussed work of Thomas Kuhn consists in the following: new paradigms are established by a crucial scientific discovery which proposes an alternative way of seeing. The paradigm refers not to the external world but to the problems associated with experiment and calculation. The new paradigm becomes "normal" (read: hegemonic) because it solves these problems by accounting for anomalies connected to the older paradigm. Thus, the realist assumptions of previous history and philosophy of science are effectively challenged by Kuhn's historically detailed investigations into the crucial moments of paradigm shifts. Paradigms are incommensurable. The evolutionary, progressive hypothesis that characterized traditional accounts, most of which were offered by scientists, is apparently denied. The mechanical worldview, overturned by a series of discoveries, particularly that ether was an untenable hypothesis (it was subjected to experimental proof and failed), leads to the overthrow of the concepts of absolute space and absolute time. If scientific discovery is at all continuous, it is united only by the critical temper of the scientific community, the procedural unity of observation/mathematics/experiment, and the thematic constancy of the z (theme) factor.

Kuhn alludes to the possibility that to account for paradigm shifts, a description of the "internal" process of discovery is necessary but insufficient. Here, Kuhn is constrained to go beyond the logical and thematic explanations for why anomalies perceived in the course of doing "normal" science become occasions for the emergence of new paradigms. Puzzles, the stuff of normal science's corrections, extensions, and improvements within the paradigm, are now perceived as serious anomalies, and this produces a "crisis." The crisis appears when the scientific community recognizes that something is profoundly wrong with the old paradigm. A new paradigm does not emerge simply because the crisis has been recognized. A startling new way of seeing must be proposed, one which enables scientists not only to see more acutely but to expand their vision, to see new things. The old paradigm must be replaced if science as an enterprise is to be preserved.

As a descriptive category, "paradigm" is full of ambiguities. Kuhn's critics have discovered at least twenty different definitions for the concept. Kuhn himself complains about the "misunderstandings" that have attended the appearance of The Structure of Scientific Revolutions. The book appeared in 1962, four years after Hanson's extensive discussion of the concept. But, in his "Second Thoughts on Paradigms," Kuhn seeks to clarify his own meaning of the concept. Here, he declares that he always meant paradigm to be located "in close proximity" to the phrase "scientific community": "A paradigm is what members of a scientific community, and they alone, share. Conversely, it is their possession of a common paradigm that constitutes a scientific community of a group of otherwise disparate men."[31] These communities have an "independent existence." Then, Kuhn introduces a substitution, the idea of "disciplinary matrix" which is the multitude of "commitments," including that of symbolic generalizations, models, and exemplars rather than analogies. These "ordered elements" become a common language that binds the community. This language consists, according to Kuhn, of a set of methodological formalisms by which the object of knowledge is represented. Of these, Kuhn awards priority to shared examples, not, as many others have determined, shared rules by which the process of investigation or the presentation of results is represented. These examples identify what is similar and what is different in the goals and the *problems* that constitute scientific activity. When scientists see the resemblance among apparently disparate goals and problems, they are seeing these elements.

We know that what Kuhn means by a disciplinary matrix is very narrow: the communities for which symbolic generalizations are not typically modes of representation do not adopt models and exemplars to signify resemblances in goals and problem-complexes, are not members of the scientific community to which Kuhn (and other historians and philosophers who have adopted physics as the exemplar of what they mean by science) refer. Thus, Kuhn really wants to identify what, in another language, may be regarded as *hegemonic* science. Its independence is modal with respect to processes of knowledge acquisition. Science, then, is a *system of representation* that becomes a series of shared commitments. Those who do not share these commitments are excluded and may not use the designation "science" to describe their activity. Had Kuhn left the definition at the categorical level, that is, had he followed Peirce by identifying knowledge with the community that has shared education, apprenticeship and institutional membership, and employment, we could infer a plurality of communities, each of which construes a "disciplinary matrix" corresponding to its traditions. Instead, he specifies the nature of this matrix in the image of physics, chemistry, molecular biology, and those social sciences that emulate natural sciences in the hopes of being admitted into its precincts.

Now, this definition contains normative features that are in dispute among philosophers and critics of science. The three key elements of scientific commitment named by Kuhn privilege the abstraction of quantitative from qualitative properties of objects and entail onstention, that is, the display of objects by means of mathematical pictures. The epistemological status of these representations is the crucial problematic of contemporary philosophy, whether following the program of Marxism or logical empiricism. Kuhn, true to the limits imposed by the historiographic community, is content to describe rather than narrate the way in which what counts as *the* scientific community is constituted.[32] The sociological elements of his description have been fairly well accepted since the time of Merton's work, which became a subdiscipline of sociology a half-century ago. The scientific community is formed by its common education, apprenticeship, and professional relationships. But when Kuhn adds what counts as valid processes of knowledge acquisition to his description, he takes a position that significantly modifies his earlier account. On the one hand, he retains historical relativism insofar as he makes no normative claims for the theories generated by his designated scientific community. On the other hand, he reproduces the conception of physics as the *model* science, introducing an ontological dimension to his description.

In order to rescue Kuhn from the charge that he has simply ontologized the enlightenment conception of normative science in his definition of paradigm, we would require a similar sociological account of how these elements were constructed. Unhappily, no such account is offered in his later reflections on the process of paradigm formation. Instead, we are asked to accept an ahistorical conception of the proximity of scientific community with paradigm, which results in ideology formation rather than clarification.

There are two things to note about this brief survey of Kuhn's contribution to the "revolution" that occurred in the theory of science in the second half of this century. First, for Kuhn, as for mainstream sociology of science, the wider context signifies the *scientific community*. Although he goes beyond the traditional view that scientific change is the work of "great men" isolated in their laboratories, he does not go so far as to link science with a conception of the social that extends beyond the precincts of science itself. Second, a new paradigm is defined as a new "worldview," not merely a collection of theories. In Kuhn's account, although these do not reflect objective reality more accurately than the older paradigm, they do represent "progress" insofar as they yield concrete advantages over the displaced science. The history of science is the history of paradigm shifts that are incommensurable, but Kuhn refuses to say that one is not "better" than the other.

Now, there are several arguments that can be offered for these positions. We may hold that science or, indeed, any discursive practice that entails the formation of a community that adopts a separate object, specialized languages, and procedures that demarcate it from others is *sui generis* autonomous. If we take the social division of labor as the scene of knowledge production and can demonstrate that any sector is incommensurable with any other (or at least that its links are truly external), the history of science is relieved of the obligation to make wider connections. Under this sign, one can adopt the position of Laclau and Mouffe for the "impossibility of the social."[33] For them, society is constituted as an abstraction that is itself dependent on theory, or, more precisely, that expresses the knowledge/power link. Thus, the Durkheimian invocation to treat society as a social fact, which means to regard the concept as axiomatic, is vehemently denied. Althusser's principle of "relative" autonomy of ideological apparatuses is similarly refused; the term "relative" connotes smuggling the "economic" or the "ideological" through the back door.

Implicitly, Kuhn, following the traditions of the history of science, allows that the metaphysical is inscribed in scientific inquiry, but science as discursive formation depends for its reproduction only on an "independent" scientific community which shares a series of meanings of (transformed) ordinary terms such as observation, experiment, and rational behavior.

With Popper, Bachelard, and Althusser, Kuhn shares the concept that paradigm shifts succeed because they solve problems better than the old paradigm and offer new problems for solution. With Bachelard, he stresses greater "precision" and persuasiveness. But then Kuhn makes a single move away from the purely rational explanation for paradigm shifts. Although reason is the necessary condition for success, it is insufficient. "Conversion" of a significant group of respected members of the scientific community is a crucial component. This conversion is followed by an active intervention into the community's professional journals, conferences, and other forums:

> At the start, a new candidate for paradigm may have few
> supporters, and on occasions the supporters' motives may be
> suspect. Nevertheless, *if they are competent*, they will improve
> it, explore its possibilities, and show what it would be like to
> belong to a community guided by it. And, as that goes on, the
> number and strength of the persuasive arguments in its favor
> will increase (if the paradigm is destined to win its fight). . . .
> Gradually the number of experiments, instruments, articles and
> books based on the paradigm will multiply. [My emphasis.][34]

Soon, only a few "elderly" holdouts remain. The fight has been won.

In Kuhn's later version of paradigm definition, his specificiation of the "disciplinary matrix" as a set of elements referring to methodological criteria for constituting the scientific community and its activity bears strong resemblance to Merton's admittedly weaker "norms and values" description of the relation of science to society. In this respect, the earlier concept of paradigm and paradigm shift provided greater latitude for the historian or sociologist of science to enter the hazardous waters in which the *text* of scientific knowledge could be linked to the *context* of power. That is, Kuhn (1962) seemed to be presenting a discourse of emancipation from the confines of logical empiricism or, indeed, from any theory of science that posited a relatively unmediated relation between the internal dynamic of discovery and its object. Kuhn (1974) makes explicit what was there in the earlier work, but hidden by the shining light of his epistemological relativism which abandoned the realist discourse within which the history and philosophy of science had been ensconced. The earlier Kuhn exhibited what Roy Wallis called "a more methodologically agnostic view of the truth claims of science"[35] than had been available in the Merton thesis. I wish to argue, however, that his retreat from that promising beginning illustrates the power of the concept of the "rationality of science." After suffering some damage within history and philosophy, the concept recovered its composure with the help of Thomas Kuhn, who, more than his predecessors, had opened the gates of doubt. There are hints that a political struggle accompanies paradigm shifts, but if the paradigm is "destined" to win, these may only accelerate or retard the length of time necessary for victory. In the end, Kuhn has supplemented the internalist account, leaving space for a politics, economics, and ideology of the scientific community, which may mediate the process of change but cannot determine its direction. Rather, according to Kuhn, what is crucial is that a group of "competent" scientists are converted and take the lead in developing the paradigm and disseminating its results to the community, which is ultimately persuaded by reason.

While strictly precluding a study of the relationship between the values and norms of the societies within which science functions and the content of scientific discovery, Merton insists that the social context determines the degree to which the autonomy of science is made possible.[36] Writing during the apex of fascism's power in Germany and the increasing occupation of continental Europe, Merton goes so far as to assert that the progress of science, which consists always in its freedom *from* political control, depends on the normative conditions of society as a whole. These conditions will determine, among other things, the system of rewards and punishments that prevail within the scientific community as well as the ability of that community to conduct critical inquiry. But reason governs the process of scientific discovery and the establishment of truth itself. Science is distinguished from other types of knowledge by its ability to adhere to rational criteria for legitimating its results. Thus, any social influence that demands that irrational criteria be employed, i.e., norms imposed on science from without, is clearly egregious.

We are witnessing the slow, discontinuous breakup of the old worldview according to which physical science offers context-free knowledge of the external world, knowledge whose certainty may be posited as a cultural ideal to which other disciplines should strive. This ideal is by no means shattered. On the contrary, its periodic revival attests to what Dewey calls the "quest for certainty," which demands a secure and unimpeachable referent and method of science. In ethics, for example, after a long record of relativism, a most influential recent work, Alisdair McIntrye's *Beyond Virtue,* argues that an ethical theory worthy of the name requires a concept of substance such as that provided by Aristotle, from which our concepts of the right may be adduced. Current physics ardently seeks the unified field theory that has eluded it since the turn of the century. Its search for the ultimate particle, or, more precisely, the ultimate form of physical matter (now said to be shaped as in the metaphor of "strings" rather than particles), reveals that scientific commitment is not merely methodological, as in Kuhn's disciplinary matrix, but metaphysical as well. Clearly, substantiality is making a comeback, a theme to which I shall return in the final chapter.

The eternal recurrence of thematic antinomies such as relative and absolute objects as well as truths is not a matter of *succession* of one as opposed to the other. Rather, with Feyerabend and modern evolutionists, I hold that dominance does not signify the disappearance of the other, only its marginalization.[37] But the coexistence, albeit unequal, of opposing theories signifying alternate worldviews amounts to a challenge to ideas of absolute truth. I would argue that it shows the relativity of theory to metatheory, the elements of which constitute a matrix of its presuppositions. Kuhn's concept of anomaly leaves room for the empirical, which, in this framework, is an indeterminate catalyst to confirmation or infirmation. "Necessity" merely describes the boundaries of possibilities that are indicated by the elements of commitment that bind the particular scientific community. To transgress this boundary usually condemns the scientist to exclusion. In a universe in which there exists a plurality of scientific communities, each with its own disciplinary matrix, and access to resources, such exclusion is not tragic. This is not the situation in physics or in the theory of science. Nevertheless, the theory of science, in the analytic mode, has been in a state of disquiet for most of the twentieth century. Encouraged by Popper's assurance that the truth if not the meaning of any proposition can be subjected to refutation and thereby the uncertainties constructed by the cultural arbitrary may be vitiated, analytic theorizing could remain satisfied with its edifying task. However, even edification proved fraught with dangers. For the commitment to problematizing the *method* of science rather than its result has created new difficulties.

The virtue of analytic philosophy, compared to more traditional modes of theorizing or narration, is its practice of close reading. In some respects, logical empiricism since Carnap becomes even unwittingly a deconstructive activity because it interrogates the validity of some of the crucial components by which scientific law is constructed, especially observation, experiment, and inference. As we have seen, such diverse theorists as Feyerabend, Hanson, Kuhn, Quine, and Dewey have all concluded that the hypothesis of epistemological and ontological *relativism* is a reasonable interpetation deriving from physical discoveries of this century. Not only has realism been severely tested, but the status of physical law has been called into question.

Recall that Feyerabend, perhaps the most radical of those nurtured in the Popperian school, argues that scientific truth is established according to a political and cultural arbitrary, i.e., the power of dominant forces within the scientific community. As with Kuhn, his fealty to science consists entirely in his own version of logical empiricism as a methodological postulate. That is, for Feyerabend and other rebellious Popperians, the criteria consists in whether a given paradigm satisfies the requirements of empirical induction, explanatory power, and falsifiability. The hegemonic paradigm is not established, in his view, by the *superiority* or even the scope of its explanation, an idea that Kuhn seems to advocate by his notion of anomaly. Rather, competing explanatory schema are adjudicated by extrascientific forces.

Feyerabend comes out in favor of sloppiness, that is, of a scientific politics that recognizes that new ideas can flourish only in a society that is constituted anarchically (he really means pluralism). To prepare for this stand, Feyerabend joins Kurt Hubner and what he understands of Marx and Lenin by insisting that statements of a scientific nature cannot be abstracted in their truth value from both their history and the context within which they have been constituted: "Science is a complex and heterogeneous historical process which contains vague and incoherent anticipations of future ideologies side by side with highly sophisticated theoretical systems and petrified forms of thought."[38] This idea, adapted from Lenin's theory of uneven development, implies that methodology, with its reductionist and ahistorical presuppositions, cannot hope to clear science of its impurities; but Lakatos and Popper argue that methodology can do so, because heterogeneity is built into science's internal structure. Although Feyerabend endorses empiricism as a component of his anarchist science, it is by no means a governor. Moreover, since scientific frameworks are incommensurable, there is no possibility of disconfirmation on the basis of already relativized evidence.

Yet, possibly responding to his own community—the theorists of science for whom legitimacy cannot be conferred upon those who

transgress the logical framework established by analytic philosophy—
Feyerabend is constrained to make his arguments within the boundaries of this discourse and to confine his references to "history" to science itself. In the context of the widely held belief, especially among
historians and philosophers, that science is more or less free of external influences, especially in the construction of theory, one must con-
sult the work of Koyre, Burtt, and Hubner, each of whom insists that
there are metaphysical foundations to physical science. Of these,
Hubner comes closest to what corresponds superficially to an externalist explanation for developments in the history of physics and astronomy. Burtt had already shown[39] that the Ptolemic system against which
Copernicus and Kepler rebelled could not be characterized as antiem-
pirical, that is, it did satisfy the condition that science be based on
observation. Nor were its theories "inadequate as explanations for
phenomena." Further, Burtt demonstrates that Ptolemic science was
capable of prediction. Moreover, in terms of the canon of common
sense, there were reasons for the empirically minded to prefer it to the
newer proposals. Burtt argues that the replacement of the old scholastic system with the Galilean theory owed its content to the Christian
idea of God the inventor and the world as a machine. "God thus ceases
to be a Supreme Good in any important sense; he is a huge mechanical
inventor, whose power is appealed to merely to account for the first
appearance of atoms, the tendency becoming more and more irresist-
ible as time goes on to lodge all further causality for whatever effects in
the atoms themselves."[40]

Hubner attributes Kepler's decision to follow Copernicus to "renaissance humanism," which allowed him to advance his bold hypothesis
that the movement of heavenly bodies could be related to concomitant
movements of things on earth. Kepler is portrayed by Hubner as a
"sleepwalker" following a culturally determined course rather than
submitting to methodological imperatives of falsifiability such as those
proposed by Popper and Carnap. Similarly, Hubner depicts the controversy over the philosophical principles underlying quantum mechanics
and relativity between Einstein and Bohr in terms of the conflict
between realism and relativism. More to the point, the nature of
"reality" is at issue. For Einstein, "reality consists of substances which
remain unaffected by the relations between substances. According to
the other axiom, that of Bohr, reality is essentially a relation between
substances and measurement . . . for Bohr measurement constitutes a
reality. For Einstein, relations are defined on the basis of substances;
for Bohr, substances are defined on the basis of relations.[41] The two
axioms are incapable of refutation in the Popperian sense. Hubner

reports the failed efforts of various philosophers and physicists to fal-sify/verify one or another of the opposed axioms or principles. Finally, he concludes that axioms are not subject to methodological refutation but express cultural/metaphysical a priori propositions. Indeed, Hubner argues, a priori principles are intrinsic to scientific discovery, and no "facts" can be adduced apart from these principles. But these are not mental structures alone. In turn, they are determined by what Hubner calls the "historical situation," which is basically the cultural conditions, or what phenomenology calls the "life-world" within which scientific discovery takes place.

Hubner argues that science is, in part, its history as well as the situation in which science is embedded. "System-ensembles," a term cor-responding roughly to paradigms, shift on the basis of anomalies that occur in the course of doing science, an activity that proceeds accord-ing to *rules* and the axiomatic principles to which these are related. Inconsistencies in this system-ensemble (understood as the relation of these terms) produce changes. The axiomatic principles are culturally, i.e., historically, situated and therefore are part of the system-en-semble. In turn, "every historical period is determined by a system-en-semble" which already contains the "situation" within which science functions.[42] Thus, "internal" inconsistencies within the system-en-semble are not the same as the conventional internalist explanations of paradigms shifts which dominate philosophy and history for science, since inscribed in the system-ensemble is the metaphysical/cultural context.

This history of science is neither externalist nor internalist precisely because it has posited an unstable unity among the components of sci-ence that includes "every" historical factor that bears on scientific development, including, presumably, the economic, political, and social world. This is, of course, a theory of science that owes a great deal to Bohr's principle that "substance" cannot be separated from relations, which occupy a position of primacy in determining what we mean by reality. These "relations" preclude the traditional subject/ob-ject split characteristic of the Kantian as well as empiricist epistemolo-gies as well as philosophical realism.

Hubner's characterization of history since the Renaissance as "the scientific-technological age" means that he holds that "today science plays the decisive role and that everything falls under its influence" just as "theology once permeated the whole fabric of life."[43] He infers that science has become a new theology understood as above the sociocultural situation because it is held "competent in all matters and [we] grant it the right to speak in behalf of everything." But his relati-

vization of scientific truth does not assign science to the realm of ideology, or, more exactly, to false consciousness. Rather, Hubner advances a view similar to what I call the "new" *sociology* of science. The
concept of system-ensemble removes the distinction between science,
technology, and economic and political structures; science is embedded in these structures, just as they are inscribed in science. The relation of determination between these ensembles is, in principle, indeterminate but depends, instead, on the historical situation. Although
Hubner does not explicitly draw these implications, it is clear that his
wider idea of science embraces the totality of the historical situation of
both past and present. In the second volume of this work I shall discuss Hubner's evaluation of the social and political consequences of
the emergence of science and technology to a preeminent position in
the contemporary world. In the present context, I want to introduce
the distinction between his theory of science and the more normative
materialist and logical empiricist positions.

Hubner is perhaps the first analytic theorist of science to acknowledge the influence of hermeneutic/dialectical accounts. In a sense, as
with critical analytic philosophy, this step is reasonable. For the crucial
methodological principle of hermeneutics is to take the "life-world" as
the proper object of inquiry, including the life-world of science. This
entails a close reading of the language as well as the processes of scientific discovery. To this must be added the crucial contribution of dialectical thought—a thing *is* its history, which implies that scientific law
is constituted by the totality of developments, including experiments
intended to demonstrate concepts quite disparate from those embodied in the outcome; the cultural, economic, and political *context* that
influences what is "seen" and how,[44] as well as what choices are made
from the observations available; and, of course, the intellectual orientations of the scientific community which establishes the validity of any
result.

Hubner's concept of system-ensemble tries to account for the heterogeneity of forces that constitute the theory, its discontinuities as
well as continuities with the past, as well as the degree to which it
anticipates the future. Reflecting upon the prevalent Marxist view that
"since the very dawn of man, technology has been undergoing an
internal process of self-development, the significant stages of which
were accompanied by social upheavals," Hubner, in concert with contemporary poststructuralism, wants to argue that science and technology are constituted by mutations, which, from the point of view of
rational criteria of predictability, are, in principle, undecidable. "Rational decisions are historical in that they are always connected to his-

torical conditions. They are determined by a situation; and therefore there exist no intrinsic rational contents which might take on the form of particular goals and assumptions," which in turn "can never make a claim to be universally valid for all times."[45]

In sum, while Hubner calls attention to the historical constitution of what is taken as scientific "fact," he posits no essential continuity between past and future, both with respect to both intention and the validity of the methods of scientific investigation. However, Hubner wishes to make clear that although the transition from myth to science during the ancient Greek era is a "mutation," mythic and scientific accounts of the world are incommensurable. The fundamental characteristic of science is its analytical procedures as opposed to the synthetic procedures of Greek myth. Modern science is relentlessly secular, that is, refuses to ascribe quality and cause to "divine primordial forms"; science is profane, myth sacred. By abolishing the sacred, science also transforms all of the categories that underlay human experience, especially substance, quality, causality. Hence, despite his pluralist intention to validate myth as an *alternative* way to understand experience by refusing to make invidious comparisons between the two, and thereby making "room" for art and sensuous experience as deriving from the mythical and retaining a justified existence in our world, they are plainly marginalized in Hubner's discourse by two critical moves: first, by his complete silence on biology or social discourse as proper objects of metascience, and by his implicit marginalization of extrascientific spheres, a form of structuration that can be traced at least to the Kantian origins of the analytic reason from which Hubner only partly separates itself.

In fact, Hubner's main argument is that experience is constituted by the theoretical and cultural a priori, and, in this sense, his work parallels that of Dewey who reached similar conclusions. Unlike Quine, who is constrained to retain experience along the boundary edges of scientific process, an article of faith that links almost all theories of science to enlightenment presuppositions, Hubner wants to place "reason," taken here as a historically contextualized mode rather than a universal a priori, as the foundation of science. Yet, having arisen by mutation from synthetic myth, science retains its status as a privileged discourse in Hubner's pantheon, and its validity resides in powers of abstraction, which becomes the principle upon which the entire edifice of Galilean physics is erected. Naturally, as we saw in Sohn-Rethel's Marxist account of the history of science, this is precisely the core of the hermeneutic/dialectical critique of modern science.

CHAPTER 10
THE SCIENCE OF
SOCIOLOGY AND THE
SOCIOLOGY OF SCIENCE

Earlier in this book, I introduced a concept of epistēmē as a way of seeing that is specific to a historical period but that is, at the same time, discontinuous in time and space. When explicating the intellectual and cultural influences in the development of quantum mechanics in the 1920s, we saw that what might be called a modernist discourse permeates parts of western and central Europe from the middle of the nineteenth century into our own time. The crucial features of this discourse are: the renunciation of foundationalism in ethics and epistemology; skepticism regarding the unified field in "nature" as an ontologically certain entity; and the positing of discontinuity as the chief characteristic of both the historical process and the physical field.[1]

The possibility of human and social sciences independent of physics and biology becomes a major problematic in the late nineteenth century. Inundated by the will to scientificity that pervades all intellectual culture (even the use of the term "science" signifies the degree to which scientific ideology remains hegemonic), those who would found sciences of history, culture, and society are faced with three nagging and apparently unsolvable puzzles. First, can human (social) relations be "reduced" to physical or biological laws? Those who followed Francis Galton and Herbert Spencer held to the primacy of biological "nature" in considerations governing human relations. Later, some American sociologists sought to employ mathematical and other methods borrowed from physics and chemistry in the study of social relations. For American social science of the first half of the twentieth century, either the intellectual orientations of a biologically rooted "organicism" projected in terms of a social "system" whose elements are functionally interrelated, or the postulate of social physics which derives generalizations from quantitatively adduced "data" culled from surveys, qualified as first steps on the long road to genuine science.

Second, midcentury leaders of American sociology such as Robert Merton and Talcott Parsons never tired of reiterating the problematic character of social knowledge, because of both its relative infancy compared to the "hard" sciences and the impossibility of replicating the experimental method in the social sphere. Controlled experiments may be conducted in psychology, but history, economics, and sociology do not lend themselves to such methods because the social cannot be fixed in space and time.

A third debate concerns the "specificity" of the social in comparison to other fields of inquiry. To what does social inquiry refer? Is there a social "system" that integrates culture and personality systems, as Parsons attempted to theorize?[2] Does the "social" consist in a particular series of problems unified only by scientific method modified to the specificity of predefined social problems? In this mode, Robert Merton, perhaps the most influential theorist on the practice of social science in the last half of the twentieth century, argued for the proposition that social theories could only be of the "middle range," by which he meant a sphere of generality whose boundary condition was that statements be subject to empirical test.[3] Thus, Merton rejects the arguments of his teacher, Parsons, that theory could make statements for which "data" did not (yet) exist; where Parsons's intellectual orientation was speculative, even as it purported to refer, in the last instance, to empirical reality, Merton insists that such speculation be suspended until the maturation of sociology as a cumulative science permits empirical or logical proofs. However, what unites both writers is their common "frame of reference," a *system of action* in which the postulate of methodological individualism may be made. That is, for the preponderant American sociological community, the social system is built inductively from units involving social actors. The action *situation* in which "ego" confronts "alter" is the primary unit of analysis.[4] Moreover, analytic methods borrowed from philosophy and mathematics are the privileged mode of theorizing for this school. Underlying the action situation is a normative or cultural system that motivates and sets boundaries of the choices actors will make. These are variable and ultimately indeterminate because the cultural system is not unified but is constituted as a typology of orientations, the choice from which depends in the last analysis on the vicissitudes of the personality system that in the Freudian transmutation is called a psychic structure and is by definition undecidable. Or, to cite Pareto, the subject matter of sociology is nonlogical actions.[5] American sociology constructs its axiomatic structure in almost explicit opposition to two alternative paradigms: Marxism, by which it understands the underlying concept that the mode of production of material life constitutes the unit of analysis, and psychologism, which reduces social action to the propensities of individuals. What Merton designates as middle-range issues are, for Marxism, derivative of the forces and relations of production. In opposition to this Marxism, Parsonian/Mertonian sociology follows Weber by insisting that "ideas" have material effectivity, particularly those that may be named intellectual and moral norms guiding action. Parsons never doubts that the economic sphere is central to human life and that, consequently, the rules governing action are instrumental/rational. But "interests" are by no means the only orientation governing action. The personality system, conceived as a wild card in action systems, introduces nonlogical considerations; actors may cathect on objects that do not correspond to their cognitively rational motives. Parsons's categories of affectivity, expressive modes of action, and the frank employment of the psychoanalytic term "cathexis" are amendations to Pareto's category of nonlogical behavior, which accounts for the indeterminacy of the choices made by actors in any given situation and may produce instability of the social system.

Parsons holds the cultural system constant and posits its integration under equilibrium conditions into the social system. Thus, even if the personality system may disrupt the reproduction of the social system, the normative order remains the countervailing structure that ensures stability. In turn, this normative system is transmitted to individuals institutionally through family, school, church, and law, in the so-called socialization processes.

Now, since Weber posited the action system as the scene of sociological investigation, theorists have recognized that social equilibrium (the term is Parsons's) remains problematic. Socialization is a process that owes its existence to the lack of a monochronic general action system. Its three components obey different regularities, or laws, and their mutual integration is always questionable, both because of the fact that their mutual integration must be reestablished in every action situation and because of their logical distinction. Yet, the probability of equilibrium in any given space/time continuum is established by the assumed complicity of the cultural system with the social system. Since individuals are formed in and through institutions that are the material repositories of the normative order, only the vicissitudes of the personality have the capacity to disrupt the social system.

For Merton, the scene of action is not the social system. Rather, actors form relations in the context of *communities*, which, at a higher level of abstraction, are linked to a larger society. Yet, because of his invocation that propositions refer to an empirical frame of reference subject to confirmation or infirmation, we are advised to say nothing of such phenomena as production *in general*, social relations *taken as a whole*, or the *totality of social structures*. The best we can do is find the homologies between distinct spheres of human action. By rigorously employing scientific methods such as modeling, the hypothetic/deductive system, and verification, we can be reasonably sure that our investigations lead to solid knowledge.

This is not to suggest that Merton and his school are empiricists as well as positivists. On the contrary. Merton makes a crucial distinction that enables the investigator to deal with the complexity of objects, which cannot be sensibly apprehended. It is that between manifest and latent functions. The manifest function of an element of a (sub) system does not exhaust its effectivity, according to Merton. What we see is not altogether what we get. The interaction of elements of a system of action have consequences that are often unexpected in terms of the empirical world as perceived on the surface. These consequences are indeterminate from the perspective of the investigator at the outset of the inquiry. Having discovered that they have shown themselves in the course of interaction, the investigator may infer that these functions are part of the system; their latency cannot be posited a priori but can be established only retrospectively.

Thus, the object of social investigation is to discover functional elements of a given system of action which do not appear to observation until the completion of the given interaction. This is surely a "speculative" element in the Mertonian paradigm and is borrowed from Durkheim's work on suicide. Recall Durkheim's view that while suicide is an apparent sign of social disequilibrium because it violates the cultural system, which commands that suicide be abhorred as an antisocial act, it may simply reinforce the solidarity of the community, which now rededicates itself to life and the tasks the community has not yet fulfilled. Further, suicide functions to spur reform in the rules governing social action if it has not been determined that the individual is deviant. In either case, suicide reintegrates the society rather than results in its dismemberment. A similar argument is made about crime which contrary to common sense (which understands it as antagonistic to social and cultural integration) is, for Durkheim, a necessary referent for law and instruments of social order. The criminal validates the system of justice; deviance turns out to be the mirror image of norms.

Writing in the 1930s and 1940s, Parsons was influenced by the burgeoning prominence of psychoanalysis, both as a therapeutic practice and as a mode of social explanation. His designation of the "personality system" as one of the three principal components of the general system of action and as constituting, therefore, an independent variable in determining the social system, was not followed by most sociologists, who stayed close to Merton's adoption of only the elements of Parsons's paradigm which remained useful for middle-range theorizing. Nevertheless, Merton joined Parsons in rejecting the Hegelian concept of "totality," which posited the eventual unity of subject and object. Nor did sociology enter the deep waters of epistemological inquiry. Instead, action systems were studied at a "distance." The postulate of action as a mode of interaction between self and other was supplemented by the notion of symbolic communication as the chief mechanism by which culture is reproduced. I would argue that despite the considerable debt sociology in its American manifestation owed to some of its European forebears, especially Weber and Durkheim, the central intellectual orientations of American sociology are pragmatism and physics. The main concurrences of social theory to the pragmatic reading of scientific method are as follows.

1. The statement that the social system is constituted primarily by the interaction of two actors and that what we mean by the "social system" is induced from the multiplicity of situations in which actors engage each other.

2. The concept of latent functions that derive from unintended consequences of social action. As with Dewey, the object, therefore, is eventual and cannot be specified a priori. All social knowledge is a project whose object is the indeterminate system of action between ego and alter.

3. The only a priori is the methodological postulate of empirical verification, a social adaptation to the experimental method.

4. The organicist presuppositions of Dewey's naturalism are embedded in the structure/functionalism of the mainstream of American sociology, as is the notion of the social system as an unstable equilibrium constituted by its elements.

5. As with Thomas Kuhn's idea of normal science, day-to-day American sociology consists in puzzle solving within the paradigm of social action as the unit of analysis. The concept of puzzle solving corresponds to the problem-solving approach of pragmatism that rejects the search for grand theory, which in physics as well as social sciences is

regarded as metaphysics. As we saw in Chapter 9, Dewey's commentary on Heisenberg's uncertainty relation is that it verified the particular nature of the object of scientific discovery. But mainstream sociology has not adopted Dewey's other postulate—that the object is constituted by inquiry, that its results are consequent of the relations between the knower and the known. The epistemological stance of sociology has been to take social action at a distance, to forget that the question asked in a survey constitutes the response and that "observation of fact" is itself problematic. To be sure, some critical schools in American sociology, such as ethnomethodology, make this problem the object of knowledge; the investigator is an element of the investigation, whose intervention is part of the field.

Sociology is mindful of Weber's admonition that fact must be separated from value, that to be a scientist means to expel the bias that may be present in sociologists' interests. It is widely believed that such methodological devices as multivariate regressions, modeling and rigorously maintaining the empirical referent to theorizing constitute a check on the value orientations of the investigator. But, with few exceptions, these methodological failsafes are relied on to validate the results, which, in relation to the frame of reference, are considered valid.

American social science, therefore, has followed the Popperian invocation of method to make knowledge "objective." Its assumptions, however, are by no means as sophisticated as the those of physical science, for which mathematical precision has become its intellectual ideal. For whereas physical science, as we have seen, has been prepared to recognize the indeterminacy of its results and the relativity of its theory, social science is still embarked on the quest for certainty.

However, in many other respects, American sociology is a discourse that constitutes an element of a postmodern epistēmē. For methodological foundationalism is by no means as secure as the epistemological variety. Further, social sciences as a set of disciplines are plagued by the plurality of paradigms that compete at both margins and center for intellectual space. For, even if the premise of methodological individualism and the rule that theories be of the middle range (and consequently subject to empirical reference and test) dominate the field, the leading paradigms within the sociological community have been unable, as in physics and chemistry, to impose their paradigm on others. That is, the positivists do not enjoy effective control over the system of rewards, including recognition of what constitutes true sociology, except in the United States.[6]

That is not to claim that no elite exists in the discipline. The leading university departments of sociology, even in Europe, East and West, are predominantly but not exclusively dominated by positivists, but by no means are they only of the quantitative variety. In social science, not only is the discipline divided by a social division of labor, that is, by field (deviance, medical, educational, community, urban, political, etc.), but there is also a technical division between theorists and investigators. Social theorists, as in physics, enjoy higher status, and, among these, mathematical modelers and other quantitative methodologists do not (yet) occupy a space of privilege. Rather, the high-status theorists are analytic, i.e., those who interrogate, critically or otherwise, the categories of the discipline.[7] Like other sciences, social sciences are preoccupied with metascience but reserve the privilege of this activity for a select few. Today, most investigative social scientists practice normal science, working in one or another positivist paradigm.

Thus, even American social science, which has an international reputation for its methodological orientation and, as Mannheim remarked more than fifty years ago, seems to have forgotten the "substantive" side of sociological theorizing, has recently discovered its own theoretical poverty.[8] With the recent infatuation with some European social theorists, such as Anthony Giddens, Michel Foucault, Pierre Bourdieu, and particularly Jurgen Habermas (whose Weberian premises make American sociologists feel they are on familiar ground when reading him), a new generation of social theorists may arise in this country.

I would not want to make too much of the parallel between the interwar rise of the United States to global supremacy and the emergence of a social science that was concerned not with the dynamics of historical transformation but with the conditions for social equilibrium; and I would insist that the epistēmē of this period exhibits features not directly attributable to the economic and political position of the United States in world affairs. Nonetheless, the effect of microinteractionist sociology, in the interwar and post-World War II periods, was to shift the paradigmatic weight from questions of historical transformation to problems of social integration. Moreover, these problems were taken in their specificity, abstracted from systemic considerations. In this respect, Merton's functionalism represents a retreat from the Parsonian *system*, which, contrary to the accusations made against it by C. Wright Mills and Alvin Gouldner, was deeply influenced by the imperative to explain the causes of social change. Even Merton, who took sociology into the less ambitious territory of "social problems," attempts in one place in his magnum opus to argue that Marxism is fully compatible with functionalist explanation.[9] Indeed, for

Merton and his associates such as Paul Lazersfeld, the major fault of Marxism is not its materialist explanation of history with its emphasis on the mode of production. Rather, like the analytic Marxists who repeat Merton's objections thirty years later, he merely wants to purge social theory of its speculative elements and place its axiomatic and propositional structure on firm empirical grounds. And this course is to correspond to the analytic philosophical conception of science offered by the Vienna circle, Carnap and Reichenbach, Hempel, and, in a different register, Karl Popper. As we shall see, the sociological study of science forms the backdrop of the effort to turn sociology into a science.

Parsons's conception of the social system retained a considerable portion of the scaffolding inherited from the German conception of the human sciences. "Interpretive understanding" (*Verstehen*) is the core of the action situation. Actors do not simply enter into relations within a given situation on the basis of their "interests" or the propensities imposed on them by their (mostly) nonlogical personality system. They interpret the behavior of the other and reflexively endow *meaning* to both their own behavior as well as the outcome of the interaction. In this sense, despite his growing distance after the 1930s from this hermeneutic vision of the human sciences (for example, his organicism mainly displayed in his anticipation of systems theory), Parsons's work may be considered continuous with the European, particularly Kantian, perspectives on the possibility of the human sciences.

The American reduction of this tradition took the form of reliance on survey research: investigators take seriously what actors *say*, especially their accounts of both situations and beliefs. However, this apparent opening to subjectively meaningful utterance is limited by its quantifiability, and interpretation is left to the investigator. Thus, under the impact of the will to (natural) scientificity, sociologists, in the name of the empirical reference and the primacy of methodology, become enslaved to the structured interview, aggregation, and statistical averaging as mechanisms of generating social knowledge. Similarly, "normal" political science is nearly entirely encompassed by survey methods. Its frankly behaviorist orientation focuses on voting, aggregated attitudes toward specific issues and events. To determine the character of American political culture, it is sufficient, for many investigators, to aggregate the results of a structured interview with a cross section of voters, segregated by class, race, gender, age, occupation, region, and such other variables as provide a simulated totality. The key task is to predict and control election outcomes on the basis of voter attitudes and beliefs about issues and events and its corollary, to reduce the margin of error by perfecting the algorithms of research.[10]

In this social paradigm, "action" is reduced to passively induced attitudes which manifest themselves in the voting booth or, alternatively, the reluctance to vote. This research aims at abstracting the relevance of the interview situation from the object of inquiry and attempts, thereby, to get back to the things themselves, at least in relation to the electoral context within which all questions are asked. Needless to say, the aspiration to accurate prediction is often frustrated in an increasingly unstable national political system which fails to account for volatility in voting behavior. In recent years, the study of American politics, with the notable exception of some theorists such as Walter Dean Burnham, has abandoned explanation in favor of description culled from aggregate surveys. In effect, there is a merger between political journalism, which in the United States reduces new analysis to data, and some branches of political science, which are content to *construct* a model of behavior generalized from survey data. Of course, construction is never innocent for several reasons:

1. The data themselves are construed within a framework of implied theories of social structure. The categories employed to select respondents not only are statistically derived from the demographic characteristics of the country but adopt an implicit conception of the social order. The lower one goes in the investigative hierarchy, the more stratification theory is taken for granted. Moreover, the stratification system is taken as an objective system in which individual membership is largely involuntary.

Yet, true to the indeterminacy of social theory, the categories employed in the survey sample are not presumed to offer predictive material to link social identification with either attitudes or behavior. The stratification system is employed primarily to ensure that the sample of individual interviews is broadly representative of the voting population or the otherwise relevant group. Analytic techniques may provide means for testing hypotheses about correlations between status attitudes and behavior, if voting for the choices offered by the political system is reflective of the latter.

2. The questions asked are constructed by the investigator. That is, the meanings imputed to action are transfigured by the interview situation into a series of structured, imposed questions that may or may not be of interest to the interviewee. Weberian conceptions of the action situation presuppose that actors have both cognitive and affective interest in interaction and that the situation is constructed, in part, by their intentions. In contrast, the interview situation is typically consti-

tuted by the interviewer's interests, whereas the interviewee may or may not perceive the encounter subjectively. Upon this shaky "unit," much political science and empirical sociology are constructed.

The tradition that American social science explicitly rejects treats the study of social phenomena from premises that can only be called philosophical. In the latter half of the nineteenth century, the question was posed in Kantian fashion: social relations exist, how are they possible? Historical and social sciences were constituted by this question. Lacking a failsafe method of verification and a controlled experimental situation within which to make observations, the social sciences are obliged to acknowledge, with Weber, that social knowledge is an interpretation among other interpretations which may also be valid. Not only is this condition produced by the infancy of these sciences; social behavior being "meaningful" is, thereby, always subject to the understanding of the investigator as well as the sometimes ambiguous self-interpretation of the actor(s). Peter Winch, accepting this unique character of social relations, has argued that Wittgenstein's philosophy of languages provides an important methodological amendment to other sociological methods: "If social relations between men exist only in and through their ideas, then, since the relations between ideas are internal relations, social relations must be species of internal relations too."[11] In Winch's model, the relations between ideas are logical relations, so the rules governing logic should govern social relations as well. Winch's position "conflicts, of course, with Karl Popper's 'postulate of methodological individualism' " whose fundamental tool is models.[12] Popper is, perhaps, the most outspoken nominalist in the philosophy of science, and his postulate of methodological individualism is still dominant in Anglo-American social science. Its origin is undoubtedly derived from the eighteenth-century economists' presupposition that economic relations spring from the actions of self-interested, rational actors. Even though Popper dispenses with the normative requirement of "rational actors," admitting that individuals may harbor nonrational expectations and attitudes, he stands with those, including Merton and Parsons, who would hold up the criterion of natural sciences to the social sciences, replicating, where possible, what they understand as the presuppositions of modern particle physics. Winch's argument is that the objects of social scientific inquiry are the logical relations between ideas that discursively constitute the social. Here Winch demarcates the social field from physics: "social action can more profitably be compared to the exchange of ideas in conversation than to the interaction of *forces* in a physical system" [my emphasis].[13] Consequently, among the objects to be investigated are language systems in use within a given social context, although these do not exhaust the material of social relations.

Thus, like Rorty's program in which philosophy becomes interpretive conversation in the European mode, sociology's objects are the regularities exhibited by the exchange of ideas between speakers starting from somewhat different premises. Winch's program is remarkably similar to that of the ethnomethodology school, which insists that the exchange of meaningful discourse is really *the* object of sociological inquiry. For both positions, relations, not individuals, are the methodological a priori and "conversation," not aggregated and reduced attitudes and expectations, are the material from which interpretive understanding is culled.

Harold Garfinkel tells us that if we want to find how relations are constituted, the "observer" disrupts everyday life to find out how it is constituted again. For without intervention, without participating in the exchange, social relations remain opaque to social inquiry.

One of the distinguishing features of Winch's and Garfinkel's critiques of positivist sociology is their insistence that the social world is constituted by actors *reflexively*. How we act entails evaluation of the consequences of our prior actions. So, rules of conduct are constantly being challenged, modified, resisted by actors for whom they are no longer self-evident. Wittgenstein regards the "given" as a mode of life which under ordinary conditions translates into rules of interaction. Following rules is what we mean by social life, because they define not only what people must do in a specific social context, but what is a mistake and under what conditions people may seek to change the rules, i.e., transform the social context of interactions. Rules may refer to the boundary conditions of exchange of ideas, the use of language, procedures governing conduct, and so on. What is clear about them is their social, not individual, character; they define what it means to be a member of a "given" community and what it takes to be excluded from it. However, one may not choose to be excluded from the language community, which is the ineluctable feature of social life.

Obviously, to know the rules implies that community members are more than habitual in their action. Rule making and rule breaking are both conscious actions and subject to evaluation. This is the sense in which every community is a discursive formation impelled by consent as well as necessity. For if necessity were the only constraint to action, social relations would be merely epiphenomena of biological relations. Production, in this model, becomes *social* reproduction, a permanent analogue to biological reproduction.

Underlying this approach to social science is the concept that individuals are inducted into the social world through language/discourse and that our mutual relations are regulated through exchange. It may be inferred that Winch's idea of a social science is really the addition of Wittgenstein's concept that discourse is the social mode of life, to Weber's theory of action, which always includes subjectively meaningful understanding.

The positivist and nominalist perspectives have become so powerful in social sciences that the Weberian program has been largely confined to a relatively marginal subdiscipline in sociology, that which addresses the social determinants of knowledge. The sociology of knowledge is by no means a *generally* accepted subdiscipline in social sciences. Or, to be more accurate, what is meant by the sociology of knowledge is often construed to refer to the ways in which knowledge *communities*, rather than what counts as scientific knowledge itself, are mediated by discourse, language use, rules of interaction, and so forth. This paradigm shift away from subjectively meaningful interaction, initiated by Merton in the late 1930s, addresses not the relation of science to ideology, since physical sciences are presumed to be ideology-proof by way of method. Merton asks questions that are quite distinct from the social constitution of knowledge; for example, the degree to which intellectual orientations may be permeated by extrascientific interests that have effectivity in the character of knowledge itself.[14] Rather, his work sidesteps the question of the social determinants of scientific knowledge and focuses instead on two ancillary issues. The first is the normative structures within a society that may foster or hinder the pursuit of pure science. His specific question is, what are the values that must be present in society that lead to support for the scientific community? These include such norms as skepticism, without which new discoveries would bow to dogmatism, and universalism, by which is meant the commitment to seeking truth that is in turn free of particular social interests. Therefore, disinterestedness does not refer to the absence of the universal aims of science as an interest, but to particular interests that might impede discovery; and, finally, "communism," that is, commitment to collective and congenial work—the view that the discoveries made by an individual contribute to the progress of the community as a whole and must be understood as a contribution to the larger project of seeking universal truth. Merton equates these values with those of democratic societies, which are the only context within which scientific truth must be pursued.

Of course, this is not necessarily true. Although his essay on the democratic prerequisites for free science was composed in the midst of the rise of fascism, it would have to be assumed that Soviet and Nazi ideology, with their (differing) doctrines of the internal relation of knowledge and interest, the Soviet concept of dialectics as a metascience unifying others and its partisanship rather than skepticism, would be inimical to scientific "progress." Although doctrinal considerations play a role in Soviet science, they are also present in the formation of public recognition of American, and West European sciences. As we have seen in chapters 8 and 9, it is not difficult to cite examples that show partisanship and interestedness as universal traits, within both society and the scientific community. No science is free, even relatively speaking, of interest, and Merton's norms, however widely held among both scientists and others, are merely one set of operators in the scientific enterprise.

Second, Merton wants to discover how rewards work in the scientific community, and more especially, how scientists are created in the education system, laboratories, and professional associations. Scientists are found to quarrel over who first discovered a particular scientific phenomenon; they race to publish their results before another investigator beats them to the punch. In short, the scientific "community" is marked by intense competitiveness, even in "big sciences" such as high energy physics and molecular biology.[15]

Careers are built by getting there first, including winning Nobel prizes with all the prestige and power that accrue to those who win them. This focus on the internal struggles within the scientific community dominates much of Merton's later work on science and constitutes the real paradigm of what became a genuine subdiscipline of sociology, the study of scientific communities. Just as sociology studies geographic, professional and demographic communities, so science becomes a legitimate community exhibiting the characteristics of any sociological object. What we want to know, according to Merton and his followers, is how scientists behave in their professional lives, what are the relations of power and privilege within the community, and how the scientific elites are formed and reproduced.

Merton's studies of the politics of the scientific community devolving around "reward" structures could provide a link with his earlier interest in the bearing of socioeconomic factors on selection of scientific problems. However, this is not the direction of his later work. Instead, Merton interprets the reward system in terms of the "institutional norms of science." Competitiveness is not the product of so-called human nature or personal ego, on the one hand; nor does

Merton again refer to the socioeconomic influence. Now, the institutional norms of originality, role performance, etc., take precedence over other influences. In effect, the scientific community becomes the context to which the social study of science refers. And it is a context that begins to resembles a discursive formation in Wittgenstein's sense more than a power/knowledge axis, in the manner of Foucault's understanding of discourse. Norms are not free-floating ideas but are integrated institutionally by being linked to professional reward structures such as recognition and career advancement.

This emphasis represents a departure from Merton's early work in which scientific studies are considered from the point of view of the sociology of knowledge. What Merton wanted to know in the 1930s was not only which value systems were compatible with the process of scientific development, but also the influence of the context within which science is produced on the forms of knowledge. His assumption, which pioneers of the sociology of knowledge such as Karl Mannheim were unwilling to make, is that science is not to be exempted from the general proposition that knowledge is mediated by the social context within which it functions, especially the intellectual orientations, the social or class interest of knowledge producers, and the value systems that accompany knowledge production. This perspective, most forcefully presented by Hessen's study of Newton's *Principia*, deeply influenced Merton's work. Both in his first book, *Science, Technology and Society in 17th Century England*, and in his defense of Hessen against G. N. Clark's critique of "The Social and Economic Roots of Newton's *Principia*" as crude economic determinism, Merton shows considerable appreciation of the central argument of Hessen's work — that technology influences the content of science, and that economic and ideological influences cannot be divorced from scientific discovery. In short, the social context considered in its totality influences the nature of scientific knowledge.

To be sure, in the mid-1930s, Merton is already developing his concern with the normative character of science and its relation to democratic culture. But his concerns embrace the influence of practical problems of navigation on fields of science such as astronomy. His corroborative case studies of transport, the military, and mining provide empirical verification for Hessen's thesis that, in Merton's words, "the range of problems investigated by 17th century scientists was appreciably influenced by the socio-economic structure of the period."[16]

Yet, Merton stops short of endorsing Hessen's insistence that the content of scientific knowledge is linked to the socioeconomic structure, but confines his concerns to the *selection* of scientific problems.

I can find no extended discussion in Merton's writings on science that explores the relations of scientific knowledge itself to the technical, economic, or value issues that surround it. Here, Merton does part company with Hessen and adopts the Mannheimian position on the sociology of knowledge: science *itself* must be exempt from the statement that knowledge is mediated by social interest. Rather, economic and technical considerations contribute to determining what problems will be studied, for, as Marx remarks in the famous *Preface to the Contribution to the Critique of Political Economy*, humans take up only those problems "for which the material conditions for their solution are already present," and this includes questions bearing on our knowledge of the external world. There is, according to Merton, a scientific ideology corresponding to different historical periods. But this ideology tells us only the general intellectual orientation of scientists toward research, even if it does not originate in the scientific community but is embedded in social structure. Thus, Merton finds that the value system we generally associate with the Enlightenment is also intrinsic to democratic societies. Science is linked to its social context through a common ideology, but if we follow Merton's work carefully, this influence is felt most strongly in the area of *method*. Such characteristics as skepticism, disinterestedness, etc., are safeguards against the influence of social interest on the content of science. So, a specific methodological ideology is the way to separate fact from value at the level of scientific theory.

Between World War II and the mid-1960s, the sociology of science is dominated by Merton's program. Studies of the scientific community, the influence of military and other social consideration on the selection of scientific problems, and the normative relation of science to the social order are the three legitimate areas of scholarship in social studies of science. Efforts to trace the social constitution of scientific facts are considered marginal to or are excluded altogether from mainstream considerations.

Merton's enduring contribution to the sociology of science, then, is his insistence that a politics of science exists, that the scientific community reflects, in deep ways, the normative order, not only its positive features that promote the search for dispassionate truth but its "sordid" aspects such as conflicts over the priority of scientific discovery. Moreover, the link between the socioeconomic structure and the selection of appropriate problems for investigation has opened the way for an enormously fruitful and detailed account of the ways that the state and large corporations have intervened in science to determine what is to be studied.

To be sure, the most important work in the politics of science has been performed by journalists such as Dan Greenberg and David Dickson, whose comprehensive studies of the close links of the scientific community with political and economic power, especially in the United States, show clearly what Merton indicated only programmatically earlier in his work.[17] Or scientists themselves, notably Joel Primack and Frank Von Hippel, have argued that even when the formal machinery exists for the participation of scientists in shaping national policy in science and technology, corporate and political establishments, which control the use of science in industry and as instruments of foreign and military policy, usually ignore their advice. Merton's own students and colleagues have focused, instead, on the internal workings of the scientific community considered as a political arena.[18] After the 1960s, American sociology of science has virtually ignored the part of Merton's program that calls for the exploration of the relation of the socioeconomic structure to the selection of scientific problems. One can only speculate about the reasons for this silence, perhaps suggesting that the selection of sociological problems is influenced by the political context as well.

Despite the ambiguity of Kuhn's *Structure of Scientific Revolutions* for a specifically social theory of science, its reception resulted in a sharp turn from the dominant paradigm. Although first published in 1962 in what retrospectively must be understood as a much longer wave of critical studies of the history and philosophy of science (see chapter 9), the full impact of Kuhn's account was not really evident until the latter years of the decade, when Imre Lakatos organized a major symposium whose focus was the contrast between the Kuhnian and Popperian images of science.[19] By the time of its publication, it was evident that the truth status of scientific facts was in contention. Nor could the experimental method be relied upon to vitiate the argument of the cultural conditioning of these facts. Significantly, in the 1970s, Merton himself joined forces with Kuhn to assist in the translation and publication of a long neglected study in the sociology of science, Ludvik Fleck's study of the origin and development of the modern concept of syphilis, originally published in 1934,[20] coincidentally the year that Karl Popper's *Logic of Scientific Discovery* made its appearance. Fleck argues that the substance of scientific knowledge is culturally and socially "conditioned," that the facts adduced are constructed historically and cannot be understood as simply the result of observation and experiment. What Fleck calls "thought collectives" (the scientific community) adopt "thought styles" (paradigms of knowledge) that enframe the ways in which observations are assimilated into explanations. Thought styles tend to be "structurally complete and closed systems of opinion" which offer enduring resistance to anything that contradicts it. Therefore, "l) a contradiction to the system appears unthinkable, 2) what does not fit into the system remains unseen, 3) alternatively, if noticed, it is kept secret, and 4) laborious efforts are made to explain an exception in terms that do not contradict the system,"[21] and the exception is reconstrued, explained, and described in a way that tends to corroborate the system. Moreover, conceptions and evidence have no logical relation since there are many possible "correct" conceptions that can be derived from the same evidence.

Fleck's work may or may not have been known to Merton in the early years of the 1930s. Whatever the case, his mid-1970s collaboration with Kuhn and others to bring Fleck's work to English readers signaled the changing mood of social inquiry into science. For Fleck's study foreshadowed the new sociology of science, which was inspired by Kuhn's refutation of the Popperian objectivist concept of scientific discovery and explicitly criticized Merton's studied avoidance of the issues of the social relations of scientific knowledge. In the early 1970s, a relatively small group of (primarily) British sociologists and anthropologists undertook, simultaneously, to critique the Mertonian conception of the sociology of science and to study the social influences on scientific knowledge. Among the strongest contributions to this critique was Ian Mitroff's attack against Merton's argument for social norms as the chief regulative device for ensuring the progress of science. On the basis of interviews with scientists, he discovered that such phenomena as organized skepticism and disinterestedness are combined with the reverse. Mitroff argues:

> that the norm of emotional neutrality is countered by a norm of emotional commitment. Thus, many scientists . . . said that strong, even 'unreasonable' commitment to one's ideas was necessary in science, because without it researchers would be unable to bring to fruition lengthy and laborious projects or to withstand the disappointments which inevitably attend the exploration of the recalcitrant empirical world. Similarly, the norm of universalism appears to be balanced by a norm of particularism. Scientists frequently regard it as perfectly acceptable to judge knowledge-claims on the basis of personal criteria.[22]

And "Mitroff produces evidence to show that the ideal of common ownership of knowledge is balanced by a norm in favor of secrecy." According to Michael Mulkey, "Mitroff's central argument, then, is that there is not one set of norms in science but two norms."[23] Of course, Merton implicitly acknowledged this argument in 1957 in his study "Priorities in Scientific Discovery," where the sordid issues were, to some extent, sorted out. However, there is no doubt that the few instances where the politics of the community are understood to contradict the normative order are viewed by Mertonians, not as evidence of dual norms, but as ways in which "scientists are human," that is, emotional deviations from recognized standards. And Merton himself insisted that "competition" rather than communism in scientific discovery is a product of another institutional norm, the commitment to progress.

But the new sociology of science does not stop with the effort to revise Merton's program. Within this group, a branch known as the Edinburgh School, associated with the work of Barry Barnes and David Bloor, advances a new, "strong" program for the discipline. In Bloor's words: "The sociologist will be concerned with beliefs which are taken for granted or institutionalized, or invested with authority by groups of men. . . . The sociology of knowledge focuses on the distribution of belief and the various factors that influence it."[24] And, for Barnes and Bloor, social interest is among the influences on belief as well as on what counts as scientific knowledge. Their major innovation is to regard science as subject to these assumptions, rebuking the arguments emanating from Mannheim that science is not socially mediated knowledge. The program addresses the question of the social causes of scientific knowledge, including sociology: "In principle, its patterns of explanation would have to be applicable to sociology itself."[25]

Bloor is careful to acknowledge the forebears of this renovation. Durkheim's call for impartiality "with respect to truth and falsity, rationality and irrationality, successor or failure" would be strictly observed. This ensures that sociology could study marginalized as well as mainstream science and would not make the a priori presupposition that mainstream scientific knowledge is always the rational choice from data. Further, Bloor wants to follow social theory's invocation to the primacy of method. In short, the sociology of science would maintain the commitment to accepted science rather than constituting itself as a counterscience.

In fact, there is considerable continuity between the "new" sociology of science and the "old" model. For the most part, the site of social studies of science remains the scientific community. However, evidence for this community is not found chiefly in journals or in professional associations but in the laboratory. The crucial study that adopts this perspective, Bruno Latour and Steve Woolgar's *Laboratory Life*,[26] follows Bloor's defense of empirical knowledge. The laboratory is selected as the site of social relations of science, not only because this is where scientific knowledge is produced, but also because it is where the interactions among scientists engaged in production may be observed. The "observability" of scientific work consists in three artifacts: (1) the written results of scientific inquiry designated as "inscriptions"; (2) scientists' conversations recorded on video- or audiotape, which provide "reasoning . . . displayed in the midst of orders of intersubjectively accountable details," that is, the order of spoken utterances by different parties in conversation; (3) the compositional order of manipulated materials at the laboratory.

These are the "data" of social scientific investigation into scientific inquiry. By paying "painstaking" attention to the details, especially the temporal ordering of materials, conversations, and inscriptions, the *informal* versus the formal procedures are revealed. This program of ethnomethodology, when applied to scientific work, provides a close reading of the labor process of knowledge production and is expected to show the degree to which a second normative set is produced by the multiplicity of interactions. It is precisely in the ordinariness of scientific work that investigators expect to discover how science is done. The assumption is not that scientists do not mean what they say, but that human praxis obeys rules that reveal the variability of scientific methods and are not present unless the most minute details are attended. Further, since the "object" in mathematical and physical inquiry but also in biochemistry is rarely "seen" but must be inferred indirectly, the process of inferential consensus becomes for some in the new sociology of science a central object of inquiry. Rather than presupposing "interest" as a component of knowledge, investigators hope by "thick" description to make visible what goes on. What goes on, in the first place, is talk, or, in Bloor's terms, negotiation as to what is "seen," what to make of it, and how to inscribe it in writing.

For some investigators, the "it" represents something of a mystery. For example, Woolgar, reflecting in 1983 on the development of this field, expresses confusion about whether his earlier study *Laboratory Life* discovered that the new reality was constituted by interactions among scientists in the Salk laboratory or whether it was *uncovered*. This issue reflects the difficulties in sorting out the constructivist and realist theories of science. Negotiation among scientists takes place to name, describe, and represent the significance of the scientific "fact." But, asks Woolgar, are the constructivists guilty of a priori judgment? Perhaps the accounts by scientists *uncover* rather than *produce* the objects?[27]

Of course, the ethnomethodological approach to scientific practice cannot accept the strong program since it presupposes a set of relationships that, although embedded in practices, are understood as relatively autonomous. That is, knowledge presupposes social contexts, which according to Edinburgh theorists include the socioeconomic structure and "interests." Garfinkel remains neutral on these questions, which, in any case, cannot be settled in advance. Nor, I may add, does the construction of the object of inquiry permit such issues to intervene in the accounts. By adopting a perspective on what constitutes a proper social scientific object, namely, the "local" production of scientific knowledge, the laboratory, and confining the "data" to conversation, the ordering of manipulated materials (what is, in Heidegger's terms, "at hand"), and inscriptions, ethnographers cannot resolve the debate between the concept of knowledge as socially constructed through interaction and its "objective" character. In Woolgar's terms, the irony of social studies of science that commit themselves to get as close as possible to the site of production is that this issue becomes undecidable.

However, more recent studies have moved further in the direction of validating the argument that science in action is science constructed by discourse, technology, and power. Sharon Traweek's study of the culture of a major research laboratory in physics argues that machinery not only mediates what is seen in the course of scientific experiment, but becomes a signifier of power both within the research community and between science and society.[28] She demonstrates by empirical inquiry the problematic character of observation. In her account, the machine constitutes, in part, the object. In turn, the machinery is built by scientists according to their own theoretical models.

Andrew Pickering's history of the quark in quantum physics shows that the theoretical models adopted determine the selection of the objects. For example, "from the experimental point of view . . . two key differences between the new (quark) physics and the old physics which preceded it were: an emphasis on the use of lepton rather than hadron beams; and, within hadron-beam physics, an emphasis on hard rather than soft scattering achieved through novel combinations of detectors."[29] As machines change, so do physical objects and vice versa. This account differs from the traditional view of machinery as a tool of experimental physics. Like Traweek, Michael Lynch, et al., Pickering is concerned to show that the construction of machines that conform to a model of physical reality themselves play a role in the construction of the object. But his account goes further: like Quine's rueful discussion of the indetermination by experience of scientific propositions, Pickering's argument is that each step in the development of the concepts underlying the quark is indeterminate from the experimental data, and that quarks remain to this day only one possible inference among other choices which would be consistent with what scientists agree is observed. I think it is safe to say that the *form* of scientific facts is socially determined and reflects "interest." Thus what counts as science today is the "useful," the quantitative, the formula. This limitation may be so narrow that within it the data *do* often dictate only one possible inference. Pickering says that even if we accept science on its own terms (and he appears afraid to do otherwise), the data are consistent with many inferences, so that the inference we make is in a sense arbitrary. But this may not be true, and perhaps it need not be true for us to say that such an inference is socially determined. The inference is forced by the narrow, socially determined boundary conditions that define what is acceptable as science. Why should we accept science on its own terms? For Pickering, the new physics is a worldview, and he disputes the scientists' own accounts of the evolution of the quark as an inescapable conclusion derived from experimental facts. Instead, Pickering asserts that these facts are actually "judgments" which are obscured by inscriptions made by scientists themselves. In other words, he argues that the presuppositions of realist theories disguise the construed character of physical reality. Having discovered this unremarkable (at least from the point of view of the new sociology of science) manner in which scientific fact is asserted, he goes further: the changes in high energy physics were culturally bound by research traditions and "shared sets of research resources" among actors. Moreover, these traditions constitute the conditions under which certain inferences, rather than others, are made from data. Pickering's account of the preferred research tradition in physics not only emphasizes the familiar, that is, types of explanation and the procedures used to verify them, but also argues that simplicity governs preference. In his account, quarks are the name assigned to (absent) phenomena that cohere with particle rather than field theories, which, in each case, offer different, although equally plausible, explanations for the same (inferred) observation. That the majority of the scientific community chose one over another is a function of scientists' preference for the tradition rather than the validity of explanation.

However, Pickering does not reach back far enough into the history of physics to find the basis of the research tradition from which the quark explanation emanates. It may not be found inside the tradition but in the ideology of science, in the differences behind field versus particle theories, simple versus complex explanations, the bias toward certainty rather than indeterminateness. But this might take the social investigator outside the laboratory and even beyond the scientific community. This might entail a critique of research tradition as a kind of ideology that guides interpretation, since the discourses of interpretation depend on but are not governed by facts adduced by experiment.

Third, Bruno Latour, whose pioneering earlier work with Woolgar asserted the social construction of scientific facts as the outcome of interactions among those performing scientific work, has made two significant amendments: that what counts as scientific knowledge achieves this status by linking itself to power—the power of leading figures in the scientific community and the state, and that of corporations that sponsor and transform research into technology.[30] In both his historical study of Pasteur's discovery of a serum for curing anthrax and his more recent study of three major contemporary discoveries, he insists that an examination of scientific discourse is necessary but insufficient to establish the mode by which experimental results are converted into accepted truth; and that the laboratory has become a new center of power and is a model for society. Latour accepts the argument that a discourse analysis of science is needed to understand how social context is embedded in practice. Indeed, his close reading of both scientific and technological discoveries reinvents a *rhetorical* approach to the analysis of texts. Where Pickering's inferences from this type of analysis remain always implicit because, following the poststructuralist understanding of language, he holds that discursive formations are identical to what we mean by the social context, Latour says explicitly that the development of knowledge as a productive force, and of science as the core knowledge determining technology in all its manifestations, changes our understanding of the significance of scientific research. The laboratory is now the late capitalist factory; its productive apparatus is no longer a reflection of social relations that lie outside it, but has become the set of internal relations that constitute what we mean by society. Pickering's slogan to epitomize his findings is that science is "opportunism in context"; Latour's is "Give me a laboratory, I will raise the world."

Latour provides the first account of science in terms of Foucault's declaration that knowledge/power is the object of inquiry, where power is constituted discursively through rhetoric, negotiation and inscription, but also by the political alliance of the scientists, the state, and capital. He offers examples of Merton's sordid politics of the scientific community, especially the competition between Watson/Crick's postulate that the structure of DNA is represented in the metaphor of a double helix and that of Linus Pauling, who offered a triple helix view. Moreover, he is among the few social investigators who has refused to accept the distinction between the power relations that constitute scientific facts and those of technology. And he draws the firm conclusion that nature is constituted in and through culture, including language, so that discourse must be taken as an object of inquiry. This will lead, no doubt, to fruitful studies of the texts of the traditions of which Pickering speaks.

But, at a deeper level than these conclusions, I must note the superficial resemblance, in some respects, between the new and the old sociology of science, or, more to the point, the degree to which the presuppositions of the positivist tradition still form the backdrop of this discourse. Although the new and the old paradigms differ as to expected findings, that is, differ as to whether the point of the study is to determine the degree to which scientific fact is constituted socially, with few exceptions each side expects to find answers in an examination of the ethnographic data. The sociology of knowledge is construed as an empirical study that fits neatly into Merton's program for middle-range theory (here, ethnomethodology parts company by bracketing the question of theory since any "explanations" contaminate the purpose for which the study of scientific work is conducted: to make visible that which is disguised by discourse). On a deeper level, the new sociology of science tends to confine its study to the laboratory, for the reasons given by Latour, which, in any case, are retrospectively adduced, and because it accepts the argument that scientific discovery is underdetermined not only by the "facts" since these are in turn constructed, but by the socioeconomic structure as well. Augustine Brannigan states the case best: "Language, as philosophers like George Herbert Mead and the phenomenologists Husserl and Heidegger have noted, not only communicates information, it constitutes and forms the social relations of its speakers."[31] His reference is to Wittgenstein's translator and a leading commentator, G. E. M. Anscombe. Discussing to the concept of causality, Anscombe argues "how we come by our primary knowledge of causality" is that "in learning to speak we learned the linguistic representation and application of a host of causal concepts."[32] Language discourse is the primary socializer into the scientific worldview. Thus, Brannigan adapts Wittgenstein's model of language as a mode of life to explain science. The process of scientific discovery is understood as meaningful action determined not by the nature of the object but by the conventions that are culturally acquired, especially in terms of the language games that constitute interaction.

However, in the final account, science is understood as Parsons and Weber understand all action—as socially meaningful. In turn, scientists inherit their presuppositions from the socializing work of language, which embodies folk and other cultural traditions. The accounts scientists make of their own discoveries, which are invariably framed in causal discourse, are seen by Brannigan as instances of folk reasoning rather than as representations of natural events.

From the perspective of traditional sociology, the main contribution of the new work in the social study of science is in no way a challenge to the basic paradigm of the sociology of knowledge: it merely insists that scientific discovery not be exempted from rules governing investigations into knowledge in which interest, culture, and power are recognized as mediations. Brannigan's treatment of science is parallel to the sociological study of religion developed in Weber's *Economy and Society*, and this is, of course, a scandal given that social science, no less than physics or biology, dedicates itself to finding Truth, defined in this context as rationally adduced knowledge that combines observation and experiment, rather than relying on intuition, revelation, or transcendence. To place science on the same plane as folk religion is surely controversial, but it must be recognized that Bloor and Brannigan, perhaps the most forthright and philosophically inclined among social theorists of science, are careful to associate themselves with the dominant paradigm in sociology, precisely in terms of its axiom that the social system is constructed from units of action, that actors interpret action-situations teleologically, that is, in accordance with their own purposes, but also with their unconscious motives that intrude in scientific discovery; and, equally important, that the cultural system, conceived here as intellectual orientations, is integrated into what counts as knowledge.

Perhaps Latour has gone the farthest in implementing Winch's program for social science. I would argue that what is lacking in the social studies of science is a way to make the power/knowledge link commensurable with the idea of science as a discursive formation. Foucault's argument that the "social" may be considered only as a series of discursive formations surely has advantages over the older base/superstructure models or theories in which action and communication are considered separately.

The knowledge/power axis, in Foucault's archaeology, implies that power relations cannot be subsumed under discourse, even if discourses always embody power. For, even in Latour's study of the emergence of Pasteur's laboratory as a new power center, this development resulted from the authority the latter's discoveries received from state/corporate power: "He (Pasteur) transfers himself and his laboratory into the mist [sic] of a world untouched by laboratory science. Beer, wine, vinegar, diseases of silk worms, antisepsy and later asepsy, had already been treated through these moves. Once more, he does the same with a new problem: anthrax."[33] In the case of anthrax, Pasteur moves his laboratory onto a farm site. "They learn from the field of veterinary, translating each item of veterinary science into their own terms so that working on their terms is also working on the field."[34] This *positioning* of the laboratory in the field of production follows the fact that "it is in laboratories that most new sources of power are generated."[35] The laboratory becomes the new social context as well as the site of knowledge production. The traditional distinction between inside and outside (the famous "externalist or internalist" accounts of the process of scientific discovery) is overcome.

Surely, Latour is right to insist that the laboratory is a site of power, constituting both text and context and also a new model for society. And it may also be said that knowledge producers in this new social paradigm *impose* a new regime upon production. But sociology still has to account for other power centers that may not be the repositories of scientific knowledge themselves, but that still command knowledge as a (crucial) resource for the reproduction of their institutional power.

It would be interesting to apply Latour's model to the study of the relation of the American military to the scientific community. The military is a discursive formation with its own language, rules of action, and priorities. Its connection to knowledge has become absolutely fundamental for the maintenance of its power, and it does not directly incorporate science institutionally, except in selected fields such as civil engineering, ordinance and strategy and tactics, and other questions bearing on organization. In relation to the development of modern weaponry, it is certainly dependent on relatively autonomous knowledge, which, even if financed by public (military) funds, remains separate. Could it be argued, as Latour appears to do, that the military is subsumed under science and technology since laboratories are now the "new" sources of power?

As we have seen in chapter 8, on Soviet science, if one accepts the postulate that the scientific/technological nexus is today the major new source of power, the capacity of scientific communities, by their assent or refusal, to determine the configuration of other discursive formations, to become the core constituent of action situations in all spheres, brings into existence new relations of domination and subordination.

Latour has adopted, even if unwittingly, STR (the Scientific-Techno-logical Revolution) as the leading postulate of a new sociology of science. To push the argument further, he wants to argue that the sociology of science is not merely a subdiscipline of social science, but has become the heart of the discipline. "Lab studies" are "the key to a sociological understanding of society itself" because "it is time for sociology of science to show sociologists and social historians how societies are displaced and reformed with and through the very contents of science."[36] The laboratory has burst the boundaries of its technical funding and the "future reservoir of political power is in the making."[37]

This is the *strongest program* for the sociology of science, but its validity is not an empirical question. At stake is the claim not only that knowledge has become the leading productive force, but that scientific knowledge contexts are displacing other power centers, making them contingent. Whereas Marx assimilates knowledge to intellectual labor, which is, in modern capitalism, subsumed under capital, especially in the labor process, Latour claims this relation is in the process of being "reversed" and argues for the materiality and effectivity of its discourse.

How does this occur? Here, Gouldner's point becomes relevant. When knowledge producers develop their own culture of critical discourse and generalize it as a social model, science is positioned to impose this social model because its method is reflexive. The "interests" underlying scientific discovery are not extrinsic to the ordinary normative and experimental structures of the community, but are embedded in both ideology and procedures. If science is organized around the principle of reflexive self-understanding which goes beyond skepticism, we can expect that its relations with other power/discursive formations cannot be reduced to its subsumption but must be termed "alliances" in which dominance and subordination are not determined in advance.

This hypothesis implies that at least on the level of the elite, scientists are conscious of the power/knowledge axis, and it is a choice rather than a relation of coercion that cements the links between physics and chemistry and the military, biology and large medical corporations, geology and the oil industry, sociology and the social policy apparatus of the state. Thus, from a subdiscipline of social science, social studies of science become a way to study and understand power, especially class relations.

Whether the scientific community constitutes itself as a class is an empirical, historical, and theoretical question. The empirical/historical dimension depends, in the last instance, on how classes are theorized. If the classes are conceptualized as self-constituted historical actors, the argument that intellectuals, particularly scientists, are a class in formation is entirely plausible. But this concept is not necessarily compatible with the Marxist expectation that rising classes are historically progressive. For the centrality of science and technology to capitalist and really existing socialist formations suggests only that the power/knowledge axis becomes a crucial problematic for the survival of old classes and ruling groups of these formations. One must allow for the possibility that the bearers of knowledge as historical agents may forge compromises with the old hegemonies, rather than acting independently to challenge that power. This possibility assumes what Marxism has been unwilling to admit: the capacity of the traditional bourgeois or even postbourgeois discourses and power centers to assimilate new power centers into their fold, or even to negotiate the terms of a settlement.

This model predicts neither contradiction nor integration, only that both are possible. It retains the indeterminacy of the postmodern epistēmē without positing complete deterritorialization in which power is so dispersed as to become unintelligible. Further, it retains the crucial idea of agency, even if no longer lodged in the old actors. Or, at least, it enlarges the range of historical actors to include not only the new social movements not directly linked to class positions but also new classes whose identity springs not from relations of ownership or even control of the means of production, even if they are crucial to production itself. The concept of the formation of the intellectuals as a class relies heavily on the relatively autonomous mode of discourse as social object, the knowledge/power axis, and the economic/social formation, without establishing in advance the priority of one over the other.

The empirical/historical question is how discursive formations position themselves in these relations, what is their political effectivity, and which alliances do they choose to make in their own interest or in relation to the (conflictual) structures that have formed them and reproduced them?

This perspective differs from the recent tendency of some, like Ernesto Laclau and Chantal Mouffe, Latour and others, to collapse social relations into discourse. In the first place, arguments for the disappearance of the social as a category independent of discursive formations tend to read the knowledge/power link into history in a way that becomes ahistorical. The historical argument, which asserts that this link has been made under specific conditions, is far more persuasive, even though Foucault's studies demonstrate, I think convincingly, that knowledge is connected to power prior to the late nineteenth century. But his studies sidestep the question of class formation. Instead, Foucault means to show that moral discourses contain a profoundly political content, and that the position of a discursive formation relative to others influences its effectivity. Laclau and Mouffe generalize this amendment to the study of social power and, in the process, abolish classes. Refusing the reduction of the social to discursive formation denies not the validity of such categories, but that only discourse encompasses all relations. Post-Foucaultian social theory has assimilated his work to a centralizing postulate that is unwarranted. Surely, economic and political relations are constituted through language, and in this sense, examinations of inscriptions, conversations, and of the position of knowledge in the power matrix are extremely important, especially because these studies are virtually absent from social sciences. That social studies of science have entered this research will prove rewarding for clarifying the way in which scientific knowledge is constituted as power.

Second, one would not wish to deny the materiality of language/knowledge, and thereby reduce either to a function of the socioeconomic structure. And it may prove to be true that the socioeconomic structure is constructed from the laboratory, the knowledge-factories upon which it increasingly depends, especially in advanced industrial societies, in both East and West. The question is whether the older socioeconomic structure, which is based on the state-capital conjuncture, dissolves into discourse? Or, whether the scientific labor process and its procedures dominate all production? We might want to adopt, provisionally, a threefold structure of social action in which the old mode of production, including the classes and institutions tied to it, retains considerable power, including its claims on knowledge production. That it is forced to give ground in the wake of the emergence of a scientific/technical social category seems unassailable.

What needs to be avoided is the tendency to displace classes by discourse and thereby to establish a new universal. For the value of Foucault's archaeology consists precisely in its antiessentialism, its ideas of dispersal, deterritorialization, and its refusal to designate, in advance, univocal motor forces that propel history. We can fruitfully adopt the concept that institutionalized knowledge is, *sui generis*, a new power center which inscribes the social without retrospectively reading this relation into the past. Instead, we may understand the position of science and its apparatus in the social formations as historically specific, as constituting a new and undeniable problematic of these formations whose consideration makes them intelligible.

This formulation avoids the tendency, characteristic of a considerable portion of contemporary social theory, to assert scientific and/or technological determinism, or, to be more exact, the determination of knowledge, over all other discourses and social relations. Rather, I want to advance the postulate that knowledge is a discursive formation that, at the same time, constitutes itself and its agents as new historical actors. That science has become indispensable for virtually all social formations in the late twentieth century and has established its hegemony on the ideological plane seems unexceptionable. That it is required to connect itself to the capital/state axis threatens to undermine its autonomy, although what we mean by these other formations cannot be separated from the will to scientificity. (The examples are almost too numerous to mention, but it is enough to recall that the concept of the scientific management of society has become the ideal of this axis. What is meant by this slogan varies according to time and place. But the desire to transform politics and economics into sciences, rather than keep them as arts, is nearly universal.)

The real question is whether the culture of critical discourse is a leading ideology of the social categories that surround science as a discourse, not the least of which are the scientists themselves. We may read Popper's corpus as an effort to establish the hegemony of his conception of critical science over the entire enterprise. In a different register, Kuhn's internal referent for paradigm shifts leans in the same direction. For if science is not subject to the fundamental influence/determination of other power centers and their discourses, we may assert the autonomy of science and its effectivity without positing its subsumption under capital, not only with respect to the selection of problems to be studied, but with respect to the contents of scientific discoveries.

If science is a discourse whose status as privileged inquiry within the social formation is historically rather than naturally constituted, its autonomy is always mediated and therefore relative to its position within the social formation of which it is a part. Its place is constantly renegotiated with other power centers, and the degree of its "freedom" is always understood in context.

CHAPTER 11
SCIENTISM OR
CRITICAL SCIENCE:
The Debates in Biology

The effort to construct a unified field theory among the sciences ultimately rests on the validity of the reduction of "life itself" and its forms
to physical and chemical categories. This effort has remained a cultural
ideal, not only for the physical sciences but also for biology and the
social sciences. As we saw in the preceding chapter, social sciences
have adopted at least three major stances with respect to reductionism. One powerful tendency in social theory wishes only to emulate
the methods of the physical sciences while holding to Durkheim's position that society, or, to be more exact, human action, is an irreducible
social fact. This logical contradiction produces a more wholehearted
embracing of physics. Accordingly, the distinction between the organic
and the inorganic becomes moot as interaction is linked to relations
among molecules, or humans are defined as animals that process interaction (or information). The mechanism of the old physics is reproduced in the human sciences.

The second tendency, identified most appropriately with Parsons,
holds that human action is an indeterminate relation among independent systems which, when synchronized properly, constitute an *organism*; but that the organism is subject to disruption or mutation
depending on the form of the relations among its constituent systems.
Parsonian systems theory, supplemented by parallel developments in
metabiology, has had an enormous influence on contemporary social
theory, notably in the work of Jurgen Habermas and neo-Marxists,

such as Immanuel Wallerstein, who have developed a picture of social relations in the metaphor of "world systems."[1] Like Parsons's model of social action, the world system is an asymmetrical unity, which has as its central category the unequal exchange of labor, commodities, and capital. These elements are subject to empirical specification, which, when reduced to sets of propositions, can be tested by the normal methods of physical sciences, particularly mathematics.

These dominant strains of social science implicitly hold to a concept of the unified field that depends on methodological uniformity rather than on ontological reductionism. However, they do contain features that, if taken seriously, presuppose either a physicalist or organicist conception of human action. Physicalism is most evident in the methodological assumption of individualism, which does not make the further specification of rational choice. Organicism is intrinsic to systems thinking, but also to the larger categories of morphological biology. Concepts such as structure and function, which are still powerful in anthropology, with their immanent teleological presuppositions, systems, and entailments, constitute an image of society in which the self-production of humans is mediated by an ecological model. In this version of social action, people do not make history as they please but are subject to systemic constraints over which they have no direct control. These constraints have features of a kind of natural order.

Both Marx and Marxism have often construed social relations in an organicist model. As I have argued, self-produced history is, for Marx, subordinate, except at key moments of historical transformation, to morphological determinations. In current terms, boundary conditions are not external to the course of action but have an internal relation of determination, at least in the last instance. Thus, Althusserian Marxism, which adapts structuralist categories to historical materialism, can discover corroboration in the works of Marx and Lenin precisely because of the hegemony that evolutionary ideology often exercised over not only Marx but social science as a whole. Even dialectical thought is profoundly indebted to organicism. Hegel is a great systems builder, albeit in the historical, evolutionary mode. His philosophy consists in the dialectical tension between history and system, time and telos. Hegel asserts that natural science, including biology, represents aspects of mind that are alienated, as if they have independent existence and obey different laws.[2] Hegel's concept of the unified field exactly reverses physicalism and biologism by arguing that return of the subject to itself takes place through its separation from nature. Accordingly, the notion of nature as objective and independent of spirit corresponds to the practical approach to nature, nature as object for appropriation. Taken in its epistemological aspect, nature confronts us as "otherness," as pure externality. From this perspective, physics appears to deal with the universal, unchanging laws of nature precisely because it unself-consciously takes nature as immediate object. This opposition is resolved by the dialectical unity in which subject and object are reunited. For Hegel, the scientific view of nature is necessarily one-sided, which does not imply falsehood. Only when the partial totalizations of natural science are negated within the subject-object totality does natural science assume its appropriate relation.

Hegel is worth quoting at length:

> The real totality of body as the infinite process in which individuality determines itself to particularity and finitude, and equally negates this and returns to itself, re-establishing itself at the end of the process as its beginning, is thus an elevation into the first ideality of Nature, but an ideality which is fulfilled and as self-related *negative* unity, has essentially developed the nature of self and become *subjective*. This accomplished, the *Idea* has entered into existence, at first immediate existence, Life.[3]

Hegel's formula for Life is Shape (form) infused with vitality, which, taken together, constitute Total Structure. Animals possess self-movement and voice but no self-consciousness, a property reserved for the human subject who reflexively recognizes nature only as a "mirror of ourselves" rather than as an external and alien object.

This is the basis of Lukács's argument that science cannot be a praxis as long as it takes reified nature as its object. Only when we understand nature as a social category can a critical science be born which, in any case, could not be based on the maxim "being is being measured." However, to argue that nature is a social category does not nullify the historical proposition that takes nature as prior to humans. As I shall insist in this chapter and in chapter 12, critical science must work with the play of these contradictory statements.

The third tendency in the human sciences, in which Lukács participates, focuses on the other side of the mirror. Dilthey, Ranke, Natorp, and, indeed, the pioneers of social theory, Simmel and Weber, abandon the dialectic of labor as the foundation of social action and, in the process, separate social relations from their place in natural history. What remains of the scientific worldview are the *methods* of natural science, but the life world is considered to be radically different from physical nature. For Weber, what is borrowed from physical science is the ideal of value freedom, which, in turn, is based on taking the object of knowledge *at a distance*, implying that the subjectivity of the investigator can be abstracted from interaction. Of course, neither neo-Marxism nor those following Husserl and Heidegger harbor such illusions. But few followed the Frankfurt school, Husserl, and Heidegger in the critique of natural science. Just as Horkheimer and Adorno showed that the fear of nature led to its domination by "man"—nature is understood here in a double sense as both the external world and the otherness of human nature—so we must admit that the idea of science as a labor process, or, more exactly, as a form of intellectual labor, has been largely jettisoned as a social object by the human sciences (except, of course, by the new sociology of science). Fearing the reduction of social relations to biological or physical laws, the human sciences have struck a deal with natural science: "Let us stipulate the duality of nature and culture, subsuming neither under the other. Let us admit that the laws of nature cannot be applied to human relations; nor does the inverse hold true."

Of course, it hasn't worked. The drive of the physical sciences to discover the unified field may founder on the eternal problems posed by relativity and quantum physics, but these obstacles have not prevented both biology and the social sciences from exercising their will to scientificity by translation of all forms of life to the interactions of atoms and molecules, or to quantitative relations which, for them, cannot describe the world of social interaction. Reductionism flourishes when the one-sidedness of nature as object of positive science is not challenged. The social sciences are imbued with the desire for legitimacy, to achieve maturity by proving to themselves and the natural science that propositions about the social world can be expressed as precise sets of objective quantitative relations. Therefore, its categories are construed in the image of empirical verifiability, a strategy which necessarily narrows the scope of inquiry to the realm of particularity. Consequently, the social is taken at a distance, and the relation of science/social science to the power/knowledge axis is rarely, if ever, questioned.

The attempt to reduce the explanation of biological phenomena to chemical and physical determinants is not new. Indeed, the search for the single material substratum of life has occupied biologists for a century, particularly since the emergence of genetics as the core branch of the discipline. The gene, which biology holds contains the source of inherited characteristics of all life, was, almost from the beginning, formulated in the terms appropriate to the search for the unified field. According to this formula, the task of science is to discover the simple, singular substance that contains and explains the multiplicity of life forms. The gene, ensconced in the cells, accounted not only for reproduction of identities, but for (most) variations in species of life. Some variations were the products of mutations, by definition "accidents" attributable to indeterminate causes which nevertheless could be reproduced in the laboratory; some mutations, such as those produced as the effects of radiation, implicitly acknowledge the independent influence of the environment on genetic structure but only under special conditions.

The emergence of new species has presented a paradox for genetics. On the one hand, evolutionary theory asserts that new species have emerged over time, sometimes replacing those that are less adaptable to changing environments. On the other hand, the Lamarckian hypothesis, that the organism responds to the environment by changing the genetic structure of future generations, is rejected. How to account for evolutionary change within the framework of genetics? "A living organism is an entity that can utilize chemicals and energy from the environment to reproduce itself, can undergo a permanent change (a mutation) which is transmitted to succeeding generations, and by accumulation of numbers of such mutations, can evolve into a distinctly new living form (a new species)"[4]—this according to a panel of the National Academy of Sciences (1970), appointed to report on the "state of the art." American genetics is overtaken by molecular biology in just one decade after the Watson-Crick discovery of the double helixical structure of DNA. But this discovery signals more than an advance in understanding the material substrate of genes. It begins the drive to reduce all problems of biology to their physical and chemical elements.

Echoing Crick, Kenneth Schaeffner puts the issue directly: "Biology is nothing more than chemistry—but chemistry nonetheless, in which the chemical systemization of chemical elements plays an important role."[5] For Schaeffner, the question is whether analogous causal sequences can be translated from the "correspondence rules" that apply from one discipline to another. In his view, all the basic propositions of genetics can be translated into chemical propositions without losing the validity of causal relations. In short, biology is reducible to physical science; it if possesses autonomy in any sense, the biological level becomes distinct only in respect to *organization*.

This view resonates of earlier work by Ludwig Bertalanffy, W. Ross Ashby, and other theorists who attempted to liken all life to system, that is, to a series of (open-ended) formal relations whose configuration constituted a *content*. The development of systems theory responds to the Watson-Crick hypothesis of the chemical/physical basis of life which makes biology truly a physical science. "Life" is a particular organization of elements into deoxyribonucleic acid, abbreviated as DNA. Against the ideology of evolutionism, molecular biology poses its reductionist ideologies of system and physicalism and has replaced references to the specific concept of organism, which retains the idea of life as an ontologically autonomous level of historical being. Needless to say, although the battle among biologists is by no means ended, reductionism has achieved hegemony in the past thirty years. The entire organism is explained in terms of molecular arrangements, the newest phase of which is the discovery of the molecular brain. In this reduction, not only intelligence but emotion becomes little more than a chemical mutation, and since mental illness is explained in terms of secretions and organization of molecules, it is subject to chemical and, on occasion, physical treatment. And, by extension, social relations become types of behavior that now lose their qualitative specification and can be understood as consequences of recombinant molecules. Thus, the specificity of the social is attacked from both sides: by those who would reduce all such relations to discursive formation, privileging language and narratives as the source of such mysteries as subjectivity; and by those who elevate the system or organization of molecules to a biological, psychological, and social first principle. In both moves, the problem of the sloppiness of biological and social sciences is resolved by abolishing them, or, more exactly, by merging them into physics and chemistry. The counterattack has been mounted by seeking to assert that the chemical organization of biological life constitutes merely a boundary condition which "is always extraneous to the process which it delimits."[6] According to Michael Polanyi, DNA cannot determine the content of the information contained in the DNA code, even if it can be likened to the "motive force" of the code:

> "Every system conveying information is under dual control, for every such system restricts and orders, in the service of conveying its information, extensive resources of particulars that would otherwise be left at random, and thereby acts as a boundary condition. In the case of DNA, this boundary condition is a blueprint for the growing organism.[7]

Therefore, meaningful information "in the DNA molecule is independent of the chemical forces within [it]."[8]

Polanyi joins the issue on the concept of integrative levels, according to which operations at a "higher level rely on the working of principles of a lower level"[9] but cannot be reduced to them. Life, then, possesses an irreducible structure.

Francisco Ayala distinguishes biology from physics and chemistry by invoking its reliance on teleological explanations, "which apply to organisms and only to them in the material world." Organisms are purposeful systems "where there exists a mechanism that enables the system to reach or to maintain a specific property despite environmental fluctuations." Ayala argues that organismic explanations "cannot be reformulated in non-teleological form without loss of explanatory content."[10] And, since "teleological explanations cannot be dispensed with in biology,"[11] Schaeffner's claim that correspondence rules can be devised to make its laws commensurable with chemistry is totally denied.

Thus, one argument against reductionism relies heavily on organisms as self-regulating systems in which form and function are articulated relatively autonomously from the environment, which includes the physical conditions of their existence. However, Levins and Lewontin make their own case against reductionism on a different plane.[12] They argue that the "past evolution of each species" determines the mechanisms of adaptation, but also that the species should not be considered as an isolated phenomenon. Species are part of population communities that interact with their ecological environment and adapt to it as well as transform it. Their point is that the community within which a species functions constitutes an irreducible level of living organism. This population ecology approach puts the issue on new ground, addressing the claims of molecular biology directly. "We must reject the molecular euphoria which has led to dismissing population, organismic, evolutionary, and ecological studies as forms of 'stamp collecting' and allowed museum collections to be neglected."[13]

It seems to me that Levins and Lewontin avoid some of the pitfalls present in the arguments of Ayala and Polanyi, which remain at the level of the single organism interacting only with its own chemical basis. By developing what might be called a biological theory of "action," that is, by making interaction between populations of organisms consisting of different species the object of biological knowledge and, further, by insisting on the effectivity of organisms on their environment as well as the inverse, the internalist/externalist dead end is avoided and the problem of reductionism put on a larger ecological canvas. The genetic structure of the organism remains, as with Polanyi, a *constraint* on the types of adaptation and reproduction available to the community and to the species, but this constraint is now part of the multiple determinations that constitute the intergenerational life of the species. Whereas Polanyi must speak of systems of dual control, Levins and Lewontin see the "whole as a contingent structure in reciprocal interaction with its own parts and with the greater whole of which it is a part,"[14] i.e., the ecosystem. The idea of "reciprocal interaction," in which so-called higher levels of the organism not only are constructed from lower levels but also determine those lower levels, recalls the development of German biology at the turn of the twentieth century. At that time, the whole organism was thought to be constituted as the object of knowledge. And, as Levins, Lewontin, and Stephen Gould have argued, the interactions of organism must be seen in the context of a double relation: to its community and to the environment which forms a contingent, open-ended system.

Taking the ecological system as the scene of biological knowledge entails the pertinence of an entirely different set of relations. No longer is the organism understood as an isolated entity and built up from its genetic material taken as the "object." The refusal of the analytic presupposition of methodological individualism and its correlate, inductive explanations that rely on the movement of the simple to the complex as a nonreversible process, suggests a different inquiry than that of molecular biology and molecular genetics. This inquiry might be described as a new organicism in which the effectivity of the DNA molecule becomes an element among several constituting the field with and within which the organism interacts. This perspective does not vitiate the contributions of those like Polanyi and Ayala who put their antireductionist arguments entirely within the presuppositions of methodological individualism. Rather, teleonomic or purposive elements become integral to ecological fields.

Ecological biology differs from the old environmentalism because it posits the effectivity of species and its community on the environment; unlike Lamarckianism, it is no longer a question of characteristics resulting from the adaptation of organism to environment becoming subject to heredity, but that the changes brought about by the activity of the organism on the context within which it functions are relevant to the emergence of new characteristics. According to this view, for example, nutritional factors, which are changed by knowledge of the relation of diet to health, affect a person's life span. The genetic code becomes not a determinant of a life span, but a boundary condition, a constraint which may limit the degree to which reciprocal relations within an ecological system may operate. But the variation between Third World human life spans and those enjoyed by populations living in environments in which fresh fruits and vegetables are part of everyday diets surely constrains the effectivity of genetic codes for explaining longevity of populations.

On the other side, the organism also has a double referent. On the one hand, its inborn characteristics limit the degree of its possibilities, both mental and physical, as well as "dictate" the constitution of some of its "needs" and wants. On the other hand, as organisms, we are self-produced. Although it is true that we inherit biological limits to our individual action, as part of wider communities we, in Marx's terms, multiply our productive powers by cooperation and conflict. (I take conflict as a necessary and positive aspect of social relations even in communities that adopt the ideology of harmony among persons as their guiding principle). Moreover, we never reach our full inherited potential either physically or intellectually because, among other things, our social relations (e.g., class, race, sex) constitute themselves as further boundary conditions that partly restrict and direct our development. These are, generally, more effective as boundaries than as biological constraints. It has been an unfortunate tendency among social determinists to discount the reciprocal relations between biological and social levels, undoubtedly because of the reduction implied by admitting that these exist. We can see in the development of social theory since the late nineteenth century a pervasive effort to establish the "social" as a fact irreducible to biological or physical categories. In recent times, the reemergence of biological and chemical/physical determinism (the combination of which I designate as physicalism) of social and psychological phenomena has only strengthened the resolve of antireductionism in the social and psychological sciences to deny the effectivity of physical elements in the construction of personality or social systems. Social and psychological theory are reacting to the "discovery" that disorders that psychoanalysis ascribed to the interaction of the psychic structure with its social environment, especially the family and sexual relations, can be explained in chemical terms and made subject to medication; the reduction of the functions of the brain to their neural boundary conditions; and even more egregiously, the reduction of human behavior to alleged biological propensities such as territoriality, innate aggression, and competitiveness. In this ideological context in which the model of the organism proposed by molecular biology becomes the generalized model for all social and psychological phenomena, i.e., the internalist reduction of all relations to a unicausal chemical substance, the increasingly embattled minority closes its mind to any variety of the interactionist model such as the concept of integrative levels, the new organicism of population biology, ecology, or teleonomic assumptions of evolutionary biology.

Nevertheless, social theory without a "field" that acknowledges the reciprocity both within the organism, and between it, its community, and its environment is unable to mount a significant critique of what has become the most universal expression of human life—the domination of nature and domination as the characteristic mode of social relations. As we have seen, the domination of nature is the crucial absence in the text of social theory since, until recently, this relation has been taken as part of the given, mystified by the edict against the reduction of the social. This historical move nullifies the earlier understanding that humans are part of natural history. What must be added is the capacity of the production and reproduction of social life to permanently alter the ecosystem within which human communities interact with other species, other life forms, and the physical environment of which we are a part.

Ecological theory teaches us that, taken together, these constitute a boundary condition on human action, the ignorance of which has marked our relations, at least since the industrializing era. As Horkheimer and Adorno, Murray Bookchin and Carolyn Merchant, among others, have argued, the refusal to acknowledge this constraint has been constituted through human domination. Orlando Patterson has demonstrated brilliantly that Blacks are made slaves through their deratiocination; their otherness is displaced as an absence of human status.[15] Elsewhere, I have insisted that male domination is intimately linked with the process by which women are identified with nature as alien and antagonistic to man. Just as the dominant ideology insists that nature must be subordinated to human needs, not only to fulfill them, but to liberate humanity from its thrall, so woman is functionally subordinated to man as mate and mother of his children.[16]

Of course, the new social movements of the past three decades are organized against domination, even as the (Western) working class identifies itself increasingly with domination despite or because of its partial eclipse as a historical actor. But only the social ecology movement has consistently possessed the vision to link the domination of nature to human domination as aspects of the same reciprocal relation. For it is clear that the absence of the biological and, more broadly, a conception of humans as part of natural history are routinely producing massive disasters. This is not the place to provide extensive documentation of the multitude of violations visited upon our ecosystem by willful human domination. In any case, most of these are well-known: the progressive thinning of the ozone layer by industrial pollution; the pertinence of nutritional, industrial, and consumption patterns to the elevation of cancer to epidemic proportions in industrial societies; the profound and long-term devastating effects on whole populations of nuclear energy and weapons and the "greenhouse effect" which may reduce our oceans to dead seas.

What is systematically denied is the fact that these crises are symptoms of the larger blindness of scientific and technological society, that the pores of our culture, especially its discourses, are surfeited by the absences and the occlusions that might save us. Thus, addressing the key issues of the social theory of science requires a radical reconstitution of its categories by analysis: the object of knowlege must be shifted to include not only language and its discursive formations or the social and economic structures constituted by the knowledge/ power axis, but also the reciprocal relations between humans and our internal, biological "nature" and the ecosystem of which we are a part. This is a paradigm shift that entails theorizing the link between psychological/biological categories, language/discourse, and social relations.

Currently, each of these inquiries succeeds by subsuming the others under its categories, e.g., Lacan's dictum that the unconscious is "structured like a language," elements of which can be borrowed from semiotics; or Althusser's theory of ideology, which borrows Lacan's categories of symbolic order and imaginary to elucidate how ideology works. These moves may make the unconscious intelligible without resorting to essentialist reductions, just as nature becomes a social category for Lukács. But while something is gained by these concatenations, we lose the sense of specificity of each realm. Language, social relations, or whatever, become the occasion for the occlusion of the wider stage within which the human drama is played out. By failing to theorize constraints, we reify "progress," growth, and other signs of the imperium as infinite and inexhaustible possibilities.

Similarly, medical science treats human disorders as, in principle, subject to objective therapies, that is, treatments that do not depend on indeterminate interactions for their effectivity, just as physics, with more confidence than it can justify, regularly announces that it is on the brink of discovering the ultimate building blocks of the universe, a testimony not so much to the reality of its perpetually imminent discovery as to its will to final knowledge (read: domination) of nature.

To argue that the biological and ecological levels constitute boundary conditions of human action does not thereby reduce social relations. Instead, it simply specifies the limits that are imposed by these conditions on social arrangements. In a word, our freedom to construe interactions "as we please" is an expression of the hubris of scientific discourses that believe that "nature" can be permanently rendered a dependent variable of human action. But the elements of nature are not infinitely elastic; they obey laws of homeostatic interaction which can only be violated only at human risk. Drawing inferences from this discovery, we learn that some of our cherished Enlightenment categories, particularly the key myth that the universe is inexhaustible and subject to human dominion, are plainly false. Even more important, they are dangerous to the survival of the human species as well as many others. I cannot rehearse here the consequences of overdevelopment on the fate of plant and animal species that have become extinct or are endangered. But the issue is not limited to the question whether this or that species will exist in the near future; more significant is the disturbances to the ecological systems of which they (and we) are a part. The violence visited upon the interactive systems that form the basis of the production and reproduction of life is undertaken, almost invariably, in the name of progress and economic growth, on the implicit assumption that the environment requires little or no consideration, that is, does not enjoy the status of *agency*.

Clearly, the ideology of unimpaired economic development is no longer acceptable. But the question cannot be restricted to slowing down or exercising "care" in plundering the earth. Such postures recognize implicitly that ecosystems are constraints on human action, but have not yet concluded that nature must be treated as an agent of the social and historical process. The concept of natural agency should be taken as more than a metaphor. It is not a question of mystifying nature as possessing *consciousness* or *will*. However, if we acknowledge that human life is only part of a larger interactive set of relations, we cannot ignore the effectivity of the two types of interactions. First, we interact with our own physiological and biological systems, which constrain our capacity to use them as fungible elements in human action. The psychic structure, including the unconscious, cannot be reduced to linguistic categories. Not only is the unconscious the repository of the biographic and historic "scars" that have formed over the traumas of social existence, but it also makes demands on us that, having been displaced in various ways, exact a price on our character. And character is a constraint on even the most nimble intelligence and body. Those who hold the psyche as the product merely of its physical and chemical components must believe, in principle, that science can discover mechanical means to cancel out the disturbances that might arise from disequilibrium in these elements. In this mode of theorizing, there is no psyche capable of exerting its own effectivity on the person. Second, those who would apply the results of the analytic processes associated with molecular biology to genetic engineering have stubbornly resisted warnings that on both ethical and ecological grounds their interventions may have dire consequences.

It would be too crude to explain such blindness simply by venality or greed, although there can be little doubt that the commercial uses of genetic engineering are lucrative for those who own the means of production of simulacra or clones, as well as those whose knowledge has been allied to economic power. The refusal of genetic engineers and their corporate allies to heed warnings may be primarily attributed to the discourse that holds nature mute, that implicitly accepts the idea that human action meets no structurally constituted resistance. Fundamentally, then, the "optimism" that argues that human intervention in nature has no serious life-threatening consequences results from the posture that accords to nature no agency and does not recognize that it is a boundary condition for action.[17]

Beyond survivalist arguments, genetic engineering poses serious ethical issues. In capitalist and state socialist societies, the ownership of most scientific and technological "products" is separated both from the producers and from the larger community. Gene splicing is a technique that may change the key characteristics of a species. As with its predecessor techniques of crossbreeding, the purposes for which such changes are intended do not necessarily correspond to the community interest. (A simple example is growing tomatoes that have harder skins so that they may be picked mechanically without suffering a high incidence of bruising. The tomato is rendered less tasty, but farm productivity is dramatically increased.)[18] Even more substantial, these changes are in the hands of private interests. Courts have determined, in some cases, that the guiding principle governing the right of corporations to own patents bearing on species-changing technologies, or the creation of entirely new species, is the vested rights of property. Some courts have determined that the products of genetic engineering may be treated like any other commodity, subject only to the normal safeguards that might apply, for example, to a potentially lethal drug. The problem of the right of private interests to own life forms unilaterally never occurs to the jurists as a barrier to genetic engineering and its private appropriation.

Two kinds of ethical arguments may be advanced against genetic engineering. The first is the religious objection that any intentional alteration of life forms, especially human life, violates God's will expressed in the natural order. If this view were accepted, any intervention in nature, such as breeding two different plant or animal species, would be opposed unless human life is privileged. However, if the question of God's will is laid aside, the ethical question turns on the right of property or any private interest to make such determinations. Since regulatory agencies have normally collaborated with private corporations or the government, which in Eastern bloc countries acts as a private interest, the appeal to the doctrine of social control over biological destiny often falls on deaf ears. Yet, when life forms are at stake, the problem of who has the power over scientific discovery is thrust to the center of the political discourse.

In Chapter 10, I insisted that the notion of the freedom of science is severely tested in contemporary life, where knowledge cannot be meaningfully separated from power, either of the state or of large private corporations. As the case of MIT's agreements with bioengineering and other private corporations shows, scientific discovery increasingly assumes no pristine existence apart from the uses to which it is intended. This does not imply that all science has become applied science. There is still some scientific work for which practical implications are not yet evident. Yet, as demonstrated by Einstein's participation in the development of the atom bomb, the emergence of molecular biology, and the relationship between solid state physics and the semi-conductor, only marginal science enjoys the luxury of complete autonomy, until applications are found for its work.

The heart of the ethical issue, then, is the degree to which social (as opposed to state) control over the results of scientific work is imposed. For the mounting evidence is that scientific communities are often compromised by their alliances. On the other hand, the need is urgent for bringing science into the public sphere of discussion, debate, and policy determination, a process which entails large-scale scientific education, an issue to which I shall return in the final chapter.

We can see how the doctrine of mute nature has worked to endanger life in the case of nuclear energy. Nuclear energy is both an ordinary industrial extraactive technology and an unusual one. Ordinary (compared with oil, coal, and mineral extraction) insofar as industrial processing makes some of the elements of geological terrain sources of energy. Unusual because the consequences of a nuclear accident are potentially disastrous not only for those in close proximity to the site of a reactor, but for those far from the incident. The reason is that the technology of nuclear energy involves harnessing highly radioactive materials, which, if they escape their containers, are dangerous to life, especially in high-density areas. The difference between radiation dispersal and ordinary pollution created by fossil fuels is that the former may affect not only those who immediately ingest it, but future generations as well because it produces mutations in genetic material.

France, whose government, political parties, trade unions, and intellectuals have been relentlessly optimistic about the safety of their nuclear industry, which in 1987 supplied about 65 percent of their energy, has also been blind to ecological warnings. But, in 1987, accidents were reported in several nuclear plants, events that prompted several important political parties (of the opposition) to demand closing them. It is hard to avoid surmising that in this most antinaturalist culture, economic and social policy follow the stricture that ecological thinking is merely one among other discursive formations whose validity may be relativized. As with the Chernobyl accident a year earlier, it is doubtful that the evidence that safety in nuclear plants may be chimerical will profoundly affect public policy in France for two reasons: the level of investment in nuclear energy may have reached a point of no return; and the ideology of the dominion of science over nature, prompted by the mind/body dichotomy, is deeply embedded in French culture.

Perhaps it is the German skepticism concerning science, which, as we saw in chapter 9, is profoundly linked to the importance of romantic naturphilosophie in the German intellectual tradition as well as in cultural life, that accounts for the rise of the Green movement in that country. In contrast, France and the United Kingdom remain deeply ensconced in positivist culture. There, socialists and economic liberals agree on the virtues of pursuing growth policies based on the exploitation of both labor and nature, and particularly the harnessing of knowledge as a productive force. But in Germany this remarkable movement makes explicit the link between the domination of nature and human domination, arguing that the "internal ecology" of social relations parallels the interactions of the ecosystems. Consequently, its program is antinuclear, feminist, and calls for limits on economic growth, while advocating urban, labor, and social policies that recognize human biology, human communities, and the larger ecosystems as (historical and moral) agents.

In the United States, events such as the Three Mile Island nuclear accident in 1979, social protest against the further development of the nuclear industry, against the widespread use of insecticides in agriculture, and the broad movement to save natural resources against real estate and industrial development, have elicited some changes in public policy, a process that was far more rapid in the 1970s than in the 1980s. The Carter administration adopted some policies that favored conservation of natural resources and the struggle against urban real estate development and won some significant victories at the local level. Since 1981, the government has rededicated itself to policies of development, arguing that these are the best ways to maintain a viable economy. Ecologists and environmentalists have lost some of the ground they had gained in the 1970s when nuclear development came to a virtual halt. There is as yet no serious electoral expression of the ecology forces in the United States, and feminist movements have not, with some exceptions, linked their struggles with those of ecologists. In the recent past, however, feminist critiques of technologies of reproduction promise to draw the two movements closer together. These technologies treat women's sexual organs as vessels to be used by "science" to reproduce human life from donors who are indifferent to the offspring. Further, genetic engineers view in vitro reproduction as an opportunity to experiment with their long cherished dream of altering the physical and mental characteristics of large sections of the population in accordance with once discredited eugenic standards.

Not only should the politics of science make explicit the knowledge/power axis that has become a new force in human affairs; it must also address the character of scientific ideology, especially its acceptance of the doctrine of nature as a passive object of human intervention. For, unless nature is treated as an agent, the presuppositions and the practices of the instrumentalization of nature dominate not only thought but power relations.

CHAPTER 12
TOWARD A NEW SOCIAL
THEORY OF SCIENCE

It remains for me to show *how* science is social relations. For even if Marcuse has already argued for the ineluctability of the relation between science, technology, and social domination, he has not provided a systematic, detailed explanation of the way in which, by virtue of its own concepts and methods, scientific practices promotes a universe in which domination of nature is linked to the domination of humans, or the *way* in which science is a form of power. The fundamental argument advanced by Horkheimer and Adorno concerning the presumptions of science and technology refers to the Enlightenment's transformation of nature into an object, the simultaneous estrangement of nature by taking it as an antagonist rather than a partner of human culture and the organization of production and the state by bureaucratic management. Thus, "rationality" itself becomes inextricably linked to domination. The division of labor is now conceived as the segmentation of tasks and their performance by qualified persons who have found their place in the order of nature and society by purely individual talents, while, conforming to the pregiven organizational chart that defines the boundaries of their authority and has the status of a natural fact.

The imputation to nature of characteristics that are nothing more than the objectivization of the table of organization of the social world has imparted to both the appearance of natural facts as well. On the one hand, nature is taken as external to human consciousness, a hallmark of the scientific attitude since the sixteenth century. On the other hand, as we have seen, humans have attempted to make their world "scientific," to organize economic, political, and social life according to a system of rationality that is understood as consistent with natural law. The point made by Critical Theory is that society has reified nature and its own relations. Although external to consciousness, we have imposed our own class-dominated, bureaucratically organized division of labor on the external world, which, in contemplation, appears at once lawlike in its structure, chaotic. For the floods, torrential rains, snows, earthquakes, as much as disease, in short, the unexpected "revolts" of nature have become phenomena to be controlled; these events define the task of science, the horizon in the quest for dominion. Just as nature is understood as subject to subsumption under human powers, so humans themselves are increasingly regarded as controllable. The "sciences of man" have been mobilized in the drive for social and psychological domination; the instincts, according to B. F. Skinner, the latest of the behavior managers, must be directed, lest the famous "war of all against all," that is, the freedom of the market-place, result in the destruction of the human species itself.[1] Of course, Skinner is just the most "humanistic" of the scientific managers of human behavior. At a more global level, governments have now recruited social science in a wide-ranging effort to manage international relations and biology to "manage" human reproduction.

For in late industrial societies, east and west, the fear of nature has been projected onto the fear of humans, a fear of their psychic structure as much as their command over the material means of destruction. And, of course, the object of economic science throughout most of the twentieth century has become the uncontrolled vicissitudes of civil society that lead to inexplicable crises or wars. Economics is now nothing but a technology masked as social theory for assisting the state, in both East and West, in managing the economy, or, to be more precise, in displacing contradictions from the relations of material production to other institutions and countries.

We have already seen that the choice of scientific inquiry is now subject to corporate and state determination, not necessarily by coercion or even material incentives of a crude type, but in virtue of the capital requirements of the apparatus of scientific work. For the costs of scientific research (at both "natural and social levels") have made it virtually impossible for the aristocratic scientists to pursue their labors in the privacy of their estates. Nor can scientific work proceed on the basis of patronage by aristocrats and prosperous business persons.

The days are over when Frederick Engels, for example, a Manchester textile employer, could slip an occasional twenty pounds to his friend Karl Marx so that the progress of the development of a science of society would not be disrupted by Marx's need to write a few pieces for the *New York Tribune* to keep family and body together. Now the patron has become a corporate foundation established as a tax dodge and to provide a bureaucracy for the conduct of research that appears independent of corporate sponsorship. The state, too, insofar as its veneer of ideological and economic neutrality has become a major support for research, is a crucial patron. And corporations, confined for the most part by their emphasis on "applied" research (a euphemism for technological development), help define the parameters of the financing of all sciences.

The state and the large corporate foundations claim that their grants are free of political influence and that they have created a mechanism for bringing members of the scientific community into decision making, to legitimate their claims to neutrality. Many scientists believe that such measures safeguard the integrity of "pure" scientific research. However, it must be noted that leading scientists and technologists are increasingly integrated into the management of the state. When Harold Brown, president of the California Institute for Technology, a major scientific and technical research and educational center, became Jimmy Carter's Secretary of Defense, it was a sign that the scientific establishment had achieved considerable status in the commanding heights of political power in America. For it is one thing to place scientists high in the councils of the National Science Foundation, a government agency that funds research, and quite another to acknowledge that science is no longer a "servant of power" (in Loren Baritz's terms).[2] Science and power are now in the process of merging, a frightening verification of the theoretical appraisal that the division between the state and civil society[3] is undergoing a collapse, if it has not already been accomplished. From the political and ideological perspective, the control of scientific research by the guardians of "normal" science ensures, in the first place, that the *choice of the field for scientific inquiry* will remain subject to the determination of those who are considered mainstream scientists. By bringing them into the apparatus of government and the corporation foundations, as well as the large corporations (which perform about one-third of all scientific research), the social system legitimates its own claim to neutrality from the conflicts of civil society and makes sure that science reproduces a certain direction of inquiry. It is true that the Department of Defense finances "pure" research, that the Department of Health and Welfare is

interested in biological sciences, not only those fields subject to immediate application, but also more long-range and theoretical areas. The
vast sums of money appropriated by the state and private foundations
each year are subject to the priorities of Congress, the executive
branch of government, and economic considerations. It is not that the
government and corporate foundation bureaucracies *consciously*
believe in the subordination of science; on the contrary, many if not
most are persuaded that the autonomy of research ought to be maintained "in the national interest." But the management of science by
leading scientists and scientific bureaucrats (those who *were* scientists,
but have become managers of research projects and institutions such
as universities that manage research) has an ideological content, insofar as science is itself a "normative" activity. (1) Theory and method, (2)
the form of result, (3) field of inquiry, and (4) the constitution of the
scientific object are categories within which norms are established and
normative activity defined. To the extent that the norms within these
categories are part of the taken-for-granted assumptions of science,
coercive control by the state and corporate apparatuses becomes more
or less unnecessary; scientific culture (the matrix of conventions of
inquiry, social and political networks among scientists, the language
and institutions of science) is usually the sufficient condition for bringing science within the apparatus politically and for reproducing ideologies in which the domination of nature, already a normatively constructed practice, is linked to the domination of humans.

Beyond the social determination of the field of inquiry, the *constitution of the scientific object of knowledge of inquiry is linked to the prevailing social and technical division of labor.* At the level of the so-
called social division of labor, science itself is constituted by various
distinct fields or practices whose demarcation has purely historical
roots. Or, the rationality that informs the division of conventional scientific fields may be shown to conform to systems of classification that
have historical origins. There is a *social logic* of the classification of the
sciences, rather than a logic that is "given" by nature.

It is by now commonplace to remark that the relation of magic, science, and philosophy is an internal, rather than external, conjunctural
phenomenon. Historians of science have pointed out that science
breaks from magic and religion and becomes a secular philosophy only
during the Athenian city-state, and breaks with philosophy during the
Renaissance period, during the development of capitalism. These
breaks are accompanied by mutations in the logic of natural and social
inquiry whose relationship to changes in the concept of rationality is

intimate. Yet, the logic of science is continuous as well. This continuity may be traced, in part, to Aristotle's invention of a binary structure for classifying *levels* of material things. Aristotle presented the system of classification according to which physics, astronomy, biology, and other sciences were arranged as an outcome of the logic of identity and exclusion. This logic proceeds from a system of binaries (the either/or) in which the exclusion of oppositions within the thing (an explicit repudiation of Heracleitus and the other pre-Socratic philosophers) is taken as the foundation. The principle of classification also contains the concepts of levels, according to which reality is arranged hierarchically in a series of mutually excluding layers of object-classes.

Aristotle deals in both a theory of "natural causes" and a theory of the "prime mover"—God. There is both a scientific and a religious element in his theory, and the religious side gets bracketed by modern science; but the logic of noncontradiction and hierarchy remains central to both stages of the history of sciences. Thus, even if we may sidestep for now the question of whether ancient and medieval sciences were dedicated to empirical observation as the ground for their theoretical formulations (as Feyerabend claims),[4] the foundation of modern science accepted much of Aristotle's classification system, because the *logic* of scientific inquiry remained largely intact until the nineteenth century. We may speak of an *Aristotelian* revolution as the genuine epistemological break between what is called science and prior forms of natural inquiry, because it is he who divides the natural as well as the social world and thus defines the scientific object as one level of that world.[5]

The concept of levels of material existence was at first integrated into "natural philosophy" only implicitly. But, by the sixteenth century, it was clear that a hierarchy of sciences had already taken shape. Mathematics, of course, was the "queen" of the sciences, because its object had the greatest degree of generality and was not confined to any one science. Physics was privileged among the so-called empirical sciences because it attempted to find the most fundamental unit of matter to which all other levels could be reduced and to discover the laws of material motion and transformation. Astronomy was similarly important but, as we have already seen, lived in the intersection of practical concerns such as navigation and the ideal of knowledge for its own sake. In a sense, it was not a separate field of science because it was subsumed, from the beginning of the modern era, by physics. Chemistry, biology, and psychology are sciences whose separation from philosophical speculation came only with the period of industrialization. The development of psychology as an independent discipline really came out of the mind/body split introduced by Cartesian meta-physics and the British empiricists. As a branch of biology, the field had been dominated by the reduction of mental phenomena to properties that could be traced to physical and chemical causes. As late as the time of Freud, the conflict between the concept of the mind as a legit-imate object of knowledge and its reduction to the instincts, which were supposed to possess a bioenergic character, made the object of knowledge ambiguous.

The division among the sciences into fields of inquiry was generated by the adaptation of the rules of formal logic to systems of classifica-tion, the concept of levels in nature, which is grounded in (1) the log-ical principle of boundaries or exclusion, where a thing is only what it is and excludes another, and (2) the principle of contradiction, which asserts the separability of a thing from anything else, or a determinate other. A and B exclude each other by definition but may be arranged in a grid as binary oppositions or vertically as an ordered discourse. The concept of hierarchy, in which these defined sciences are arranged in vertical order, is presented in Aristotle, and in the modern era, legiti-mated by the social organization of sciences as if these are a property of nature.

The concept of levels in vertical ascending or descending order pre-supposes the formal rationality of Aristotelian philosophy and cannot be taken as an observation from nature. In turn, this type of rationality, challenged in the modern period by Vico and Hegel,[6] has remained hegemonic in the constitution of the fields of scientific inquiry, deter-mining the relative weight, both within the scientific community and in society at large, of the various disciplines. I call these divisions *social* because they correspond to various branches of industry under capi-talism, which are similarly historically formed—as industry is formed. In place of steel, automobiles, electronics, retail, public sector, etc., science is organized into various fields that are considered as separate discourses corresponding to separate levels of reality.

But science is also organized according to a technical division of labor in two ways: first, the distinctions between pure and applied sci-ences, and second, between these and "technology," appear to be objectively rooted in the nature of the objects. Scientific workers in these three fields acquire a common education only at the beginning of their training; occupational nomenclature as well as prescribed cur-riculum vary depending on which of the fields is pursued. The applied scientist may develop new types of cosmetics, synthetic fibers, or arti-ficial fertilizers, but not concern himself/herself with the physical chemistry of matter except insofar as it relates to the development of a product. The character of the scientific object here is socially determined, more specifically, has to do with the difference between the pure researcher, who is engaged in making discoveries about the "nature" of the material world, apparently with no ulterior purpose beyond the discovery itself, and the applied researcher, who is part of the system of rational-purposive action: knowledge is gained about the material world, indeed, the object of knowledge is clearly defined in terms of its commercial, industrial, or social uses. Society has placed an ideological premium on the acquisition of knowledge for its own sake and an economic premium on the practical, i.e., technical outcomes of knowledge. In this respect, the technical division of labor masks the fact that both types of knowledge depend on historical and social context—particularly the way in which the relation of the state, the economy, and the ideological forms divides scientific labor. The parameters of the object of knowledge are socially prescribed depending upon this context, and are given significance by their respective institutional frameworks. Thus, discoveries made in each sphere that do not conform to the teleological presuppositions of that sphere may be suppressed. Even though the pure scientist may discover a practical use for his/her theory, such applications would tend to be "forgotten," laid aside, regarded as either inappropriate or trivial because intervention may not bring prestige with the community of pure science. Conversely, the industrial laboratory is not considered a place for "pure science." The applied scientists' work must yield a marketable product, that is, a product which can yield at least the average rate of profit.

The technical division of labor *appears* to be normal, a division of branches of scientific labor, just as any other two branches of the same industry seem naturally divided according to types of tasks. But this logic masks the hierarchical character of the divisions; the ideological outcome is to preserve the notion of a pure science unsullied by the marketplace. At the same time, the difference between "pure" and "applied" science retains the concept of the neutrality of at least one branch of scientific research.

The organization of scientific research takes place on the same basis as the organization of industrial labor. At the pinnacle of a research project or institution there is a manager; in the university, the site of much of what is called pure science, this person is a professor and is considered a working scientist and teacher. But, as a manager, this individual has a specific set of responsibilities that correspond to those of any other superintendent: raising funds for the research, hiring and firing professional and technical staff, and equally important, maintaining relations with the administration of the university, the funding source, and the scientific community at large. He is an administrator/ politician whose skills range from those of persuasion to repression. Often, he becomes a manager as a result of his stature in the field of inquiry, usually gained years before his academic ascendancy. Thus, he becomes the logical person for the managerial function because he belongs to the social network called the scientific community and the academic community. His ability to maneuver within bureaucracies is made possible by years of sitting on committees, including review panels for the National Science Foundation and other agencies where he has gained contacts. When he writes a research proposal, scientific bureaucrats and university administrators are inclined to accept it because they know him to be a *normal* scientist who is also an effective negotiator and operator.[7]

I have already remarked that scientific research is no longer the province of individuals; it is now a vast social, economic, and political enterprise. In the first place, it is located in mainstream institutions of society: the universities, corporations, and "independent" institutes, often connected to both schools and business. The nature of the research process in both the pure and applied branches is connected to the use of high technology equipment amounting to several million dollars for even modest projects. Often the equipment is nontransferable, that is, cannot be used for other projects because it is designed for specific experiments, by scientists working on the project. It is not unusual for many projects to be funded as joint ventures among several institutions who are similarly connected to government agencies and the scientific establishment.

Each project is organized as a job hierarchy below the principal investigator, or manager of the project. Research scientists fall into two distinct groups within universities: the academic faculty, whose major responsibilty is to work on the project and whose ancillary task is teaching, and a growing number of researchers hired for the individual project, possessing no academic standing. These are usually postdoctoral fellows or migrant employees who are hired on contract and who perform their duties at the direction of the faculty members or directly under the principal investigator. Often their activities do not differ from those of other researchers on the university faculty, but even if they are included as authors when the results of the project are published, this may not be significant for their careers.[8]

The homology between migrant construction or farm workers and many scientific researchers does not refer to the type of work they perform, but to their *social position* within the scientific project. Even if highly paid, they are never integrated into the "scientific community";

their relationship to the institutions of primary research is always tenuous because they are only part of the "team" for the life of the contract. Thus, their influence is usually confined to the single project. They are often assigned specific tasks within a much larger investigation, which, however complex and based on their sophistication about the major issues in the field, is largely controlled by those with stable ties to the scientific establishment, including those institutions charged with the bulk of research.

Graduate students in physics, chemistry, and biology occupy an ambiguous position with respect to research. Typically, their work is linked to their doctoral dissertation. In turn, the choice of both field of inquiry and object is defined by their thesis director, who may be either an academic member of a research team or a manager. The function of the doctoral study in the sciences, as in the social sciences and humanities, remains primarily to constitute a *rite of passage* into the field. But the natural sciences, and increasingly the social sciences, possess more power than other disciplines to ensure that the candidate has internalized their ideological precepts. Since the process of knowledge acquisition is far more collective, the research project is the scene for the reproduction of the scientific laborer as much as the production of knowledge. The graduate student is obliged to accept a thesis director who not only advises on intellectual issues connected to the research topic, but selects the topic, supervises the methodology of inquiry, and manages the process itself. The graduate student is often completely under the domination of faculty, which not only approves the results, but intervenes at every step of the research to ensure that the intellectual canons of the scientific community are observed.[9]

Thus, the process of paradigm maintenance is built into the training of the graduate student in the sciences, since the student is already a regular member of the research team and the dissertation consists in little more than a report of results of one segment of the project, reconceptualized in academic terms so as to meet the requirements of the university. This merging of the academic process in the sciences with the research functions of the university serves to undermine any critical function of the academy at its foundations, because the social organization of scientific labor at the micro level conflates learning with performance. The concept of critical distance associated with the acquisition of theoretical knowledge is foreshortened by the increasing tendency of social and natural scientific graduate study to be integrated with research projects that are commissioned by the government and by private foundations or are indirectly (or directly) associated with product development for private industry. In many sciences where the experimental activity depends on the availability of fairly expensive equipment, the learning process for students takes place within the framework of research projects even if the research takes place on the university site. Access to up-to-date equipment is limited to those who participate as members of the research team.

This rule is equally true for the social sciences which increasingly employ computers for calculation and data gathering. As quantitative methods become more "normal" in the social sciences in comparison to older qualitative research methodologies such as ethnography, the whole enterprise of social scientific research, in both universities and private or government research institutions, is bound up with norms of natural science. Although it is still possible for students to enter the social sciences without being committed to its positivist objectives, it is increasingly difficult to complete graduate degrees in most major universities without participating in funded research projects. It is not only that the student is obliged to learn certain methods that imply scientific ideologies or paradigms. The basis upon which degrees are granted is increasingly bound up with such research as is currently funded by the government or private foundations. In sociology, economics, and political science, most funded research contains explicit or implicit policy outcomes. The mobilization of the social sciences for purposes of helping to shape and evalute social policy has become characteristic of research since World War II. The research is linked with instrumental ends that become disturbing for the social sciences, and much of the research is fragmented, narrow, and, in the last analysis, remains part of system maintenance. It is not uncommon for large grants to be made for social research on drug abuse, alcoholism, child abuse, and criminal behavior. These areas are selected by funding agencies on a relatively arbitrary basis: the topic has become politically significant; the "public is clamoring" for solutions. Federal, state, and local governments are under pressure to "do something." The task of research is either to *legitimate* a series of preselected responses by providing a "scientific" veneer for policies, or to help guide political and social agencies in the selection of options.

Very little of this applied research has the time, the money, or the inclination to challenge the presuppositions of these topics, although individual researchers sometimes become critics of the categories and institutions associated with these practices. For example, can the causes of crime be discovered on the basis of data that focus on individuals, or even local environments such as neighborhoods? Or, to be more far-reaching, does crime have a cause? Or, do we have an adequate definition of what we mean by crime, that is, is it a discontinuous set of behaviors and attitudes that can be separated out from the complex of activities that make up daily life within society?

Social research devoted to discovering the causes for criminal behavior often selects the wrong object (the individual), makes the assumption that crime can be abstracted from a larger series, and that the medical model (treatment) may be applied to its cure. Thus, much of the debate in what is called in sociology "deviance" centers on options that are circumscribed by these viewpoints. Students who select "deviance" as their area of sociological study are provided with funds to complete their dissertation through their participation in projects supported by government agencies. In many cases, their adviser is principal investigator for the project and, as in physics and biology, sets the parameter for the student's research; the student is then assigned to a particular segment of the project. Without this opportunity to work on the project, the student would be obliged to teach in undergraduate programs in order to earn sufficient money to write. Since many departments of sociology are oriented nearly completely to policy research, this student might not find an adviser, a necessary requisite for the dissertation. And, even if such a person could be found from among senior faculty, the burdens of being a teaching assistant or an adjunct for several courses might prove too difficult to allow the student to do much writing and research. The induction of social scientists into paradigm maintenance, while not yet as overwhelming as in the natural sciences, is becoming increasingly common.

The managers, of course, are responsible for another, equally central function of research. They define or supervise the *definition of the object of knowledge*. Their position in the specialized field of inquiry is such as to enable them to determine what is an appropriate object of investigation as well as a fundable field of inquiry. This object is often technically divided, not only by the broad divisions such as between chemistry and physics, but even more narrowly. Within each scientific discipline, which defines the *level* of the scientific object, there are subdisciplines that define the field within the level. Examples of such subdisciplines are quantum theory, high energy physics, and cosmology, and even these have further divisions. Some scientists are trying to determine the properties of causality in relation to these fields, others to determine the properties of specific metals, particles, or waves, to predict their behavior and control their action in a field.

From the cognitive perspective of scientists, these fields appear to be a logical set that is generated by the progressive march of science.

One "problem" is solved and another arises as a side effect of the solution. As we gain more knowledge, so many scientists believe, we are able to pose new questions that are now capable of solution by virtue of our prior knowledge. But even if this tells one part of the story, it remains an incomplete account. The division of scientific labor into sets of problems to be solved already implies that such categories of "prediction and control" form the presupposition of science. Its *purity* is mediated by the rational-purposive character of the scientific labor process. Problem-solving aproaches to science are forms of fragmentation of the object. The subject, the scientist, has inserted herself/himself into the field of observation from the outset. Thus, science is a form of purposive action that proceeds from a rationality that is conventional, but that appears to have the force of natural fact. The progressive subdivisions within each level, or scientific discipline, already embodies the history of scientific interventions into nature. Nature has now been defined and subdivided in accordance with the social will of the scientific community as well as the corporations and the elite sciences, is organized within the boundaries of the social and technical division of labor analogous to any industrial enterprise. The organization of scientific labor is not a socially neutral process. And the determination of the object of knowledge presupposes this division of labor, the technical parameters of which correspond to those of any other branch of the labor process. Just as the assembly line is organized according to definite rules, preeminently that work is divided into a series of discrete, repetitive operations, and the function of management appears merely that of coordination of these fragmented tasks, so scientific labor is organized in terms of an object of knowledge that remains discrete and fragmented and is given significance only by its managers.

If Althusser is correct that theories, experimental method, and technique constitute the mechanism of scientific knowledge—its internal apparatus, which becomes the criterion, in the last instance, of the truth of its discoveries—then one of the crucial tasks for an argument that wishes to oppose the separation of science, ideology, and social relations is to show the ideological character of the *experimental method*. The Copernican revolution has been supposed by most writers to consist in the first place in taking account of the observation of nature in the development of scientific theory. Its *sufficient* contribution to the history of science has been traditionally understood as the insistence that a scientific proposition can be verified (falsified) only by means of experiment. However, the experiment is an ambiguous activity.

The presupposition of all science, according to Karl Popper,[10] is that no proposition may be said to be scientific unless it can be falsified by observation or experiment. Thus, no statements about the material or social world may be taken seriously if it cannot be reduced to a form that may permit the methods of experimental science to operate. This a priori is part of the apparatus of modern science, is closely linked to what contemporary sciences mean by theory: a theory is a proposition that can be verified (falsified) by experiment/observation.

Now, "observation" is not the same as "experiment" in the commonsense use of these terms. Observation usually implies the reception by the senses of external phenomena. But, for science, the term "observation" is linked to experiment; the scientist observes effects of experiments, or to be more typical, records results of experiments in numerical form. A theory is merely a hypothesis that can be quantified by means of observing effects recorded in numerical form. In these terms, the scientific observation of human behavior, no less than the observation of subatomic particles, is concerned with effects that are presumed to be accurate reflections of the intrinsic properties of the object.

The ambiguity of the experimental method becomes apparent when we remember that its preconditions include decisions about how the object of knowledge is to be constituted, not only by its classification according to levels and fields, but also in the setting up by the scientist of the *boundaries* of the experimental field. It is commonplace among scientists to reduce the number of variables to be observed to the *least number possible* because the ability to make sense of the observations is understood to be a function of this reduction. Thus, the second condition, after its definition, is to delimit the observational field by decontextualizing the object so as to facilitate the project of predicting and controlling behavior. I do not wish to dispute this procedure for the moment, merely to point out that such decontextualization implies that the category of interpretability has already changed the concept of observation to conform to the requirement of prediction and control. What is observed is the effects of an experiment on an object under certain conditions determined by the scientist and his/her apparatus. Of course, the investigator knows that decontextualization limits the claims resulting from the experiment. He/she often tries to model the environment within which the organism or object interacts within the framework of experiment. But the requirement for at least *statistical probability* for prediction and control restricts the degree to which the experimental method may be taken as an "objective" mechanism of knowledge.

Experiment is a type of human activity that is an *intervention* by the investigative team into "objective" processes. The forms of intervention entail the establishment of the context within which the object is observed; the machinery employed to produce the effects from which inferences are made according to theoretical presuppositions and from the rational-purposive basis of all modern knowledge. Werner Heisenberg's indeterminacy principle attempted to preserve the objectivity of scientific theory by inserting the observer into the observation field and arguing for a science that recognized the problematic character of the results and included these in its calculations, procedures, and results. The instrument of observation must be taken into account in the measurement of the object as well as in the determination of its position, according to Heisenberg.

The philosophical implications of this argument have disturbed scientists and philosophers because claims of science to precise calculation of the movement of particles, and the integrity of the experimental method itself have been challenged by the revelation that the object of knowledge may be our relation with the object, not the thing itself. As I have shown in chapter 9, most physicists as well as philosophers have now been obliged to acknowledge that nature offers more than one option, depending on the theoretical framework in which the experiment is conducted. There is said to be a "degree of truth" in these determinate options, now called interference of possibilities. The framework adopted by the scientist to understand the results of experiments will make the result itself.

Largely because of the development of quantum theory, modern theoretical physics, at least, has been forced to work with uncertainties. The experiment is one valid means of knowing the external world, *provided* we understand that knowledge depends on its theoretical presuppositions, which are taken as no more than possible explanations for the characteristics of matter.[11]

The crisis in science has occurred because of the challenge posed by Heisenberg and others to its positivist assumptions, particularly the notion that the relation of the observer to the observed is unproblematic. When Heisenberg reduced to mathematical language the simple notion that what is observed depends on the apparatus of observation, that is, depends on experiment, the scientificity of science was thrown into question. In the 1930s and 1940s, some, like James Jeans and Arthur

Eddington, took these discoveries to mean that nature was, at the bottom, a mental construct, or at least was unknowable. Others understood Heisenberg's principle as an invocation to further limit the variables, to further narrow their field of observation, or to calculate the impact of the mechanism of experiment and make their results more probabilistic.

Since we can only calculate effects and infer causes, control and predict behavior by constituting the object of knowledge theoretically and changing it in the process of our intervention, I argue that the so-called laws of nature are better described as *laws of science*. Scientific theory describes the relation of humans to the object of knowledge, not the objects themselves, taken at a distance. Further, our knowledge of effects is always mediated by the logic of scientific discovery, e.g., by its concepts of causality as a part of the apparatus of discovery.

Linear causality assumes that the relation of cause and effect can be expressed as a function of temporal succession. Owing to recent developments in quantum mechanics, we can postulate that it is possible to know the effects of absent causes; that is, speaking metaphorically, effects may anticipate causes so that our perception of them may precede the physical occurrence of a "cause." The hypothesis that challenges our conventional conception of linear time and causality and that asserts the possibility of time's reversal also raises the question of the degree to which the concept of "time's arrow" is inherent in all scientific theory. If these experiments are successful, the conclusions about the way time as "clock-time" has been constituted historically will be open to question. We will have "proved" by means of experiment what has long been suspected by philosophers, literary and social critics: that time is, in part, a conventional construction, its segmentation into hours and minutes a product of the need for industrial discipline, for rational organization of social labor in the early bourgeois epoch. To be sure, we may discover that time's arrow is inverted and that the evolutionary assumption of the progressive nature of time possesses only a "degree of truth" insofar as it is *one possible option* in answer to the question about the complexity of temporality. But so was the older theory of time and causation "experimentally" verified.

Since our knowledge of the external world is gained principally by theoretical construction and experiment, and since the purpose of experiment is not the verification of hypotheses in an abstract sense but the prediction and control of external phenomena, the masked basis of experiment is domination of nature and of humans. Prediction and control are presupposed in the language of science–mathematics and statistics. Mathematics is understood by scientists to be necessary to remove the ambiguities of ordinary language, to gain a degree of precision in the description of results and the inferences about the internal construction of the objects of knowledge. There is an "intrinsic uncertainty in the meaning of words," in Heisenberg's view,[12] a conclusion that already prompted Aristotle's solution to this problem by investigating the formal structure of language and "conclusions and deductions" independent of the content. Aristotle thought that formal logic would lay the "solid basis for scientific thinking."[13] That "solid basis" could be obtained only when form was abstracted from content and when the relations of these forms were expressed symbolically. The separation of quantity from quality is instrinsic to the constitution of the physical and chemical object in the reduction of all observations to their mathematical, that is, symbolic, form. I have argued that the logic of these relations has historically obeyed rules that are not ideologically innocent: they are linked to the attempt to achieve dominion over nature in order to satisfy human ends. In turn, these ends are mediated by social relations, by the further mediation provided by the scientific community, which operates according to rules of evidence and canons of truth that are part of the legacy of Western logical thought.

The implications of modern developments in theoretical physics have called into question the logical tradition because new ideas such as "coexistent states" and "uncertainty of possibilities"[14] seem incongruous with the loss of contradiction and mutual exclusion. Heisenberg has discovered that the logical principles that underly mathematical symbolization are no longer in a relation of simple correspondence to nature. The concepts derive from the theoretical ambiguity of physics, its noncorrespondence with nature, which Heisenberg has called the "limits of the correlation" between the older languages of classical physics and the new concepts soluble only if

> one confines the language to the description of facts, i.e., experimental results. However, if one wishes to speak of the atomic particles themselves, one must either use the mathematical scheme as the only supplement to natural language or one must combine it with a language that made use of a modified logic or of no well defined logic at all.[15]

When science digs into the areas of "potentialities or possibilities," it has left the world of experimentally adduced facts or things, and the old logic must be abandoned.

If natural language is filled with meanings that defy precision and thus make control and prediction difficult, the language of control, mathematics, contains within it a limit on science. The questions within scientific theory that imply a logic that defies the formal categories of space and time and that have renounced, however implicitly, the rational-purposive, i.e., experimental, basis of investigation, lead to the discovery that ordinary language, precisely because of its undecidability, is more suited to these issues. Experimental methods, the results of which are formulated in purely quantitative terms, may not be abandoned but must be placed in a subordinate position to speculative science frankly concerned with a reconceptualization of all the fundamental presuppositions of the old science.

Despite his intentions to preserve the heuristic value of experimental, mathematical science, Heisenberg opened a Pandora's box: he began the process by which physical science must become reflexive, that is, must recognize its own presuppositions as interested, that is, as ideological. A scientific practice that forces research to subordinate itself to a normative a priori of prediction and control is prevented from reformulating the object of knowledge in ways that may be suggested by the anomalies arising from its own methodology. The presence of measurement in the observational field, that is, the objectivity of mediation, the challenge to time's forward direction, which questions conventional theories of cause and effect as a temporal series, the self-limiting character of quantitative a prioris to research—these may not result in the formulation of a new paradigm as long as science remains entwined with the requirements of domination. Such questions will remain on the margins of normal science, relegated to the purgatory of projects that are poorly funded, or worse, regarded as secondary priorities in the organization of scientific research, subordinate to defense requirements, or the construction of bigger and more efficient machinery to "treat" conditions such as heart disease and cancer that have become a concomitant of industrialization. Of course, the establishment of priorities for scientific research is common to all nation-states that manage science and technology. These priorities embody the economic and political programs of these states, are formulated in terms of technical problems that require solution for the benefit of the entire society. But, as certain branches of science stumble on anomalies that cannot be fitted into the existing *ideological* presuppositions of science, not merely the particular paradigm, the question of the internal processes by which new paradigms are formulated is thrown into doubt.

Take the question of medical science. In the United States, millions of dollars every year are appropriated by government and private corporations to find "treatments" for diseases whose "causes" are related to the entire matrix of social existence within our society. The object of knowledge is defined as the human body, not principally the quality of social life. Research institutions employ a virtual army of scientists, engineers, and technicians to discover chemical treatments for these diseases. The experimental method dictates that medical treatments be tested on animals under laboratory conditions. Researchers also test hypotheses about the causes of the diseases under the same conditions. Causal factors stemming from the "external" environment, such as the relationship of cigarettes and industrial pollution to cancer and the relationship of stress produced by insecurity and tension arising from working conditions to heart disease, are acknowledged but studiously avoided as the object of knowledge. The scientific object is the *effects* of these conditions on the human body, which, in the last instance, constitutes the limit on scientific inquiry related to medicine. For medical science is confined to the study of the human body, even if practitioners may take external factors into account. Thus, the question of defining the causes and cures of infirmities finds its parameters within the characteristics of the body itself. As a result, medical science, more or less willfully, defines its tasks in terms of ameliorating the effects of a social system that produces disease. In effect, within a system where rationality is determined, in part, by the criterion of capital accumulation, the rationality of scientific research on disease cannot be in conflict with the larger rationality without risk.

Even late capitalist and state socialist science has been required to study the degree to which the rationality of capital accumulation affects the physical environment. But, because of the social divisions within the sciences, these questions are not the province of medical research. If studies of the impact on the biosphere of the use of some types of spray cans show that these devices should be banned, government regulation may prevent the entire species from becoming extinct. Yet, medical science may not adequately study the degree to which repetitive labor performed at a killing pace should be stopped to prevent heart disease, or whether certain materials that pollute the work environment with carcinogens, such as some types of plastics, should be declared illegal. Funds for medical research dealing with the prevention of disease by social means are far less available than funds for research treating the individual as the object of knowledge. Medical research in the United States is confined almost wholly to discovering treatments for diseases already presumed inevitable. Occupational health-related research comes regularly under severe political attack by large corporations that may be forced to change labor processes and equipment, both with respect to what is produced and with respect to how production is conducted. Managers of medical research projects internalize the attacks by avoiding such subjects as may be considered unlikely by funding sources, by media, and finally by their own professional communities.

Nevertheless, some preventive medical research goes on, and its results are published and sometimes disseminated beyond the networks of professional science. But environmentally related research of all types occupies the margins of medical science. It is not considered "sexy," except when its results show that specialization and segmentation have gone too far, that it is sometimes in capital's interest to pull back, either because popular movements have exerted political pressure or because the legitimacy of the state and the corporations has been undermined, as was the case with the food scandals of the early twentieth century.

The case of medical science illustrates the ideological content of science, the degree to which science is constituted by social relations. First, the object of knowledge perpetuates the division between the body, the mind, and the external environment. Second, the technical division of labor within medicine segments parts of the body so that research is often directed to discrete characteristics of each part that may be susceptible to disease, without considering the relationship between these and the body taken as a whole. There are exceptions to the extreme rationalization of the body into fragments for the purposes of study; but, for the most part, medical science accepts the invocation to reduce the number of variables within its experimental methodology to a minimum. This reduction of variability, of course, becomes ideological insofar as it induces an epistemological perspective which does not conform to the way in which the human being acts in the world, but which sets internalist boundaries that conform to the precepts of normal scientific inquiry. Thus, the chemistry of the body is a legitimate object; the work environment is not.

Third, medical science, by dealing with effects of absent causes, produces treatments that are often detrimental to the rest of the body. The "side effects" of medication are understood by specialists as unintended consequences of certain remedies. Since the object has been so narrowly construed, chemical treatments often produce more problems than they resolve. And, since the mind/body split has, for the most part, been preserved in medical research, or, to be more precise, the problems of the mind are increasingly reduced to the body, mental illness is treated as a problem of chemical imbalances, skewed metabolism, nutritional aberrations, or some other type of physical mutation. For a brief period after World War II, when ego psychology became fashionable among middle strata groups, the reverse tendency was observed. For numerous physicians as well as social workers, teachers, and nonprofessionals, many diseases could be traced to "psychosomatic" causes. The psyche was believed to produce everything from the common cold to heart disease. But this psychological reductionism was just the other side of physical reductionism.

Of course, there is a great deal of validity in both perspectives. The division of labor within science has blinded each to the dialectical relation of the body to the mind, their indissolubility. Most medical research cannot conceptualize the notion of a body-subject, that is, a body that acts in the world and has both conscious/unconscious sides which constitute physical as well as mental functions. Even if such "unscientific" conceptions could be formulated, research that began from this relationship would encounter difficulties gaining support, because of the difficulty in reducing such conceptions to discrete experiments and in quantifying the results.[16]

Yet, the consequence of our understanding of pathology resulting from the mind/body split, the methods of normal contemporary science, and the reductionist ideologies has been to perpetuate the "mysteries of the organism." Despite the fact that billions of dollars are allocated for research that seeks to develop elaborate chemical treatments, which may disturb the internal relations within the body, or to construct machines to deal with symptoms, medical science is imprisoned by its own ideology. It still operates on the basis of linear causality and has been unable to find "cures" for diseases that have no simple origin. Yet, there is no prospect for immediate change: large-scale research will continue to search within the fragmented body for the secrets of disease, for the one vaccine or chemical cure for our maladies, in the hopes of avoiding social or preventive medicines.[17]

Of course, as critics have shown,[18] the expansion of a misdirected medical research establishment is predicated upon the political pressure of large drug and machinery corporations, in alliance with hospital associations, research institutes, and the dominant forces in the medical profession who control these medical institutions. Yet, it would be a mistake to attribute *all* the problems of scientific research to the power of these institutions. Political and economic power must be linked to the preproduction of an ideology of science, in this case ideological presuppositions for the discovery of causes and treatments of diseases. For, even if "profit" was taken out of medicine, it does not follow that science would immediately turn to a more integrated, dialectical view of its object. As the experience of those countries in which the private ownership of capital has been abolished shows, the transformation of science and technology entails the transformation of the epistemological as well as the material foundations of science.

It is time to summarize the discussion of the relation of science and ideology. I shall follow it with a discussion of the problem of the truth in relation to ideology. My central thesis that science may not be considered a separate discourse from ideology depends on the following propositions:

1. The concept of the science/ideology antinomy is itself ideological because it fails to comprehend that all knowledge is a form of social relations and is discursively constituted. Within late capitalism and state socialism, these relations are organized according to a division of labor (principally the division between intellectual and manual labor). The rational-purposive basis of social production under both capitalism and state socialism means that science is a labor process as well as an ideology whose truth claims are entwined with the interest of domination. This ideological function is revealed in at least five ways:

a. The choice of the object of knowledge or inquiry is determined by the complex of economic and political alliances made between scientific institutions, corporations, and the state. These have a virtual monopoly of the means of scientific production and dictate, more or less completely, those projects that may be supported. This monopoly does not preclude funding occasional projects that depart, in one way or another (although never in toto), from the canon of normal science. But these are always relegated to exceptions, whose function is to legitimate normal, incremental science. Contrary to Kuhn's ascription of paradigmatic change to internal developments within science, particularly the relation of experimental results to accepted scientific laws, since science as well as technology are entwined with the relations of power, these changes will always be constrained and configured by social and political influences.

b. But although the end of the alliance of science with corporate and state institutions and technology is a necessary condition for an emancipatory scientific theory and practice, it is not a sufficient condition. As Marcuse has reminded us, after centuries of reproduction of society according to division of classes and the division of labor, we have internalized, even introjected, domination.[19] Specifically, this means that the *object of knowledge* has been constituted according to the social division of scientific labor (physics, chemistry, biology, psychology, and human sciences), and these categories are considered separate objective levels of natural and social reality, rather than necessary but partial abstractions from the totality of social existence.

Second, the object is constituted by a technical division of labor, where specialization of tasks within a field determines the perception of the investigator, such that a fragment is defined as the object, taken out of its natural context for the purposes of study. The result of this fragmentation is to impute characteristics to an object, discovered under certain circumstances but taken as intrinsic to the object.

c. Ways of knowing in contemporary science are discursively produced. The experimental method is not an "observation of nature" free of presuppositions. It assumes (i) that the abstraction of the object from its natural context is unproblematic, (ii) the intervention of the observer into the observed both by measurements that produce certain "effects" from which inferences are made, and by producing "causes" of the effects, such as by applying electrical currents of certain frequencies and magnitudes in physics, by inoculating animals to produce symptoms from which causes are inferred, and so on. The assumption of *intervention* is only part of the reflexivity of science to the extent that it tries to "correct" for its ingression. But the experimental method must be recognized as informed by its presupposition of intention, that is, the control of nature and humans so that their action may be predicted. Its claim to value neutrality is vitiated by the degree to which behavioral hypotheses or symptomatic readings are endemic to scientific inquiry.

d. *The form of the results* of science is historically located in the formal logic of Aristotle, which tried to understand forms of nature irrespective of their particular content. Galileo's doctrine that the "book of nature" was written in the language of mathematics was the logical culmination of this attempt to reduce nature to a form that permitted its predictability and control. But mathematics as the hegemonic form of scientific discourse constitutes a boundary to science. To the extent that knowledge is inextricably linked to language discourse, it generates a set, an internal unity that excludes those questions that require natural language and do not rationalize anomalies in terms that reduce *signification* to their technological dimensions. The separation of quantity from quality is linked to the divorce between means and ends. For example, a society wishing to "use" the results of science to serve class-determined or class-mediated human ends requires science to take a form in which it may be infinitely fungible. Even if the ostensible purposes of research are not instrumental in the immediate uses of results, the quantification of results lends technical significance to the research. Under present conditions, such technical uses are almost always connected to social and natural domination or profit.

e. The development of conventions that are legitimated and reproduced within the scientific community determine what is called truth,
and what is acceptable as science. These conventions are not separate
from the paradigms of scientific knowledge, which may be seen as
both the presuppositions and the outcome of the systemic requirement that certain traditions of inquiry, methods, and forms of results
be followed. The observance of the conventions is just the other side
of the rituals and credentials that are preconditions for entrance into
the scientific community. Thus, no less than any other specialization in
the division of labor, the scientific community regulates entrance
requirements, conditions for membership maintenance, and power
relations within the community. Since the entire machine of scientific
investigation presupposes the social relations of the prevailing mode
of production, the project of transforming science must be consonant
with the project of transforming social relations.

2. The struggle for the transformation of ideology into critical science is a project in our period which is made possible by recent philosophical work. It proceeds on the foundation that the critique of all
presuppositions of science and ideology must be the only absolute
principle of science. Thus, a reflexive science is radically opposed to
"normal" science that takes its ethical, epistemological, and methodological foundations for granted. Althusser's Freudian metaphors have
validity insofar as change can take place only when what occurs behind
our backs is brought to our conscious view, that is, when science takes
itself as its object.

Since the turn of the century, when it became apparent that capitalism had undergone a fundamental transformation in its structure and
material/ideological practices, those who wished to preserve Marxism
as a living socialist philosophy and critique of capitalist society have
tried to comprehend the character of the changes and to discover their
implications for Marxism itself. The Frankfurt School, which tried to
integrate Freud and Weber with a Marxist framework, made a serious
effort to challenge the theoretical presuppositions of Marxism in the
light of historical changes.[20] More recently, others have claimed that
the historical conditions that produced Marxism have been surpassed
by the integration of the working class into late capitalist society. As I
have shown above in my critique of Habermas, such positions do not
constitute a critique of Marxism from within but abolish its foundations
by constituting a new scientific object. At the same time, whether in its
French manifestations (the new philosophers) or its German expression, post-Marxism, since it cannot identify the motive force of history
because it has abandoned the category altogether, is left with a new

positivism. For those who have abandoned agency as a vital historical category, there are no conditions for the transformation of society. All critical theory may do is refine the categories of the eternal, reified present and wait for the "will" of humans to take care of the rest. But Gouldner is right to point to the historicity of Marxism, especially its nineteenth-century insistence on the autonomy of science. For a critical theory cannot be critical of itself if its highest aspiration remains to achieve scientificity in the traditional sense. It is this aspect of Marxism that has generated the post-Marxist antinomies as much as the changed social conditions within late capitalism and the deformations of state socialist societies.

Historical agents are not necessarily personified in particular social categories, for example, industrial workers, scientists, or women. To approach the question in this way ignores the critique made by contemporary philosophy and social theory of Marxism, or, to be more exact, merely replaces workers with women, the Third World, or whatever. What is at issue here is the discovery that agency is produced discursively and that people attach themselves to certain discourse because these offer a vision of an alternative to the system of production and reproduction of social life. I hope to show in a future work that the emergent agency today defines itself against the prevailing rationality, not within it, and this characteristic precludes the substitution of one group of personified agents for another, or the reform of the management system governing the production of either knowledge or things. For what is at stake in the struggle over historicity is the challenge to the dual system of domination over nature and humans, of which the prevailing division of labor is a crucial element, but also the entire system of social production and appropriation.

In short, the discourse we need to elaborate is already present in the various critiques of Marxism, liberalism, and conventional science.

3. Science and ideology have the same object—the material world and social relations that have been produced by social labor considered historically. All relations with this world are mediated by material and social structures and the concept of unmediated relations that may produce knowledges that are independent of the social processes by which they were acquired is the ideology of science that draws its inspiration from the bourgeois protest against feudalism. Since the relations of science, magic, and religion are internal to each other because they all purport to offer adequate explanations for natural and social phenomena, it is rank ethnocentrism to claim that one may be privileged over the others without specifying the social-historical setting which under capitalism tends to subsume all discourse under its

system of rational-purposive action. Within this framework, modern science becomes a partner of industrialization, whose social consequences have both liberated us from the brute struggle with nature, but only partially, and imprisoned us in a logic of domination and degradation.

4. If science is ideological, what is truth? My argument leads to the conclusion that truth is the critical exposition of the relations of humans to nature within a developing, historically mediated, context. Within this critical project, the form of appearance of social relations is a unidimensional rationality that implies particular conceptions of space and time, causality, and construes the object of knowledge both discursively and socially. Thus, the laws of science *are* the laws of nature, providing one specifies the system of rationality within which they are discovered. Their reified "natural" appearance remains opaque without the weapons of criticism: we can discover the external world as a product of the collective labor of centuries, not by observation, but by the construction of a series of concepts that are contradictory to the certainty of the senses that only report the surfaces. I am not claiming that critical science exists. Normal science, particularly the study of microphenomena, has also contained a critical dimension but is not, itself, critical, especially of its own presuppositions, for example, the discovery of the atom, which was theorized long before it was "observed," and the hypothesis of the psychic structure, whose existence could only be inferred from a "reading" of symptoms. In the study of society, Marx's theory of surplus value is, above all, a logical discovery; examples given to illustrate it are mostly "made up." This does not imply that Marx failed to do detailed historical and empirical economic research. My point is that Marx's theoretical framework was constituted by the relation of history to its conceptual foundations in dialectical philosophy.

But normal science is enslaved by its positivist presuppositions. Its *structure* of inquiry as much as its picture of the world makes it *positive* and *descriptive* rather than critical. For science and technology within its current framework to challenge its foundations would be tantamount to its self-immolation because it would be required to separate itself from the economic, social, and political conditions for its work as much as its epistemology.

5. What we call science and what we call ideology are distinguished *in practice* by the degree to which the acceptance of the ideology of science exempts it from being called ideology. "Ideology" is the general form of all human thought because social relations coded as discourse not only mediate but constitute thought itself—the forms of

discourse refer to the material world in ways that are demarcated from each other yet all appear as truth. The various moments of discourse appear as common sense. Ideology, in general, takes the form of discursive hegemony. The most general sense in which hegemony is exercised is the internalization of an entire conception of the material and social world by the population of a given social formation. The most powerful form of hegemony is inscribed in science, which is constituted both by its paradigms at the theoretical level and by the complex of institutions and material practices that are reproduced as self-focused and self-evident facts. Scientific ideology refers to a totality of forms of conceptualization that not only are reproduced as knowledge of the material world but take on the appearance of the material world, such that the opacity of things has the force of nature. The laws of science are perceived as properties of nature, bereft of their material and historical determinations and mediations.

Some Theses on Science and Technology

1. There are three key aspects of capitalism as a social formation: the universalization of the commodity form over all production; that is, production for exchange and for profit becomes the dominant rationale for investment of labor and money capital; the Industrial Revolution, in which machines become the central productive force, together with the technologies of the organization of labor; and urbanization, which is usually distinguished from medieval or ancient cities which were centers for regional and world trade in food. The modern city, now a site of industrial production, becomes the space of accumulation of capital whose main form is *fixed* capital, or machines and buildings.

Although scientific and technological knowledge play a crucial role in the development of capitalism, it is not until the mid-nineteenth century that intellectual labor (whose sole possession is knowledge) becomes a major productive force, partly displacing both craft and unskilled types of manual labor. In his *Grundrisse*, Marx already takes notice of this shift, which is still nascent even in England. The first half of the twentieth century is marked by the ascendancy of knowledge (if not of a new intellectual class) as the premier productive force. This transformation is manifest not only in the rise of the engineer as a key player in production, but by the advent of *scientific* management which seeks to both subordinate the manual labor force to its direction, and to impose systematic methods of control and organization over the state and the corporations. Management is an *ideology* to the degree that it seeks power on the basis of a new rationality that claims the mantle of science in affairs outside the laboratory and other traditional sites of scientific work; it is also a potentially competing community that challenges the hegemony of capital and the state, or, to be more exact, wants to impose its unique discourse on these hitherto autonomous sites.

By the first half of the twentieth century, neither science and technology nor management, the two most important forms of intellectual labor, had as yet captured more than the workplace, more than the sphere of production of things. Still to be conquered are the military and other state institutions, the crucial economic institutions of capital, principally financial agents and, more globally, cultural life.

2. The past thirty years have been characterized by the emergence of technology as the discursive formation that constitutes the dominant space of dispersion in industrial society. This entails massive impositions in (nearly) all social relations or discursive formations. Here, I define "discursive formation," following Foucault, as *a group of statements that form a unity*. These statements include the formation of the knowledge object, the rules for investigation as well as the specific claims about the objects. To say that the discourse of technology is dispersed everywhere does not signifiy a kind of technological determinism in the older meaning of the term; what is commonly understood as "technology," namely, machines or tools, do *not* constitute the determining element of social life. I am not arguing that means constitute ends in the modern world, or that the forces of production determine social relations. Instead, I understand technology as a constituted totality that drives production; more broadly, it is a type of rationality that (in Heidegger's felicitous phrase) "enframes" scientific thought and also constitutes a *sensorium*, a field of perception. Further, technology is not merely an extension of human powers, "mediating" our relation to the external world, or nature, but has penetrated human character structure. That is, one may not sharply distinguish "emotions," "feelings," and other terms of interiority from the technological sensorium, just as science has become virtually identical with technology. This statement implies that we are *wired*, that the mass communication media, perhaps the most powerful technological achievement of this century, are neither extensions of collective human powers, nor an awesome otherness standing against us, but occupy the space of social life such that no relations—those within the psychic structure, or between individuals, or among and between collectivities—can escape the enframing of technology.

3. Technology is not an epistēmē alone, although it is surely entwined with all forms of knowledge, including language and art; it is the discourse that modifies, when it does not entirely shape, objects as well as the rules of knowledge formation.

4. Thus, to speak of technology is to speak of culture; for the objects of the social world are not only constituted socially, in the sense that "natural" objects have been so severely mediated that to speak of physics, biology, etc., as direct causal agents in the social matrix appears naive. More to the point, technology as discourse defines social construction.

5. In turn, natural objects are also socially constructed. It is not a question of whether these natural objects, or, to be more precise, the objects of natural scientific knowledge, exist independently of the act of knowing. This question is answered by the assumption of "real" time as opposed to the presupposition, common among neo-Kantians, that time always has a referent, that temporality is therefore a relative, not an unconditioned, category. Surely, the earth evolved long before life on earth. The question is whether objects of natural scientific knowledge are constituted outside the social field. If this is possible, we can assume that science or art may develop procedures that effectively neutralize the effects emanating from the means by which we produce knowledge/art. Performance art may be such an attempt. The artist foregoes the use of tools—brushes, chisels, cameras, as well as the raw materials (paper, stone, film) upon which are inscribed shapes, colors, lines. Performance art, like more conventional theater, attempts to restore to the body its autonomous space. Communication is no longer mediated by things. This might satisfy the antitechnological impulse were it not for the body itself, which is already incorporated into the technological sensorium. Movement is never natural; it is enframed in technology.

6. Science invented the algorithm of reproducible, testable experiments to screen out the social world, including the prejudices of the investigator, relations of domination and subordination in the laboratory, the pressures of the scientific community to conform to established paradigms, and so on. In short, the progress of science is presumed guaranteed by the requirement that any statement about the object be falsifiable by standard experimentation (Popper). However, if it can be shown that the experimental method presupposes a technological telos, the effort to achieve value neutrality in the sciences is unachievable unless this telos is, itself, considered value neutral.

7. The fact is, science and technology have been constituted as discursive formations, which, by definition, exclude the social and cul-

tural world as relevant influences in knowledge production. Gaston Bachelard and Louis Althusser go so far as to claim that science is constituted by its separation from common sense (even though Stephen Toulmin defines science as "organized common sense.") If organization may be related to technology as one of its modes of existence, science becomes a kind of technology. Popper's definition, cited earlier, is to specify the criticism through organization as the metarule for achieving scientific truth, if by that term we mean knowledge uncontaminated by ideology. This is nearly identical to Bachelard's program for verification, although Bachelard adds the requirement that criticism extend to the axiomatic foundations of science, not merely to theories operating within a given paradigm. But, unlike Bachelard, Popper in his later work abandons the traditional epistemological assumption that knowledge is constituted through the interaction of objects to be known and the knower. For Popper, while this process may be entirely accurate as a description of how knowledge is gained, the "third world" of theories and propositions, which he calls epistemologies, stands independent of both the objects and the subjects who know them. Knowledge takes on an ontological existence and thereby itself becomes *objective*. This is, for Popper, the most reliable of the three worlds because its existence no longer depends on the fallible processes of acquisition. The theories have been tested, and the results have become autonomous. Clearly, Popper hopes to avoid the messiness of epistemological questions. By separating the processes of the production of scientific knowledge from their structure, Popper wants to remove behaviorist or psychological aspects from judgments about the contents of scientific knowledge. But, as David Bloor has pointed out in his critique of Frege, psychology is not really at issue when addressing the production of knowledge. More to the point is the question of social determinants, mediations, etc., of scientific knowledge. What we want to know is whether the products of the knowledge process, the representations called theories, propositions, and statements, can be shown to be independent of discourses of the social, cultural, ideological. We want to ask whether the products of science, notwithstanding the social processes that combine to construct them, are true. That is, can the thesis of realism be maintained with respect to scientific knowledge? The issue is not whether theories refer to an object. The question is whether that object is free of social construction, which, even if admitted, leaves no traces.

Bloor focuses his argument on mathematics, which Frege, following Kant, regards as not an abstraction from objects but as derived from pure intuition. Popper, following Brouwer's suggestion, disputes this by asserting that mathematics uses discursive, particularly logical, thought.[21] He has not escaped Kant's invocation of the mental sources of mathematical knowledge. Bloor's argument is that while Frege is right to reject the contention that the products of mathematics are derived by abstraction from material objects (such as pebbles), he also rejects the idea that arithmetic has no source outside mind. Bloor substitutes social for psychological and "material" determination of matematical knowledge. He adduces several examples culled from the literature on the philosophy of mathematics to show that many ideas correspond not to scientific necessity but to cultural influences. In this, he follows Spengler's and Wittgenstein's concept that different cultures will produce different mathematical concepts. We can show that evolving concepts in the theory of numbers as well as other branches of mathematics vary according to influences already present in a culture. Taking the evidence of Evans-Prichard's famous study of the Azande, Bloor disputes the idea that there can be only one logic. He argues that the refusal of the Azande to acknowledge the contradiction in their culture between the assumption that all are witches and the fact that this is not the case is not due to "a lack of theoretical interest" in the subject of logical contradictions. Rather, according to Bloor, the "logic" of the Azande in deflecting the contradiction is no different from "Western" logic that regularly denies that bomber pilots are murderers even though a murderer is defined as someone who deliberately kills someone else. From the point of view of the society within which pilots function, they are seen as persons who perform their duty out of loyalty to the government. From the point of view of the victim, however, they are murderers since the victim experiences only the danger of death without the referent of duty to government. Bloor invokes this example to demonstrate that social interest mediates logic, and that the assumption that there is a "pure" logic separate from culture is false.

We don't have to stop with Bloor's perspectivist refutation of the notion of objective knowledge. Scientific experiment may be shown to derive from a specific conception of "value," that of intervention into nature as the road to reliable knowledge. Or we might cite the problematic object of scientific knowledge, that the construction of the object has both cultural and political presuppositions. The point is that neither logic nor mathematics escapes the "contamination" of the social. What remains to be discovered is the nature of that contamination, its relation to the truth value of scientific propositions, and, most particularly, the role of science and technology in the social world.

8. Whatever our solution to the problems posed in Thesis 7, we are faced with the fact that science and technology remain subjects for intense discussion and disputation. These debates must be explained, for, since the nineteenth century, the advances in scientific knowledge have been supposed to eliminate the need for metaphysics, including a metaphysics whose field of discourse is science. Science should be able to "speak for itself" if its objects and methods preempt metacommentaries. All we need do, according to the most optimistic advocates of this position, is *do science*. Problems posed in the course of scientific inquiry/discovery should be subject only to solutions arrived at by those qualified to perform scientific activity. It is true that after Kant, with the exception of ethics, all philosophy is a discourse on science.[22] Moreover, with the advent of logical empiricism and logical atomism, schools associated with the rise of what is sometimes called nonclassical physics—the physics associated with relativity theory and quantum mechanics, although by no means identical with them—problems in philosophy existed only to clarify the actual results of scientific inquiry at the level of language and discourse.

The distinction between philosophy, long viewed as a speculative inquiry, and natural science, in which speculation is strictly limited by scientific method to preexperimental hypothesis, has become increasingly blurred. For, just as philosophers have confined their work to interpreting scientific outcomes, scientists have felt constrained to become philosophers to make sense of their own work. Apart from a few relatively isolated figures (Alfred North Whitehead comes to mind), philosophy has been transformed into metascience, illuminating the most general principles that are said to derive from scientific practices.[23] Philosophy is no longer magisterial in its ambition to be the most general of the sciences; philosophers either have become metamathematicians insofar as logical inquiry defines their intervention, or have tried to enter the still unresolved disputes within science.

Speculative philosophy refuses to go away, despite its intentions, because physics, mathematics, and biology, the three sciences that have generated the controversy, are themselves rent. Nonclassical physics continues to make discoveries, but the actors cannot agree on what they have found. Or, to be fair, they are unable to render clearly the significance of the products of scientific discovery because the framework of understanding is in contention.

This framework is by no means self-evident: in the torrent of commentaries on the significance of quantum mechanics and relativity theory, one may discern wide differences, even among those most responsible for the discoveries. This is another way of saying that science proves incapable of speaking for itself with a unified voice. This is not particularly new. Since the so-called Copernican revolution, both philosophers and scientists have felt the need to interpret the results of inquiry.

9. Their central interpretations are these:

a. Scientific theories are descriptions dictated by the real world. Taken together, they constitute a picture of the world as it actually exists "in nature." The quotation marks are used because there is a question concerning what we mean by nature. Since Copernican/Newtonian science insists that objects and relations must be explained by purely natural causes, the conception of nature means simply the object of scientific knowledge taken in its totality. For these purposes, we need not address whether the real is identical with something called "material" reality. For realists who have gone beyond the naive implications of materialism, the main requirement is that a theory refers to something outside itself. Referentiality is regarded as the *sine qua non* of theorizing.

b. There are no laws of nature, only laws of science. In this way of seeing, science is defined by the rules of inquiry, mediated by the level of development of (i) its theoretical apparatus and (ii) the material apparatus, i.e., technologies of inquiry. Since these mediations inevitably overdetermine what we mean by fact (that is, facts are always both theory and technology dependent), no individual fact can be falsified/verified by means of experiment without taking into account its referent(s) which constitute themselves as traditions that have become conventions of inquiry.

c. The significance of quantum mechanics and relativity theory is that scientific theories refer to the relation between the "field" of inquiry and the tools of measurement, on the one hand, and to subject/object relations, on the other. The first formulation takes the position that theory is "objective" knowledge in Popper's mode; the second reformulates the proposition in terms of interaction and mutual determination, in the manner of hermeneutic/dialectical analysis. In either case, the intervention of the "observer" is inescapable in the constitution of the field (I use the term "field" rather than "object" because "facts" refer to relations and relations of relations, rather than things). The second and third interpretations assume that uncertainty, which results from overdetermination, is both the consequence of shifting referentiality of the field and the referent of the social mediated by the subjects who engage in the process of scientific inquiry. That scientific facts are constituted socially as well as technologically infers that science is an *intervention* in both senses. This means that the object is constituted as a relation between the ends of scientific inquiry and its means. As Fleck demonstrates, facts are produced, not discovered.

10. To the degree that human sciences aspire to scientificity, they have felt obliged to parallel the epistemological and methodological norms of natural science. But what these are is in dispute within the human sciences. Weber adopted the probabilistic, uncertain stance of his contemporaries in theoretical physics. His theory of unintended consequences and his reflexive methodological comments were enframed by the same epistēmē that gave rise to relativity. In his *Protestant Ethic*, he protested against historical materialism's a priorism in accounting for both periodicity and change. Weber contended that each historical situation manifested its unique determinations depending on the conjuncture of discourses. Thus, he introduces the "principle" of historical specificity, which would argue, in terms of relativity theory, for the significance of the signifier as well as the referent in determining the nature of social relations. At the same time, Weber holds to the view that rationalization is the distinctive characteristic of modern scientific/technological society, including its thought forms. As Mannheim points out, this is a universalistic, quantitatively determined thought mode and is identified with the Enlightenment. In consequence, *science* is posed against magic and religion, which become marginalized in the bourgeois world.

Mannheim argues that the Counter-Enlightenment, or romanticism, is a deeply conservative movement. But its insistence on the return of the suppressed cannot succeed in resuscitating the "irrational" dimension and making a place for it in the modern world without, at the same time, unwittingly rationalizing this dimension. For, to consider the so-called irrational and expose the elements of its reason as *system* is already to submit to the scientific worldview, since the process of legitimation nearly always entails demonstrating that the "other" conform, at least in a great degree, to science. In this connection, one may cite the experience of parapsychology, psychoanalysis, and other suspect sciences. In both of these cases, the controversy surrounding their findings turns on whether they can be shown to conform to the canon of experimental method and deductive/inductive reason. Consequently, their advocates are obliged to argue that the discovery of the "irrational" can take place only by means of normal, enlightenment science. Only a few of their proponents, especially of psychoanalysis, insist on the methodological validity of hermeneutic/dialectic procedures of investigation.

Durkheim introduces a classical conception of society in his insistence that it be treated as a fact, that is, as indivisible object. Rules for sociological method are construed *objectively*, with the investigator standing outside the field of vision.

This dispute has set the terms of theoretical development in the social sciences since the turn of the twentieth century. Needless to say, "normal" Anglo-American social science follows the objectivist account of Durkheim rather than the hermeneutic or dialectical accounts of Weberian and phenomenological paradigms, whether neo-Marxist or not. On the other hand, recent social theory in France, Italy, and Germany roughly follows the prescriptions of relativity and quantum theory in their respective insistences on temporal discontinuities, spatial indeterminacy, and historical construction of discursive formations in social phenomena.

11. If this is so, we can reconstruct the concept of the unified field, that elusive goal of the sciences, in another way. The relation among physical, life, and human sciences is not one of determination, even in the last instance, of the physical and biological over the social. Rather, each achieves unities within macrospheres and microspheres but not necessarily between them. Further, there is an epistēmē that spans different discourses within a specific historical era.

12. Anthony Giddens has proposed that ideology be disconnected from the "philosophy of science" in favor of a concept of ideology considered "as a positive term, meaning something like an all-embracing and encyclopedic form of knowledge." Yet, he goes on to urge that ideology "should be reformulated in relation to a theory of *power* and *domination*."[24] I take this to be another sense in which the term "ideology" may be employed. A third is to define it as lived experience, the unmediated formulation of judgment—practical, "useful," interested, but surely not "false consciousness" or "bad" science in contrast to truth.

We have three meanings of ideology. The first, against which Giddens poses ideology/power/domination, is entwined with the problem of demarcation, how to distinguish science (truth) from nonscience (myth, religion, etc). I entirely agree that this *should* be a nonproblem. But it is not so simple: science and technology legitimate their privileged place by claiming to be the single source of reliable knowledge; their power/domination emanates from establishing demarcation. An enterprise that wishes to call attention to the character of science as a constituted discourse must address the problem of the elements that make up science—not only its "ideology" in the sense of worldview, but the neutrality of its practice in relation to issues of power/domination. If my thesis is right that, as Giddens says, the mode of signification is "incorporated within systems of domination to sanction their continuance," then science cannot be exempted from this study.

13. Science is a language of power, and those who bear its legitimate claims, i.e., those who are involved in the ownership and control of its processes and results, have become a distinctive social category equipped with a distinctive ideology and political program in the postwar world. The relation of science to the state is still one of subordination in both capitalist and state socialist formations, but this relation is now under attack by knowledge communities which increasingly perceive, even if they have not yet theorized, the elements of their autonomy. That science communities routinely declare their neutrality on political matters, especially on questions affecting the content of scientific knowledge, demonstrates not only the character of scientific and professional ideology, but a studied naïveté concerning the implications of accepting resources made available by the state for research and their own role in establishing priorities.

At the level of the economy, it has been apparent for the last century that science is central for the processes of economic reproduction, the manifestation of which, as the recent development of the high-temperature superconductor demonstrates, is the intimate link between the motivation for scientific discovery and the desire for technological application. A similar example is provided by the current AIDS crisis with respect to medical research. From a small group of underfunded scientists, AIDS is rapidly becoming the hottest property in the medical field as the state mobilizes scientists to deal with the crisis. As with military-related research, state priorities have the force of inexorability: the recalcitrant independent scientist had better be prepared to sacrifice her/his career or work in some subfield for which only marginal support is required.

The scientist wants his/her work to be thought of as both esoteric and socially useful in the long run. Esoteric work carries status probably because it symbolizes freedom—in the form of distance from the dictatorial marketplace. It is more difficult to get a position as a theorist in an esoteric subfield where only modest support is required (mainly just a salary; perhaps some computer time too) than it is to get a position in a subfield more immediately connected to technology, where heavy support is required. Compare solid-state physics and cosmology, for example. One third of the 11,000 physics graduate students in the United States are in the single subfield of solid state physics, and all of them will be able to get jobs in that subfield. Even though there are only a handful of cosmologists, they will have a harder time finding jobs.

14. In turn, we may not theorize the character of the state without understanding that science is the discourse of the late capitalist and the "socialist" state. Science is rapidly displacing, as dominant discourse, the old ideologies of the liberal state—chief among them, possessive individualism, which was based on the dominance of the market in the economy and the conduct of political affairs. This development coincides with the consolidation of bureaucratic power, whose rationality parallels that of science, or, to be more precise, is the ostensible social form of scientific rationality. Therefore, an alternative science would have to imagine, as a condition of its emergence, an alternative rationality which would not be based on domination.

NOTES

NOTES

Chapter 1. Science and Technology as Hegemony

1. *The New York Times*, September 12, 1986. The same theme was repeated by the press after the 508.33-point drop in the stock market on October 18, 1987.

2. Herbert Simon, *Models of Man* (New York: Wiley, 1957).

3. Ward Morehouse and M. Arun Subramanian, *The Bhopal Tragedy*, a report for the Citizens Commission on Bhopal (New York: New York Council on International and Public Affairs, 1986).

4. Edmund Husserl, *Crisis of European Sciences* and *Transcendental Phenomenology*, trans. with an introduction by David Carr (Evanston: Northwestern University Press, 1970), p. 290.

5. Max Horkheimer and Theodor Adorno, *Dialectic of the Enlightenment*, trans. John Cumming (New York: Seabury Press, 1972).

6. Herbert Marcuse, *One Dimensional Man* (Boston: Beacon Press, 1964), p. 231.

7. Gaston Bachelard, *The New Scientific Spirit*, trans. Arthur Goldhammer (Boston: Beacon Press, 1984).

8. Karl Popper, *The Logic of Scientific Discovery* (1934) (New York: Science Editions, 1961) and *Conjectures and Refutations: The Growth of Scientific Knowledge* (1962) (New York: Harper Torchbooks edition, 1968). In both works, published almost thirty years apart, Popper stresses that his philosophy is concerned chiefly with the problem of demarcation between science anad metaphysics, disputing the positivist's statement that metaphysics is "meaningless" but not its uses in other than scientific discourse.

9. Bachelard, *The New Scientific Spirit*, p. 165–66.

10. Sigmund Freud, *Civilization and Its Discontents*, trans. James Strachey (New York: Norton, 1961).

11. Although not the only exception, Bachelard is thoroughgoing in his insistence that science and technique are mutually dependent in the production of verifiable knowledge. "For science, then, the qualities of reality are functions of our rational meth-

ods. In order to establish a scientific fact, it is necessary to implement a coherent technique." (*The New Scientific Spirit*, p. 171.)

12. Raymond Callahan, *Education and the Cult of Efficiency* (Chicago: University of Chicago Press, 1962).

13. In the United States, which does not have a strong feudal tradition, modern industrial capitalism was not forged over a corpse of a powerful Catholic Church whose explanations of the natural world were closely linked to the preservation of the old order. In the United States, the church grew up alongside capitalist agriculture and industry, and many of its laity were themselves scientists and inventors.

14. Marcuse, *One Dimensional Man*, p. 33.

15. Quoted in Freud, *Civilization and Its Discontents*, p. 21.

16. Ibid., pp. 21–23.

17. So far, however, theocracies of the Middle East have subordinated ideology to private considerations and have refrained from condemning Western societies.

18. Douglas Futuyama, *Science on Trial: The Case for Evolution* (New York: Pantheon Books, 1983).

19. Ibid., p. 12.

20. Ibid., pp. 12–13.

21. Alexandre Koyre, *Newtonian Studies* (Chicago: University of Chicago Press, 1965), p. 6. Koyre, who militantly opposes "social" explanations for scientific law, nevertheless holds that "the use and growth of experimental science is not the source, but, on the contrary, the result of the new *theoretical*, that is, the new *metaphysical* approach to nature that forms the content of the scientific revolution of the seventeenth century" (ibid.).

22. Stephen J. Gould, "D'arcy Thompson and the Science of Form," in *Topics in the Philosophy of Biology*, ed. Marjorie Grene and Everett Mendelsohn, Boston, Studies in the Philosophy of Biology, vol. 27 (Dordrecht: Reidel, 1976), p. 97, footnote 59.

23. François Jacob, *The Logic of Life: The History of Heredity*, trans. Betty E. Spillman (New York: Pantheon Books, 1973), pp. 8–9.

24. Francis Crick, *Life Itself* (New York: Simon and Shuster, 1981), p. 53.

25. Ibid., pp. 49–50.

26. Ibid., p. 37.

27. Paul Davies, *Other Worlds* (New York: Simon and Shuster, 1980), pp. 42–43.

28. David Bohm, *Wholeness and the Implicate Order* (London, Boston, and Henley: Routledge and Kegan Paul, 1980), pp. 65–110.

29. Bruno Latour, "Give me a Laboratory . . . ," in *Science Observed*, ed. Karin D. Knorr-Cetina and Michael Mulkey (London and Los Angeles: Sage Publications, 1983), pp. 141–70.

30. For a superb discussion of this phenomenon, see Arthur Kroker, *Technology and the Canadian Mind* (New York: St. Martins Press, 1986).

31. Murray Bookchin, *The Modern Crisis* (Philadelphia: New Society Publishers, 1980), p. 50. The essay "What Is Social Ecology?" in this collection is probably the best short statement of Bookchin's standpoint.

32. Ibid., p. 55.

33. Simone de Beauvoir, *The Second Sex*, trans. and ed. H. M. Parshley (New York: Knopf, 1952); Charlotte Perkins Gilman, *Women and Economics* (1889) (New York: Harper and Row, 1966).

34. For a superb historical description of women's work, see especially Susan Strasser, *Never Done: A History of American Housework* (New York: Pantheon Books, 1982).

35. Evelyn Fox Keller, *Reflections on Gender and Science* (New Haven: Yale University Press, 1985), chapter 9.

36. Ibid., p. 172–73.

37. Ibid., chapter 1. Here, Keller comes close to a gendered reading of philosophy and science but backs off.

38. Carolyn Merchant, *The Death of Nature* (New York: Harper and Row, 1980).

39. Nancy Hartsock, *Money, Sex and Power* (New York: Longman, 1983), especially chapter 10.

40. Ibid., p. 241.

41. Ibid., p. 242.

42. Ibid., chapter 3.

43. Merchant, *Death of Nature*, p. 6

44. Thomas Kuhn, *Structure of Scientific Revolutions* (Chicago: University of Chicago Press, 1962), Norwood Russell Hanson, *Patterns of Discovery* (London and New York: Cambridge University Press, 1958).

45. Kuhn mentions Marxist accounts of the history of science on a few occasions in his published work, in the course of discussing the external-internal explanations for the course of scientific discovery. Although he carefully distinguishes the Marxist position, which focuses on economic/technological influences, from those like Koyre who insist on the metaphysical basis of modern science—the ideas that are brought in from the outside—his own account is not at all dependent on what he calls the "Merton thesis." Robert Merton, Kuhn says, relies in part on the extensive Marxist historiography of Puritanism to explain seventeenth-century science. In no place that I know of does Kuhn go beyond explaining ideas by other ideas or even devote a large part of an essay to Marxist historiography of science. This example, from the scholar who is not only a preeminent historian of science but a major philosopher as well, has been followed by those who have tried to stand on his and Koyre's shoulders (see, for example, Barbara Shapiro, *Probability and Certainty in Seventeenth-Century England*, Princeton: Princeton University Press, 1983).

46. Paul Forman, "Weimar Culture, Causality and Quantum Theory, 1918–1927," in *Historical Studies in Physical Sciences* (Philadelphia: University of Pennsylvania Press, 1971).

47. Robert Merton's canonical *The Sociology of Science* (Chicago: University of Chicago Press, 1973) dominated this field for decades and, in some respects, still does.

48. A fuller discussion of the sociology of science is reserved for chapter 10. For an excellent introduction to the point of view of this group (taking into account some differences among them), see Michael Mulkey, *Science and the Sociology of Knowledge* (London: Allen and Unwin, 1979); also David Bloor, *Knowledge and Social Imagery* (London: Routledge and Kegan Paul, 1978); Bruno Latour and Steve Woolgar, *Laboratory Life* (London and Los Angeles: Sage Publications, 1979), and Knorr-Cetina and Mulkey, *Science Observed* (see note 29).

Chapter 2. Marx 1: Science as Social Relations

1. Karl Marx, Preface to *A Contribution to the Critique of Political Economy*, in Karl Marx, *Selected Works*, vol. 1, ed. C. P. Dutt (New York: International Publishers, n.d.), pp. 356–57.

2. "Marxism" is here understood as the mainstream tradition stemming from Engels's reading of historical materialism. As we shall see, there is reason to hold that Marx, in the 1870s, subscribed to many of the same views.

3. Frederick Engels, *Anti-Duhring*, trans. Emile Burns (New York: International Publishers, 1939).

4. Karl Marx, *Grundrisse*, trans. with a Foreword by Martin Nicolaus (London: Penguin Books, 1973), p. 409.

5. Ibid.

6. Karl Marx, "Theses on Feuerbach," in *Early Writings*, ed. David Fernbach (New York: Vintage Books, 1975).

7. Karl Marx and Frederick Engels, *The German Ideology* (Moscow: Progress Publishers, 1964), p. 32.

8. Ibid., p. 32.

9. Ibid., p. 31

10. Marx and Engels, *Selected Correspondence* (Moscow: Foreign Language Publishers, n.d.), p. 40.

11. Marx and Engels, *The Poverty of Philosophy* (New York: International Publishers, n.d.), p. 95.

12. Georg Lukács, "What Is Orthodox Marxism?" in *History and Class Consciousness* (London: Merlin Books, 1971).

13. Engels to Kautsky in Marx and Engels, *Selected Correspondence*, p. 445.

14. Marx and Engels, *The German Ideology*, p. 41.

15. Ibid., p. 32.

16. Ibid., p. 46.

17. Ibid.

⟶ 18. Ibid.

19. Albrecht Wellmer, *Critical Theory of Society* (New York: Herder and Herder, 1970).

20. Samuel Lilley, *Men, Machines and History* (New York: International Publishers, 1966), p. 55.

21. Karl Marx, *Capital*, trans. Samuel Moore and Edward Aveling (New York: Modern Library, n.d.), vol. 1, p. 364: "If, then, the control of the capitalist is this two-fold in content, by reason of the two-fold nature of the process of production itself—which on the one hand is a social process for producing use value—on the other, capital's process for creating surplus value—in form that is purely despotic."

22. Karl Marx, *Capital* (New York: Modern Library, n.d.), vol. 1, p. 363.

23. Ibid., p. 363-64.

24. Ibid., pp. 363-64.

25. Ibid.

26. Ibid., p. 364.

27. Ibid., p. 359.

28. Ibid., p. 375.

29. Karl Marx, *Capital* (London: Penguin Books, 1975), p. 470. I use this edition here because the translation conveys Marx's meaning more pointedly.

30. Marx and Engels, *Selected Correspondence*, p. 17.

31. David Montgomery, "Job Control in the Steel Industry in the 19th Century," unpublished paper. See also David Montgomery, *Workers Control in America* (New York: Cambridge University Press, 1979).

32. Ibid.

33. See Daniel Nelson, *Managers and Workers: Origins of the New Factory System in the United States, 1880-1920* (Madison: University of Wisconsin Press 1975), especially chapters 3 and 4. See also F. W. Taylor, *Principles of Scientific Management* (New York: Norton, 1967). Taylor wished to "prove that the best management is a true science, resting on clearly defined laws, rules, and principles, as a foundation".

Chapter 3. Marx 2: The Scientific Theory of Society

1. David S. Landes, *Unbounded Prometheus* (London: Cambridge University Press, 1969).

2. Engels to Schmidt and Bloch in Marx and Engels, *Selected Correspondence* (Moscow: Foreign Language Publishers, 1953), pp. 493–500.

3. Karl Marx, article in the *New York Daily Tribune*, August 8, 1953. Quoted in Landes, *Unbounded Prometheus*, p. 39.

4. Eric Hobsbawm, *Industry and Empire* (London: Penguin Books, 1973), p. 54.

5. H. J. Habbakuk, *American and British Technology in the 19th Century* (London: Cambridge University Press, 1967). See also Marx to Annekov, *Selected Correspondence*, p. 44.

6. We might observe the same dynamic in Europe's and Japan's technological superiority to the United States in such industries as car production and steel after World War II.

7. Louis Hacker, *The Course of American Economic Growth and Development* (New York: Wiley, 1970).

8. Edward Kirkland, *Economic History of the United States* (New York: Holt, Rhinehart and Winston, 1947), p. 44.

9. Ibid.

10. Samuel Lilley, *Men, Machines and History* (New York: International Publishers, 1966).

11. Karl Marx, *Capital*, vol. 1, pp. 382–83.

12. Engels to Starkenburg, *Selected Correspondence*, pp. 548–49.

13. Marx, *Capital*, p. 407.

14. Ibid., p. 408.

15. Ibid., p. 414.

16. Ibid., p. 417.

17. Ibid., p. 421.

18. Ibid., pp. 476–77.

19. Ibid., p. 417.

20. Ibid., p. 504.

21. Karl Marx, *Grundrisse*, trans. Martin Nicholaus (London: Penguin Books, 1973), p. 701.

22. Ibid., pp. 704–5.

23. Ibid., p. 706.

24. Ibid.

25. Ibid., p. 711.

26. Ibid., p. 712.

27. Karl Marx, "Results of the Immediate Process of Production," Appendix to *Capital* (Penguin Books edition), p. 1024.

28. The issue is not resolved by ascribing scientific development to the accumulated knowledge of society; advanced capitalism may be demarcated precisely by its polarization of intellectual and manual labor in the production process.

29. Karl Marx, *Capital* (Penguin Books edition), p. 616.

30. Ibid., p. 617.

31. Marx and Engels, *Selected Correspondence*, p. 229.

32. Karl Marx, *Capital*, pp. 423–24.

33. Karl Marx, Preface to *A Contribution to the Critique of Political Economy*, in Karl Marx *Selected Works*, ed. C. P. Dult (New York: International Publishers, n.d.), pp. 356–57.

34. Christopher Caudwell, *The Crisis in Physics* (London: John Lane, The Bodley Head, 1939; reprinted, 1949, 1950).

35. Victor S. Clark, *History of American Manufacturers* (New York: McGraw-Hill, 1929), vol. 2, p. 388.

36. J. W. Oliver, *History of American Technic* (New York: Norton, 1956), p. 321.

37. Ibid., p. 322.

38. Ibid. We shall see in chapter 10 that contemporary sociology of science takes this as more than metaphor; rather, the *merger* of science with technology is embodied in the form of production.

39. Ibid., p. 355.

40. T. K. Derry and T. I. Williams, *A Short History of Technology* (London and New York: Oxford University Press, 1960), p. 611.

41. Caudwell, *The Crisis in Physics*, p. 21.

42. These processes are meticulously described in Harley Shaiken, *Work Transformed: Automation and Labor in the Computer Age* (New York: Holt, Rhinehart and Winston, 1984).

Chapter 4. Engels and the Return to Epistemology

1. Georg Lukács, *History and Class Consciousness* (London: Merlin Books, 1971), p. 3.

2. Ibid.

3. Frederick Engels, *Anti-Duhring*, 2nd ed., trans. Emile Burns (New York: International Publishers, 1939), pp. 18–19.

4. Samuel Lilley, *Men, Machines and History* (New York: International Publishers, 1966), pp. 89–90.

5. Christopher Caudwell, *The Crisis in Physics* (London: John Lane, The Bodley Head, 1939; reprinted, 1949, 1950).

6. Lukács, *History and Class Consciousness*, pp. 132–33.

7. Ibid., p. 9.

8. Ibid., p. 7.

9. Ibid., p. 10.

10. Ibid.

11. Georg Lukács, "A Chapter on Labor from the Posthumous Ontology," *New Hungarian Quarterly* 45 (1972), p. 11.

12. Ibid.

13. Karl Kautsky, *The Socialist Republic* (New York: Labor News, 1904), p. 34.

14. Ibid., p. 3.

15. Ibid., p. 47.

16. This emphasis is especially evident in Georg Lukács, *Lenin* (1924) (London: New Left Books, 1973).

17. Lukács, *History and Class Consciousness*, p. 208.

18. Ibid., p. 310.

19. George Plekhanov, "The Development of the Monist View of History" (1895), in *Selected Philosophical Works*, vol. 1, trans. A. Rothstein and R. Dixon (Moscow: Progress Publishers, 1974), pp. 664ff. In this book, it is clear how much Plekhanov relies on

Engels's conception of the primacy of the struggle between humans and nature through the development of tools as the core of historical materialism.

20. Plekhanov, *Selected Philosophical Works*, vol. 1, pp. 49–106.

21. Plekhanov, "For the Sixtieth Anniversary of Hegel's Death," in *Selected Philosophical Works*, vol. 1, p. 428.

22. Plekhanov, "Joseph Dietzgen," in *Selected Philosophical Works*, vol. 3, trans. R. Dixon (Moscow: Progress Publishers, 1976), p. 101.

23. Plekhanov, "Materialismus Militans," in *Selected Philosophical Works*, vol. 3, p. 221.

24. Ibid., pp. 223–24.

25. Ibid., p. 226.

26. Lenin, *Philosophical Notebooks*, trans. Clement Dutt, in *Collected Works*, vol. 38 (Moscow: Progress Publishers, 1976).

27. Ibid., p. 179. This preoccupation with the theory of knowledge in contrast to Marx's historical method is already prefigured in Plekhanov's "Materialismus Militans," which consists of three long "letters" attacking Mach and one of his leading Russian followers, the bolshevik Bogdanov.

28. Ibid., pp. 187–88. This quote is not atypical of Lenin's reading of Hegel through Engels's and Plekhanov's eyes; nor is the implied dualism of the distinction between nature and the "purposive activity of men" cited in note 29.

29. Ibid., p. 188.

30. Ibid., p. 188.

31. Ibid., p. 357.

32. Ibid., p. 358.

33. Ibid., pp. 160–61.

34. For a recent account, see Loren Graham, *Science and Philosophy in the Soviet Union* (New York: Knopf, 1972), pp. 425–27.

35. Karl R. Popper, *Conjectures and Refutations: The Growth of Scientific Knowledge* (1962) (New York: Harper & Row, 1968), p. 37.

36. Ibid., p. 49.

37. I. P. Pavlov, *Selected Works* (Moscow: Foreign Language Publishing House, 1955).

38. Graham, *Science and Philosophy*, p. 357.

39. Ibid.

40. Paul Ricoeur, *Freud and Philosophy*, trans. Denis Savage (New Haven and London: Yale University Press, 1970).

41. For an interesting attempt to develop the theme of individual history as an element in the constitution of human psychological activity, see Otto Fenichel, "Remarks on Erich Fromm's *Escape from Freedom*," in Otto Fenichel, *Collected Papers*, Second Series (New York: Norton, 1954).

42. Joseph Stalin, *Marxism and Linguistics* (New York: New Century Publishers, 1951).

43. A good account of the workers' opposition is Beatrice Farnsworth, *Alexandra Kollantai: Socialism, Feminism and the Bolshevik Revolution* (Palo Alto: Stanford University Press, 1980), chapter 7; see also Carmen Sirianni, *Workers Control and Socialist Democracy: The Soviet Experience* (London: Verso Editions, 1982), for a discussion that places workers' control in the wider context of the problem of democracy in a revolutionary society.

44. Caudwell, *The Crisis in Physics*, p. 5.

45. Ibid., p. 11.

46. Ibid., p. 53.

47. Ibid. Faithful to the dialectical materialist doctrine that scientific knowledge is both objective and relative to the social context within which it is produced, this "distortion" becomes a catalyst for the resolution of the antimonies only at a higher level.

48. Ibid., p. 18.

49. Ibid., p. 60. Caudwell adds, "All goes well till a point is reached where practice with its specialized theory has in each department so contradicted the general unformulated theory of science as a whole that in fact the whole philosophy of mechanism explodes" (ibid.). The resemblance of Caudwell's theory of paradigm shift to that of Thomas Kuhn's is striking.

Chapter 5. The Frankfurt School: Science and Technology as Ideology

1. For a direct attack on Engels's scientism, see Sidney Hook, *Towards an Understanding of Karl Marx* (New York: John Day, 1933), pp. 25–34. Hook focuses on Second International interpretations of Marxism as a *science* that can be separated from its revolutionary intention. At the same time, Hook celebrates Marx's "experimentalism" and "empiricism." See especially chapter 9. See also Henri Lefebvre, *Dialectical Materialism* (London: Johnathan Cope, 1968), pp. 141–44.

2. Karl Marx and Frederick Engels, *The German Ideology*, (Moscow: Progress Publishers, 1964), p. 31.

3. José Ortega y Gasset, *The Dehumanization of Art and Other Writings on Art and Culture*, trans. W. R. Trask (New York: Doubleday Anchor Books, 1956).

4. Herbert Marcuse, *One Dimensional Man* (Boston: Beacon Press, 1964).

5. In France, under the influence of Bachelard, Marxist scientificity takes not a positivistic form, but a distinctly metaphysical character. For it is Bachelard in concert with Koyre who insists that science has metaphysical presuppositions, the truth of which can be verified by experiment, but not dispensed with.

6. This *will* constitutes a central category of my discourse, so I want to discuss it a little more fully. Engels's Marxism shares with other paradigms the claim to be truth, independent of its historicity. Thus, even if relativity theory has displaced much of the Newtonian worldview, the latter retains its character as science owing to the form in which its propositions are framed, a form which permits their "premises" to be "verified in a purely empirical way" (*The German Ideology*.)

7. Claude Lévi-Strauss, "The Science of the Concrete," in *Structural Anthropology* (New York: Anchor Books, 1969).

8. Of course, Marcuse's book is written fifteen years after Horkheimer and Adorno returned to Frankfurt to reestablish the institute. But much of its impetus derives from "Some Social Implications of Technology," first published in 1941 in the Institute's *Studies in Philosophy and Social Science* and reprinted in Andrew Arato and E. Gebhardt, eds., *The Essential Frankfurt School Reader* (New York: Urizen Books, 1978).

9. Max Horkheimer and Theodor Adorno, *Dialectic of Enlightenment* (New York: Herder and Herder, 1972), p. 4.

10. Ibid., p. 24.

11. Herbert Marcuse, *One Dimensional Man*, p. 154.

12. Max Horkheimer, *Eclipse of Reason* (New York: Oxford University Press, 1947), p. 81.

13. Horkheimer and Adorno, *Dialectic of Enlightenment*, p. 25.

14. Ibid., p. 21.

15. Marcuse uses the thesis of genetic transformation metaphorically, of course. But the underlying tacit meaning of this metaphor verges on the literal.

16. See Horkheimer, *Eclipse of Reason*, chapter 1.

17. Ibid., p. 159.

18. Ibid., p. 159.

19. Ibid., p. 158.

20. Paul Willis, *Learning to Labor: How Working Class Kids Get Working Class Jobs* (New York: Columbia University Press, 1981).

21. Marcuse, *One Dimensional Man*, pp. 230–35. Marcuse's ultimate endorsement of science and technology rests on his belief that, given the proper political conditions, these instruments of domination can be finally neutralized. More, for Marcuse, science may be criticized for "transutilitarian" ends. In the last instance, Marcuse endorses the Enlightenment as *possibility* and concludes, with orthodox Marxism, that the texts of science are capable of separation from the context that produces them.

22. Ibid., p. 251.

23. Ibid., p. 153.

24. Ibid., p. 155.

25. C. F. von Weizsacker, *The History of Nature*, p. 71, cited in Marcuse, *One Dimensional Man*, p. 155.

26. Mancur Olson, *The Logic of Collective Action* (New York: Schocken Books, 1965).

27. Lucio Colletti, *Marxism and Hegel* (London: Verso Books, 1973), p. 175.

28. Ibid., pp. 174–75.

29. Indeed the key text, apart from *History and Class Consciousness*, for the Frankfurt critique of science and technology is Husserl's *Crisis of European Sciences* published late in his life as a kind of critical summa of the immanent anti-Kantianism of one of the leading Kantian philosophies of our time.

30. Galvano Della Volpe, *Logic as a Positive Science* (London: New Left Books, 1980).

31. Alfred Sohn-Rethel, *Intellectual and Manual Labor* (London: Macmillan, 1978), p. 18.

32. Ibid., p. 57.

33. Ibid., p. 66.

34. Ibid., p. 72.

35. Compare Sohn-Rethel, *Intellectual and Manual Labor*, p. 106 with Lynn White, *Medieval Technology and Social Change* (New York: Oxford University Press, 1962).

36. Bertrand Russell, *Human Knowledge*, Quoted in Sohn-Rethel, *Intellectual and Manual Labor*, p. 130.

37. Sohn-Rethel, *Intellectual and Manual Labor*, p. 179.

38. Ibid.

Chapter 6. Habermas: The Retreat from the Critique

1. Max Weber, *Economy and Society* (Berkeley: University of California Press, 1978), vol. 1, part 1.

2. Karl Mannheim, *Essays in the Sociology of Culture* (London: Routledge and Kegan Paul, 1962), part 2: "The Problem of the Intelligentsia."

3. Alvin Gouldner, *The Future of Intellectuals and the Rise of the New Class* (New York: Seabury Press, 1979).

4. See especially Herbert Marcuse, "The Affirmative Character of Culture," in *Negations* (Boston: Beacon Press, 1969).

5. James Madison, Alexander Hamilton, and John Jay, *Federalist Papers*, no. 24.

6. Karl Popper, *The Logic of Scientific Discovery* (London: Hutchinson, 1959).

7. See, e.g., J. D. Bernal, *Science in History* (London: Lawrence and Wishart, 1958).

8. Max Horkheimer and Theodor Adorno, *Dialectic of the Enlightenment* (New York: Seabury Press, 1972).

9. E. J. Djiksterhuis, *The Mechanization of the World Picture* (New York: Oxford University Press, 1961).

10. By "epistēmē," I mean a conception of the world that inscribes itself in the institutions, material practices, and social relations of an epoch.

11. Thomas Hobbes, *Leviathan* (London: Penguin Books, 1965).

12. Bernal, *Science in History*.

13. Dan Greenberg, *The Politics of Pure Science* (New York and Cleveland: World Publishing, 1967).

14. V. I. Lenin, *Materialism and Empirio-Criticism*, in *Collected Works*, vol. 14 (Moscow: Progress Publishing, 1967).

15. Louis Althusser, *Reading Capital* (London: New Left Books, 1970), chapter 1; "Contradiction and Overdetermination," in *For Marx* (New York: Vintage Books, 1970).

16. Rudolph Barho, *The Alternative* (London: New Left Books, 1979).

17. Karl Mannheim, *Ideology and Utopia* (London: Routledge and Kegan Paul, 1936).

18. Georg Lukács, "Class Consciousness," in *History and Class Consciousness* (London: Merlin Books, 1971).

19. See Althusser, *For Marx*.

20. Thomas Kuhn, *The Structure of Scientific Revolutions* (Chicago: University of Chicago Press, 1967).

21. Althusser has stated this case most succinctly in *Reading Capital*, chapter 5.

22. For recent examples, see especially the work of Eastern European philosophers in the 1960s. A representative work of this group is Ivan Petrovic, *Marx in the mid-20th Century* (New York: Doubleday Anchor Books, 1970); also Karel Kosik, *Dialectic of the Concrete* (The Hague: Mouton, 1976).

23. Habermas, "Technology and Science as 'Ideology,' " in *Toward a Rational Society* (Boston: Beacon Press, 1970).

24. Herbert Marcuse, *One Dimensional Man*, quoted by Habermas, "Technology and Science," p. 86.

25. Habermas, "Technology and Science," p. 99.

26. For a discussion of this issue, see my *Crisis in Historical Materialism* (New York: Praeger Publishers, 1981).

27. For some recent studies, see G. David Gartman, *Auto Slavery* (New Brunswick: Rutgers University Press, 1985); Dan Clawson, *Bureaucracy and the Labor Process* (New York: Monthly Review Press, 1982); Huw Beynon, *Working for Ford* (London: Routledge and Kegan Paul, 1974).

Chapter 7. Marxism as a Positive Science

1. See especially Althusser's *For Marx* (New York: Vintage Books, 1970).

2. Althusser, *Reading Capital* (London: New Left Books, 1970), p. 67.

3. Ibid., p. 4.

4. Ibid., pp. 52–53.

5. Ibid., p. 67.

6. Thomas Kuhn, *The Structure of Scientific Revolutions* (Chicago: University of Chicago Press, 1967).

7. Emile Durkheim, *Rules of Sociological Method* (New York: Free Press, 1959), p. 27.

8. Althusser, *Reading Capital*.

9. See Louis Althusser, "Freud and Lacan," in *Lenin and Philosophy* (London: New York Left Books, 1971).

10. John Mepham, "Theory of Ideology in Capital," *Working Papers in Cultural Studies* 66 (autumn 1974).

11. Ibid., p. 107.

12. Sigmund Freud, *Civilization and Its Discontents* (New York: Norton, 1965).

13. Paul Feyerabend, *Against Method* (London: New Left Books, 1976).

14. Kuhn, *Structure of Scientific Revolutions*.

15. Charles S. Pierce, "The Fixation of Belief," in *Selected Writings of C. S. Pierce*, ed. Justin Buehler (New York: Dover Edition, 1955).

16. Lukács, "Reification and the Consciousness of the Proletariat," in *History and Class Consciousness*.

17. See Jürgen Habermas, *Knowledge and Human Interests* (Boston: Beacon Press, 1971), pp. 62–63, where by insisting that Marx's mistake was to throw together "work and interaction," human sciences were merged with natural science. For Habermas, the critique of ideology consists in the effort to show that the methods of natural sciences are inappropriate for the human sciences since the latter have to do exclusively with communicative actions which are inherently self-reflexive, whereas the former are (appropriately) forms of instrumental actions. Since ideology refers to class relations and Habermas tries to show that these have been subsumed under the "general" interest entailed by production, ideology as a social category is no longer adequate to descriptions of the social world.

18. Galvano Della Volpe, *Logic as a Positive Science* (London: New Left Books, 1980).

19. Colletti, *Marxism and Hegel* (London: Verso Books, 1973), p. 199, even though "Marxism is not an epistemology in any fundamental sense."

20. Ibid., p. 227.

21. Jean-Paul Sartre, *Critique of Dialectical Reason*, trans. Alan Sheridan-Smith (London: New Left Books, 1976), p. 20.

22. Jonathan Rée, "Glossary," in ibid., p. 827.

23. It is interesting to compare this aspiration with Brecht's, who "saw himself as a theoretical Galileo whose task was to bring drama into line with Einstein" (Ronald Hayman, *Brecht: A Biography*, New York: Oxford University Press, 1983, p. 138). The Galilean model was a theater without motivation. Rather, "fate's no longer an integral power but more like a field of force" (ibid.; this quote is Brecht's own).

24. John Roemer, *Analytic Foundations of Marxian Economic Theory* (London and Boston: Cambridge University Press, 1981); Jon Elster, *Making Sense of Marx* (London and Boston: Cambridge University Press, 1985); Jon Elster, *Logic and Society* (New York: Wiley, 1978); Eric Olin Wright, *Classes* (London: New Left Books, 1985).

25. G. A. Cohen, *Karl Marx's Theory of History: A Defense* (London and New York: Cambridge University Press, 1978).

26. Ibid., p. 26.

27. I would also add: the production of trucks is not merely production of use values outside a specific social context. Their *mass production* as alternatives to trains or any other means of transport is not devoid of premises that are linked to economic and political decisions. In the same vein, roads are not roads; they are instances of historically situated social relations which imply a whole regime of *how* humans interact with nature.

28. Cohen, *Karl Marx's Theory of History*, p. 152.

29. Or, to be more precise, Cohen, echoing Marx, holds that scarcity is the a priori condition for history itself. For without the naked, shivering human production, growth

would not occur. At the same time, the concept of scarcity itself is not really historicized. When scarcity is evoked as a political and ideological weapon in contemporary late capitalist societies, its historicity is never more evident.

30. Cohen, *Karl Marx's Theory of History*, p. 153.

31. Nevertheless, I do not contest that Cohen's is a plausible interpretation of Marx and the mainstream of the Marxist tradition, even if his citations from Marx himself are employed to buttress this interpretation. Missing, of course, is Marx's own ambiguity on many of the issues.

32. In the course of rendering a generally enthusiastic account of the research project that culminated in the book *The Authoritarian Personality*, Adorno acknowledges "we had to water our wine a bit. It seems to be the defect of every form of empirical sociology that it must choose between the reliability and the profundity of its findings. It is difficult for me to avoid the suspicion that the increasing precision of methods in empirical sociology, however impeccable the arguments for them might be, often restrains scientific productivity." Theodor Adorno, "A European Scholar in America," in *The Intellectual Migration*, ed. Donald Fleming and Bernard Barlyn (Cambridge, Mass: Harvard University Press, 1969), p. 366.

33. See Alain Touraine, *The May Movement* (New York: Random House, 1971); Henri Lefebvre, *The Explosion* (New York: Monthly Review Press, 1969); Fredy Perlman, *Student-Worker Committees* (pamphlet) (Detroit: Black and Red, 1968).

34. Antonio Gramsci, *Prison Notebooks* (New York: International Publishers, 1971), pp. 376–77. One example, "the thesis which asserts that men become conscious of fundamental conflicts on the level of ideology is not psychological or moralistic in character, but structural and epistemological; and they form the habit of considering politics, and hence history, as a continuous marché de dupes, a competition in conjuring and sleight-of-hand." Page 164 also in his discussion of the relation of economic structures to ideology, pp. 168–69.

35. Ibid., pp. 244–45.

36. Ibid., p. 468, where Gramsci argues that all scientific theories "are superstructures." "According to the theory of praxis (Marxism), it is evident that it is not atomic theory that explains human history but the other way about, in other words, that atomic theory and all scientific hypotheses and opinions are superstructures." Compare this formulation to that of G. A. Cohen, for whom science is meaningful only to the extent that it is integrated with productive forces.

37. For the view of the autonomy of science from technology, see David Landes, *Unbounded Prometheus* (London: Cambridge University Press, 1969). Of course, Engels was a strong proponent of the view that science is a generalization from the practical problems generated by craft practices and economic circumstances, especially trade in the late Middle Ages. See B. Hessen, "The Social and Economic Roots of Newton's Principia," in *Science at the Crossroads* (London: Kniga, 1931).

Chapter 8. Soviet Science: The Scientific and Technological Revolution

1. Zhores Medvedev, *Soviet Science* (New York: Norton, 1978), chapter 2.

2. David Joravsky, *Soviet Marxism and Natural Science* (London: Routledge and Kegan Paul, 1961), p. 66.

3. Ibid.

4. George Konrad and Ivan Szelenyi, *The Intellectuals on the Road to Class Power* (New York: Harcourt Brace Jovanovich, 1979).

5. See Ferenc Feher, Agnes Heller, and George Markus, *Dictatorship over the Needs* (Oxford: Basil Blackwell, 1983).

6. Alvin Gouldner, *The Future of Intellectuals and the Rise of the New Class* (New York: Seabury Press, 1979).

7. Kendall Bailes, *Technology and Society under Lenin and Stalin* (Princeton: Princeton University Press, 1978).

8. For a discussion of the difference between the old "extensive" regime of capital accumulation and the "intensive" regime introduced in the wake of mechanization, see Michel Aglietta, *A Theory of Capitalist Regulation* (London: New Left Books, 1979), pp. 52–61. The intensive regime signifies the historical stage when increasing surplus value by means of lengthening the working day reaches its limit. Now workers' productivity is increased by the introduction of machinery, technologies of organization of the labor process, and so on—in short, by holding working hours constant.

9. Antonio Gramsci, "The Intellectuals," in *Prison Notebooks* (New York: International Publishers, 1971), pp. 5–23.

10. Andre Gorz, "Technology, Technicians and Class Struggle," in *Division of Labor*, ed. Andre Gorz (London: Harvester Press, 1976).

11. Thorsten Veblen, *Engineers and the Price System* (New York: B. W. Huebsch, 1921).

12. V. A. Fok, "Quantum Physics and Philosophical Problems," in *Lenin and Modern Natural Science*, ed. M. E. Omelyanovsky, trans. S. Syrovatkin (Moscow: Progress Publishers, 1948). In this essay, Fok acknowledges that Einstein's theory of relativity and the Heisenberg uncertainty principle constitute corrections to the "absolutism" of classical mechanics but in no way challenge the fundamental objectivity of physical phenomena or the ability to manage adequate descriptions of them. Probability is part of description, not a statement about the indeterminate nature of the objective, material reality under investigation.

13. George D. Holliday, *Technology Transfer to the USSR: 1928–1937 and 1966–1975: The Role of Western Technology in Soviet Economic Development* (Boulder: Westview Press, 1979), p. 46.

14. P. N. Fedoseyev, "Social Significance of the Scientific Technological Revolution," in *Scientific and Technological Revolution: Social Aspects* (London and Los Angeles: Sage, 1977), p. 88.

15. Ralf Dahrendorf, "Observations on Science and Technology in a Changing Socioeconomic Climate," in *Scientific and Technological Revolution*, p. 74.

16. P. N. Fedoseyev, "Lenin's Ideas and the Methodology of Contemporary Science," in *Lenin and Modern Natural Science*.

17. V. I. Lenin, *Materialism and Empirio-Criticism*, in *Collected Works* (Moscow: Progress Publishers, 1967), vol. 14, p. 249.

18. *Civilization at the Crossroads*, ed. Radovan Richta (Sydney: Australian Left Review, 1967), p. 235.

19. Ibid., p. 22.

20. *The Scientific and Technological Revolution: Social Effects and Prospects* (Moscow: Progress Publishers, 1972).

21. V. G. Afnasyev, *The Scientific Management of Society* (Moscow: Progress Publishers, 1971) is an explication of the centrality of systems theory to STR ideology.

22. Severy Bialer, *The Soviet Paradox* (New York: Knopf, 1986), p. 27.

23. Ibid., p. 29.

24. More exactly, he was trained as a lawyer, but has never practiced. His entire career is within the Communist Party.

25. Joravsky, *Soviet Marxism and Natural Science*, part 3.

26. B. Hessen, "Social and Economic Roots," in *Science at the Crossroads* (London: Kniga, 1931), pp. 36–37.

27. Richard Levins and Richard Lewontin, *The Dialectical Biologist* (Cambridge, Mass.: Harvard University Press, 1986).

28. Jerome Ravetz, "Tragedy in the History of Science," in *Changing Perspectives in the History of Science*, ed. Robert Young (London: Heinemann, 1971), pp. 216–20.

29. Arthur Koestler, *The Case of the Midwife Toad* (New York: Random House, 1972).

30. Ibid., p. 146.

31. E. A. Burtt, *The Metaphysical Foundations of Natural Science* (New York: Doubleday, 1924; reprinted, 1931, 1954).

32. A. D. Alexandrov, "Space and Time in Modern Physics in the light of Lenin's Philosophical Ideas," in *Lenin and Modern Natural Science*, p. 239.

33. V. A. Fok, "Quantum Physics and Philosophical Problems."

34. M. E. Omelyanovsky, *Dialectics in Modern Physics* (Moscow: Progress Publishers, 1983), p. 13.

Chapter 9. The Breakup of Certainty: Philosophy of Modern Physics

1. C. F. von Weizsacker, *The Unity of Nature* (New York: Farrar, Straus and Giroux, 1980), p. 183.

2. Ibid., p. 185.

3. W. V. O. Quine, "Two Dogmas of Empiricism," in *Can Theories Be Refuted? Essays on the Duhem-Quine Hypothesis*, ed. Sandra G. Harding (Dordrecht: Reidel, 1976), p. 45.

4. Ibid., pp. 59–60.

5. W. V. O. Quine, *The Ways of Paradox and other Essays* (Cambridge, Mass.: Harvard University Press, 1966, 1976).

6. Quine, "Two Dogmas," p. 58.

7. Ibid., p. 61.

8. Ibid., p. 63.

9. Richard Rorty, *Consequences of Pragmatism* (Minneapolis: University of Minnesota Press, 1982), pp. xviii–xix.

10. John Dewey, *The Quest for Certainty*, in *Later Works*, vol. 4 (Carbondale: Southern Illinois University Press, 1984).

11. Ibid., p. 136. "That is, it is an outcome to directed experimental questions."

12. Ibid., p. 80. "Modern experimental science is an art to control."

13. Ibid., p. 132.

14. Thomas Kuhn, *The Essential Tension* (Chicago: University of Chicago Press, 1977), pp. 105–6.

15. N. R. Hanson, *Patterns of Discovery* (London: Cambridge University Press, 1958).

16. Gerard Holton, *Thematic Origins of Scientific Thought: Kepler to Einstein* (Cambridge, Mass.: Harvard University Press, 1973).

17. Paul Forman, "Weimar Culture, Causality and Quantum Theory," in *Historical Studies in the Physical Sciences* (Philadelphia: University of Pennsylvania Press, 1971), p. 62.

18. Ibid., p. 28.

19. Oswald Spengler, *Decline of the West* (New York: Knopf, 1932), p. 167; quoted in Forman, "Weimar Culture," pp. 31–32.

20. Ibid., p. 28.

21. Ibid.

22. See Georg Lukács, *The Destruction of Reason* (London: Merlin Books, 1980), especially chapter 2. Compare Forman's contemptuous dismissal of Spengler's theory of science to Lukács's. For both, the underlying vitalism of Bergson and Spengler are, in Lukács's words, "worthless from a philosophical viewpoint."

23. Holton, *Thematic Origins of Scientific Thought*, pp. 219–21.

24. Carl Shorske, *Fin de Siecle Vienna* (New York: Knopf, 1982). See also Siegfried Kracauer, *From Caligary to Hitler* (Princeton: Princeton University Press, 1963).

25. Abraham Pais, *Subtle Is the Lord* (New York: Oxford University Press, 1982), p. 486. Pais's superb biography stresses Einstein's philosophical realism in opposition to the realism of the Copenhagen school.

26. Charles Sonders Peirce, *Collected Papers* (Cambridge, Mass.: Harvard University Press, 1965), paragraph 158, p. 65.

27. Henri Bergson, *Creative Evolution* (New York: Modern Library Edition, n.d.) chapter 3.

28. Holton, *Thematic Origins of Scientific Thought*, pp. 21–43.

29. Ludwig Wittgenstein, *Philosophical Investigations*, trans. G. E. M. Anscombe (New York: Macmillan, 1953, 1966, 1968).

30. Ibid., paragraph 241, p. 88.

31. Thomas Kuhn, *Essential Tension*, p. 294.

32. The distinction between describe and narrate is Lukács's. It refers to his method of canon formation in literature. Novels that "describe" (e.g., the naturalism of Zola) are really positivist in the manner prescribed by Auguste Comte. Narration is a self-reflexive method that recognizes that the writer is making choices from among the elements of the pure. And these choices ineluctibly constitute interpretation that is linked to interest.

33. Ernesto Laclau and Chantal Mouffe, *Hegemony and Socialist Strategy* (London: Verso Books, 1985).

34. Thomas Kuhn, *Essential Tension*, p. 220.

35. Roy Wallis, ed., *On the Margins of Science* (Keele: University of Keele, 1979), pp. 5–6.

36. Robert Merton, *The Sociology of Science* (Chicago: University of Chicago Press, 1973), especially chapter 12, "Science and the Social Order" (1938).

37. Paul Feyerabend, *Against Method* (London: New Left Books, 1976).

38. Ibid., p. 146.

39. E. A. Burtt, *The Metaphysical Foundations of Modern Science* (New York: Doubleday, 1924; reprinted, 1931, 1954).

40. Ibid., p. 90.

41. Kurt Hubner, *Critique of Scientific Reason* (Chicago: University of Chicago Press, 1983), p. 75.

42. Ibid., p. 115.

43. Ibid., p. 105.

44. One is tempted to substitute "constructs" for "influences." I have avoided the temptation because I do not want to engage in the interminable question of the distinction between "reality" and "knowledge." What is clear is that "reality" is inconceivable without its historical dimension, which includes its social, economic, and epistemic elements.

45. Hubner, *Critique of Scientific Reason*, p. 227.

Chapter 10. The Science of Sociology and the Sociology of Science

1. Even Marx spoke of "laws of tendency" rather than laws of development in a manner ascribed to him by some of his followers. Yet, there is no doubt of the strong movement within Marxism that parallels classical physics, evolutionary biology and 18th century economics in their quest for "laws of motion" of nature and society.

2. Talcott Parsons, *The Social System* (Glencoe: Free Press, 1951).

3. Robert Merton, *Social Theory and Social Structure*, enlarged edition (Glencoe: Free Press, 1968), part 1.

4. In this sense, much of American sociology is really a kind of social psychology, since it uses the presuppositions of methodological individualism. For a classical, influential work in this mode which, however, remains indebted to Hegel's dialectical phenomenology, see George Herbert Mead, *Mind, Self and Society* (Chicago: University of Chicago Press, 1934).

5. Vilfredo Pareto, *The Mind and Society*, ed. A. Livingstone (London, 1935). The influence of the work on Parsons is evident, not only in his *Structure of Social Action* (Glencoe, Ill.: Free Press, 1937), which deals explicitly with Pareto's sociology, but in the *Social System* (New York: Free Press, 1961), where the personality system becomes the crucial variable in the constitution of the social system.

6. Even in the United States, reward systems in the social scientific communities have not been unified on what constitutes "excellence" or status. For example, the 1980s has been marked by a polarization in sociology between an increasingly quantitative tendency modeling itself on the natural sciences and a more emphatic qualitative tendency often led by theorists and ethnographers who have contended as much for political power with the American Sociological Association as they have for intellectual hegemony.

7. Pierre Bourdieu, Anthony Giddens, and Jürgen Habermas have offered roads away from quantifications not only to continental but also to Anglo-American social theorists. They reinforce an older, indigenous American tradition of which Parsons, Daniel Bell, and, more recently, Jeffrey Alexander are examples.

8. Karl Mannheim, "American Sociology," in *Essays on Sociology and Social Psychology*, ed. Paul Keckemeti (London: Routledge and Kegan Paul, 1953). Referring to a compendium edited by Stuart A. Rice, *Methods in Social Science*, Mannheim remarks: "We must admit a very marked and painful disproportion between the vastness of the scientific machinery employed and the value of the ultimate results" (p. 187). The rest of his review discusses the permissive penchant of American Sociology to (1) "limit its scope" to social problems linked to social policy, and (2) a "mistrust of 'philosophy' or metaphysics" which prevents it from addressing "general" or "basic" questions. This essay, which originally appeared in the *American Journal of Sociology* (vol. 38, no. 2) (1932), is, sadly, as fresh and relevant today as it was more than fifty-five years ago.

9. See Merton, *Social Theory and Social Structure*, pp. 516–17.

10. It may safely be argued that the whole of American political science as a discipline is dominated by these concerns. This is especially true of the study of American politics, where "participation" and political behavior, the key objects of knowledge, are really euphemisms for voter surveys. There are, of course, exceptions, and the sections of American political science devoted to political theory and comparative politics are relatively free of such methodological compulsions.

11. Peter Winch, *The Idea of a Social Science* (London: Routledge and Kegan Paul, 1958), p. 123.

12. Ibid., pp. 126–27.

13. Ibid., p. 128.

14. Robert Merton, *The Sociology of Science* (Chicago: University of Chicago Press, 1972).

15. Ibid., see especially "Priorities in Scientific Discovery."

16. Ibid., p. 207. See the chapter "Interactions of Science and the Military."

17. For a recent work, see David Dickson, *The New Politics of Science* (New York: Pantheon Books, 1984). This is a nearly exhaustive treatment of science's relation to military and corporate power in the United States, implicating the universities not as dispassionate institutions of pure research but as a major purveyor of "knowledge as commodity" (chapter 2).

18. Joel Primack and Frank von Hippel, *Advice and Dissent: Scientists in the Political Arena* (New York: Basic Books, 1974). The argument here is for "public interest science," a program in which scientists can share control of political decisions. Compare Dickson's call for a broad program of "democratic control" over science and technology based on the dissemination of scientific information and literacy.

19. Imre Lakatos, ed., *Criticism and the Growth of Scientific Knowledge* (London: Cambridge University Press, 1969).

20. Ludwig Fleck, *The Genesis of a Scientific Fact* (Chicago: University of Chicago Press, 1976).

21. Ibid., p. 27.

22. Michael Mulkey, *Science and the Sociology of Knowledge* (London: Allen and Unwin, 1979), p. 67.

23. Ibid.

24. David Bloor, *Knowledge and Social Imagery* (London: Routledge and Kegan Paul, 1976), pp. 2–3.

25. Ibid., p. 5.

26. Bruno Latour and Steve Woolgar, *Laboratory Life* (London and Los Angeles: Sage Publications, 1979).

27. Steve Woolgar, "Irony in the Social Study of Science," in *Science Observed*, ed. Karin D. Knorr-Cetina and Michael Mulkay (London, Beverly Hills, and New Delhi: Sage Publications, 1983). The reference to "uncovering" objects is Heidegger's distinction made in *The Question Concerning Technology*.

28. Sharon Traweek, "Culture and the Organization of a Particle Physics Community," Ph.D. dissertation, Stanford University, 1983.

29. Andrew Pickering, *Constructing Quarks: A Sociological History of Particle Physics* (Chicago: University of Chicago Press, 1984), p. 30.

30. Bruno Latour, *Science in Action* (Cambridge, Mass.: Harvard University Press, 1986) and "Give me a Laboratory," in *Science Observed*.

31. Augustine Brannigan, *The Social Basis of Scientific Discoveries* (London: Cambridge University Press, 1981), p. 68.

32. Ibid., pp. 68–69.

33. Latour, "Give me a Laboratory," p. 144.

34. Ibid., p. 145.

35. Ibid., p. 160.

36. Ibid., p. 160.

37. Ibid., p. 168. The term "future reservoir" implies that science has not yet exercised political power in its own name. Yet, with Latour, I would argue that this is clearly on the political agenda.

Chapter 11. Scientism or Critical Science: The Debates in Biology

1. Immanuel Wallerstein, *The Modern World System* (New York: Academic Press, 1978). For a methodological critique of this work, see Stanley Aronowitz, "A Metatheoretical Critique of Immanuel Wallerstein's *Modern World System*," *Theory and Society* 10 (1981), pp. 503–20.

2. G. W. F. Hegel, *Philosophy of Nature*, trans. A. V. Miller (London: Oxford University Press, 1970).

3. Ibid., p. 356.

4. *Biology and the Future of Man*, ed. Philip Handler (New York: Oxford University Press, 1970), p. 7.

5. Kenneth Schaeffner, "The Watson-Crick Model and Reductionism," in *Topics in the Philosophy of Biology*, ed. Marjorie Grene and Everett Mendelsohn (Dordrecht and Boston: Reidel, 1976), pp. 102–3.

6. Michael Polanyi, "Life's Irreducible Structure," in Grene and Mendelsohn, *Topics in the Philosophy of Biology*, p. 130.

7. Ibid., pp. 132–33.

8. Ibid., p. 132.

9. Ibid., p. 134.

10. Francisco Ayala, *Biology as an Autonomous Science*, in Grene and Mendelsohn, *Topics in the Philosophy of Biology*, p. 314.

11. Ibid., p. 319.

12. Richard Levins and Richard Lewontin, *The Dialectical Biologist* (Cambridge, Mass.: Harvard University Press, 1986).

13. Ibid., p. 288.

14. Ibid.

15. Orlando Patterson, *Slavery and Social Death* (Cambridge, Mass.: Harvard University Press, 1984).

16. For a fuller discussion of this relation, see Stanley Aronowitz, *Crisis in Historical Materialism* (New York: Praeger Publishers, 1981), chapter 3. The ontologization of nurturance as a natural biological propensity of mothering is, curiously, not confined to male images of women, but has been adopted by one wing of feminism as evidence for the "natural" superiority of women. In cultural feminist theory, traits of nurturance are situated not in the historical conditions of women's oppression or in the social division of labor, but in their biological makeup. However, this is only the other side of the doctrine of biologically rooted subordination.

17. Witness the reluctance of even some of the most severe critics of bioengineering in the United States to declare that bioengineering as a product of molecular biology is deeply rooted in the technological imperative rather than in a rational conception of human interest. For a case in point, see Jack Doyle, *Altered Harvest* (New York: Penguin Books, 1986). After a thorough and often brilliant exploration of the actual and potentially disastrous implications of bioengineering in agriculture for human survival, Doyle's last chapter is appropriately titled "In Search of Neutral Science."

18. Doyle provides an even more dramatic example. In 1974, much of the Midwest corn crop was destroyed by ordinary breeding procedures that made the corn susceptible to a fungus. Breeding by means of bioengineering is likely to magnify the problem many fold.

Chapter 12. Toward a New Social Theory of Science

1. B. F. Skinner, *Beyond Freedom and Dignity* (New York: Norton, 1971).

2. Loren Baritz, *The Servants of Power: A History of the Use of Social Science in Industry* (Middletown, Conn.: Wesleyan University Press, 1960).

3. I use "civil society" in Gramsci's sense, "the ensemble of organisms commonly called 'private' " that "correspond[s] to . . . the function of 'hegemony' which the dominant group exercises throughout society" (Gramsci, *Prison Notebooks*, New York: International Publishers, 1971). Clearly, I hold science as a crucial element of hegemony, but it is also, increasingly, integrated with the "dominant" group rather than constituting a counterhegemony or a neutral set of practices.

4. Paul Feyerabend, *Against Method* (London: New Left Books, 1976).

5. See especially "Categorical," in Richard McKeon, ed., *Basic Works of Aristotle* (New York: Random House, 1941).

6. See *The New Science of Giavonni Bettista Vico*, trans. Thomas Bergin and Max A. Fisch (Ithaca: Cornell University Press, 1968); Hegel, Preface to *Phenomenology of the Spirit* (London: Oxford University Press, 1980).

7. I use the male pronoun deliberately to signify the prevalence of men in top scientific managerial positions. See Evelyn Fox Keller, *Reflections on Gender and Science* (New Haven: Yale University Press, 1985).

8. The division between the theorists and experimenters has been ably described by Sharon Traweek. Theorists are the higher status scientists whose work is, partially, *interpretation* of experimental results.

9. This is especially true for major research universities.

10. Karl Popper, *The Logic of Scientific Discovery* (New York: Science Editions, 1961).

11. Werner Heisenberg, *Physics and Philosophy* (New York: Harper and Row, 1958), p. 9: "we know the object of scientific knowledge only by the speculative means of axiomatic theoretic construction or postulation: Newton's suggestion that the physicist can deduce theoretical concepts from the experimental data being false." This theory, in common with the Duhem-Quine hypothesis discussed in chapter 9, has, nevertheless, escaped most natural and social scientists as well as their philosophical acolytes.

12. Ibid., p. 169.

13. Ibid. For Heisenberg, Aristotle "creates the basis for scientific thinking."

14. Ibid., p. 180.

15. Ibid., pp. 183–84.

16. See Maurice Merleau-Ponty, *Phenomenology of Perception* (London: Routledge and Kegan Paul, 1964).

17. We can observe this phenomenon graphically played out in the current AIDS crisis. While the years of pressure for preventive measures, rather than for chemical treatments exclusively, have yielded some new educational initiatives by state and local governments, instrumental science is concentrating on finding out how the virus operates so that a serum can be discovered. When the ecological issues are addressed, discussion is conflated with new calls for sexual abstinence.

18. See, e.g., Vincent Navarro, *Medicine and Capitalism* (Baltimore: Johns Hopkins Press, 1978).

19. Herbert Marcuse, *Eros and Civilization* (Boston: Beacon Press, 1955).

20. Herbert Marcuse, *One Dimensional Man* (Boston: Beacon Press, 1964).

21. Karl Popper, *Objective Knowledge* (London: Oxford University Press, 1972).

22. And even most contemporary ethical discussion labors to make its discourse scientific, although there are significant dissents from this effort.

23. See Alfred North Whitehead, *Process and Reality* (New York: The Free Press, 1929, 1969). Whitehead, who with Bertrand Russell helped establish the basis of philosophy for a theory of science at the turn of the twentieth century, argues in the Preface for speculative philosophy as a legitimate activity. Moreover, within the framework of an ontology, he argues for the primacy of "relatedness" and "becoming" as categories over "quality." Yet, there is no doubt that Whitehead's speculative cosmology claims to be an interpretation consistent with the latest developments of science.

24. Anthony Giddens, "Four Theses on Ideology," *Canadian Journal of Political and Social Theory* 7, no. 1–2 (Winter/Spring 1983), pp. 18–21.

INDEX

INDEX

Action theory: in biology, 307; and human action, 301, 310
Adorno, Theodor, 7, 126-27, 139, 152, 183, 193, 304, 317
Agriculture: and political economy, 45
Air travel: as privileged technology, 5
Alexandrov, A. D., 212
Althusser, Louis, 124-25, 126, 141, 142, 147, 154, 159, 169-84 *passim*, 193, 197, 199
Analytic/synthetic statements, 242-44
Animals: and consciousness, 303
Anomalies (Science), 244, 251, 266
Anti-Duhring, 91-93
Aristotle, 12, 13, 27, 266, 321, 338
Artificial intelligence, 4
Assembly line, 53, 83
Astronomy: and the supernatural, 17
Authority, Method of, 140

Bachelard, Gaston, 8, 183
Bacon, Francis, 23, 183
Bailes, Kendall, 214
Balibar, Etienne, 124-25
Barho, Rudolph, 155
Beauvoir, Simone de, 22
Benjamin, Walter, 22, 134
Bergson, 140

Bioengineering. *See* Genetic engineering
Biology: and action theory, 307, 311; versus chemistry, 305, 306, 307; ecological, 308; and Hegel, 302; nineteenth-century German, 308; versus physics, 306, 307; and reductionism, 304, 306; and social relations, 309
Black, Max, 28
Block, Ernst, 127
Bogdanov, A., 106
Bohm, David, 17
Bohm-Bawerk, Eugen, 138
Bohr, Niels, 17, 240, 241, 245, 255
Bolshevik Revolution, 100, 101, 116, 117
Bookchin, Murray, 21-22
Brezhnev, Leonid, 212
Bukharin, Nicholai, 198

Callahan, Raymond, 9-10
Capital, 42, 48, 56, 70, 72, 73, 102, 141
Capitalism, 93-94, 130-31, 156, 162; and bourgeois worldview, 41, 65; and development of science, 39-40, 47, 352; formation of, 342; and Marxism, 39, 42, 48, 339; and physical environment, 334; and production, 50, 53, 65, 71, 74-75; and the

proletariat, 102-4; and scientific knowledge, 41, 67, 85; and technology, 38, 39, 68, 85, 117; and U.S. manufacturing, 54-55
Carnap, Rudolph, 240, 244, 279
Carnegie, Andrew, 79
Caudwell, Christopher, 95-96, 117-19 passim, 127
Causality, 13, 17, 95, 99, 100, 109, 119, 331
CCD. See Culture of critical discourse
Chemistry, 79, 80-81, 305, 306
Civilization at the Crossroads, 218
Class relations, 46, 48-49, 309; and scientific community, 298; in U.S. building industry, 57. See also Social relations
Cohen, Gerry, 185-88, 190-91
Colletti, Lucio, 124-25, 126, 139-42 passim, 180-81, 183, 184
Communist Manifesto, The, 36, 48, 73
Communities, 275
Computers, 87; age of, 3-4; in auto industry, 84; in social sciences, 326; and stock market, 3
Concrete totality, 97-98, 103
Contribution to the Critique of Political Economy, A, 43-45, 85
Copernicus, Nicolaus, 149, 173
Corporations: and scientific research, 15, 16, 20, 293, 314, 318-20, 337
Creationism, 12
Crick, Francis, 14
Crisis in Physics, The, 95-96, 119
Critical Theory, 6-7, 129-30, 318, 340; of Frankfurt School, 122; and nature, 127; of science, 139-41, 145, 191; and technology, 132
Critique of Hegel's Philosophy of the State, 150
"Critique of the Gotha Program," 201
Culture of critical discourse, 208-9
Curie, Marie, 23

Daly, Mary, 25, 27
Darwin, Charles, 21, 105
Davies, Paul, 17
Defense budget, 10, 20
Della Volpe, Galvano, 124-25, 139-41 passim, 181-84, 191
Descartes, René, 8, 240

Dewey, John, 245-47, 248
Dialectic of the Enlightenment, 127, 183
Dialectics, 73-74, 91-93, 96, 98, 99, 106, 108, 118, 121-22, 251, 302
Dialectics of Nature, 92, 107
Dirac, Paul, 119
Durkheim, Emile, 41, 134-35, 172

Eclipse of Reason, 127
Ecological systems, 311, 312
Ecological theory. See Social ecology
Economic and Philosophic Manuscripts of 1844, The, 42
Economic development: and Kautsky, 101
Edinburgh School, 30, 289, 291
18th Brumaire, The, 36
Einstein, Albert, 17, 118, 240, 245, 255, 268, 314
Electrical industry, 81-82
Elster, John, 138, 185-86
Empiricism, 41, 239, 243, 246, 247, 275. See also Science, empirical
Empiricomonism, 106
Engels, Friedrich, 30, 37, 91, 92, 96-97, 105-6, 107, 108, 116, 119, 120, 125, 182, 319
Engels Contra Marx, 91
Enlightenment, The, 158, 173; counter-, 349; criticism of, 7; and evil, 253; and Marcuse, 11; and nature, 128, 149, 152; and premodernity, 27; value system of, 286, 312, 317
Enlightenment science, 22, 253, 349
Epistēmē, 29, 41, 152, 256, 272, 278, 298, 344, 364
Epistemology. See Knowledge, theory of
Erickson, Erik, 23
Essays on the History of Materialism, 107
Essentialism, 13
Ethics, 239, 266, 313-14; and foundationalism, 272; of science, 16, 313
Ethnomethodology, 277, 291, 294
Evolution, 12, 13, 257, 266, 307

F scale, 193
Fedoseyev, 215-17
Feher, Ferenc, 207
Feminism: and ecology, 315-16; effect of, on science and technology, 18, 22, 23,

27; and epistemology, 25; overview of, 22-23; and view of nature, 24
Feyerabend, Paul, 176
Fin-de-siècle Vienna: political milieu of, 254
Final cause, 12, 13
Fok, V., 212
Ford, Henry, 130
Forman, Paul, 252-53
Foucault, Jean Bernard Léon, 19-20, 293
Foucault, Michael, 189
Frankfurt Institute for Social Research, 127, 128
Frankfurt School, 7, 33, 183, 186, 191, 304, 339; and alternatives to technology, 122; and critique of science and technology, 126-33; and ideology, 126, 147; and Marxism, 165; and scientific neutrality, 132
Freud, Sigmund, 113-15, 174-75
Fundamentalist theology, 12
Futuyama, Douglas, 12-13

Galileo, 140, 144, 149, 151, 177, 184
Geisteswissenshaften, 254
Genetic engineering, 6, 15-16, 313, 316
German Ideology, The, 42-45, 48, 70
Gilbreth, Lillian, 23
Gilman, Charlotte Perkins, 22
Goethe, Johann Wolfgang von, 11
Gorz, Andre, 211
Gould, Stephen J., 13
Gouldner, Alvin, 147, 208-9
Government: and scientific research, 8-10, 296, 318-20, 337
Graham, Loren, 113
Gramsci, Antonio, 124, 147, 196-99, 210
Green movement, 315
Greenberg, Dan, 154
Griffin, Susan, 25, 27
Grundrisse, 68, 70, 73, 213, 342

Habermas, Jürgen, 159-69 passim, 181, 191-92
Haldane, J. B. S., 116
Hanson, Norwood, 245, 249-50, 260
Hartsock, Nancy, 24-26
Hegel, Georg Wilhelm Friedrich, 40, 42, 100-101, 106, 107-8, 109, 110, 124, 125, 141, 181, 182, 184, 185, 186, 192, 302-3

Heidegger, Martin, 7, 183, 304
Heisenberg, Werner, 17, 95, 119, 240, 241, 245, 250, 330-31
Heller, Agnes, 207
Hermeneutics, 270, 271
History and Class Consciousness, 91, 97, 140, 144
Hobbes, Thomas, 153
Horkheimer, Max, 7, 126, 127, 131, 139, 152, 304, 317
Hubner, Kurt, 267, 268-71
Human body: defined by modern medicine, 14-15, 65-66
Human error: and technology, 4
Human relations. See Social relations
Husserl, Edmund, 7, 256, 304

Ideology, 41, 43, 60, 83, 350; defined, 146; Gramsci on, 196-99; and Hegelians, 43; of nature, 19, 310; and Popper, 147; and scientific knowledge, 70, 292, 300, 337; and scientific theory, 148, 239, 286, 339, 340-42; Soviet and Nazi, 284
India: and British rule, 61
Individuality, 131-32, 137, 138
Industrial/military complex: and scientific community, 20
Industrial revolution: defined, 55; development of, 60; in England, 61-62, 63; and Marx, 66; in U.S., 62, 63
Information processor: defined, 4
Intellectual and Manual Labor, 142-44
Intention, 246
Interpretation of Dreams, 114

Jay, Martin, 184
Jevons, William, 138
Joravsky, David, 202-3

Kalinin, 202
Kama truck plant (Soviet Union), 188
Kant, Immanuel, 108, 124, 141, 172, 181, 183, 184, 192, 239, 242
Kantian thought, 7, 256
Kapital, Das. See Capital
Karl Marx's Theory of History, 188
Kautsky, Karl, 95, 101, 102-3, 104, 107, 141, 155

Keller, Evelyn Fox, 23
Kenny, Sister Elizabeth, 23
Kepler, Johannes, 149
Keynes, John M., 138
Khrushchev, Nikita, 159
Kirkland, Edward, 62
Knowledge, theory of, 17-18, 24-25, 41,
    239, 240, 262, 300, 321, 345, 349; and
    empiricism, 41, 246; and feminism,
    24-26; and foundationalism, 272; and
    Hegel, 108; and male sexuality, 23, 24;
    and Plekhanov, 107; and reflection
    theory, 95; in science, 323, 328, 338,
    346; in sociology, 283, 289, 291, 327;
    in Soviet and Nazi ideology, 284
Knowledge: versus power, 20, 38, 49, 72,
    189, 293, 295, 297, 298-99, 304, 314, 316
Kollontai, Alexandra, 207
Konrad, George, 206
Korsch, Karl, 100-1
Koyre, Alexandre, 13, 16-17
Kuhn, Thomas, 29, 119, 158, 171, 176,
    249, 251, 260, 261-64, 276, 287

Labor: abolishment of, 101; degradation
    of, 67-68, 71-72, 77, 79; dialectic of,
    76-77, 303; domination of, 80, 317;
    emancipation of, 76; and
    manufacturing, 53-54, 56-57, 58, 81-83;
    Marx's view of, 49-50, 53, 56, 67, 70,
    85-86; as mediation between humans
    and nature, 99; and production,
    46-47, 53-54, 65, 71, 74, 79;
    rationalization of, 58, 64; scarcity in
    nineteenth century U.S., 62; sexual
    division of, 24, 49, 86; technical
    division of, 323, 328, 336, 338
Laboratory Life, 289, 290
Lamarck, Jean-Baptiste Pierre Antoine de,
    115
Landes, David, 60
Laplace, Marquis de, 13
Latour, Bruno, 19, 20, 30, 289, 293
Lebenphilosophie, 253, 256
Lederer, Emile, 211
Lefebvre, Henri, 196
Leibniz, Gottfried Wilhelm von, 172
Leisure: Marx's view of, 70-71, 75, 86, 87

Lenin, Vladimir Ilyich, 94-95, 96, 104-5,
    107-8, 109, 110, 116, 117, 119, 141, 154,
    201
Lévi-Strauss, Claude, 125-26, 184
Life (Biology): defined, 306
Lilley, Samuel, 93-94
Logic as a Positive Science, 182
Logic of Scientific Discovery, 259
Logical empiricism, 242, 259, 262, 267
Lukács, Georg, 39, 91-101, 103-4, 110, 113,
    119, 127, 131, 140, 147, 156, 179, 180,
    183, 197, 254, 258
Luxemburg, Rosa, 104
Lysenko, Trofim, 115, 116, 203-4

McClintock, Barbara, 23
Mach, Ernst, 106
Madison, James, 150
Making Sense of Marx, 185
Male domination, 310; of labor, 24, 49;
    of science, 23-24, 27-28
Malthus, Thomas, 14
Management: as domination, 77, 80; as
    ideology, 343; in science and
    technology, 69, 323-34
Mannheim, Karl, 147, 155, 211
Marcuse, Herbert, 7, 10-11, 126, 127, 129,
    133, 134, 138, 139, 159-60, 163-64, 165,
    317
Marr, N.Y., 116
Marshall, Alfred, 138
Marx, Karl, 30, 36, 37, 92, 97-98, 101, 105,
    319
Marxism: Althusserian, 302; and
    machinery, 187-88; and necessary
    versus sufficient conditions, 36;
    orthodox, 35, 45, 61, 72; Plekhanov
    on, 105-6; as science, 31, 91, 93,
    94-95, 123, 142, 151, 169-200; and
    scientific theory, 29, 30-34, 37, 70, 72,
    78, 253-54; and social theory, 278-79,
    297, 302, 339; and surplus value, 341;
    Wellmer on, 48. See also Second
    International Marxism
Mass society, 122, 137
Materialism and Empirio-Criticism, 94,
    108, 109
Mathematics, 332, 338; and business, 9;
    defined, 8; and human sciences, 248,
    302; ontological status of, 240; and

physical knowledge, 259; and seventeenth-century machinery, 64; truths of, 243, 256, 257
Mechanism, 91, 92
Medical science: and definition of human body, 14, 65-66; and diagnosis/treatment, 14, 311, 333-34, 335-36; research, 334-35, 351
Medvedev, Zheres, 201, 203
Mendel-Morganism, 116
Menger, Karl, 138
Mepham, John, 174, 179
Merchant, Carolyn, 39, 254, 257
Merton, Robert, 29, 30, 273-76, 278, 283-86
Metaphysics, 189, 239, 248; and physics, 266; versus scientific philosophy, 250, 256
Mills, C. Wright, 138
Mining industry: and machine use, 64
Molecular biology, 15, 16, 18, 305, 313. See also Genetic engineering
Mutation: in science, 270
Mysticism: versus reason, 8; in science, 28, 271

National Institute for Health (U.S.), 189
Natural science: and dialectics, 92-93; and Marxism, 95, 100; and the study of society, 98; versus technology, 150
Naturalism, 245, 248, 259, 276
Nature: and bourgeois epoch, 200; and culture, 294; domination of, 77, 78, 304, 318; and Hegel, 303; and human relations, 272, 312; ideology of, 19; images of, 24; in philosophy of science, 247, 259, 302-3; randomness of, 17, 257; scientific view of, 24, 40, 78, 303, 317, 328, 331, 341, 348
Neo-Kantians, 106-7, 111, 140
Newton, Isaac, 94, 117, 118-19, 149, 158
Newtonian mechanics, 246
Nietzsche, Friedrich Wilhelm, 256
Nuclear energy, 4-5, 314; and plant accidents, 315; and social hierarchy, 5

Olson, Mancur, 138
One Dimensional Man, 127, 129, 192
Ontology, 99
Ortega y Gasset, José, 122

Paradigm shift: in philosophy of science, 260-61, 263, 264, 267; in physics, 240, 245, 250, 251, 265, 291, 332; in social sciences, 278, 283, 311
Parsons, Talcott, 273, 279, 301
"Part Played by Labor in the Transition from Ape to Man, The," 105
Pasteur, Louis, 19, 189, 293, 295-96
Patterns of Discourse, 249
Pavlov, Ivan Petrovich, 112-13, 115, 203
Peirce, Charles Sanders, 140, 171, 177, 256-57
Performance art, 344
Perlman, Fredy, 196
Phaedrus, 23
Phenomenology of Spirit, 100-101
Philosophical idealism, 247
Philosophy: analytic crisis, 245; foundationalism in, 245; poststructuralism in, 255, 270; role of, according to Engels, 92; and science, 239, 321. See also Science, philosophy of
Philosophy of Poverty, 44
Philosophy of Science. See Science, philosophy of
Physicists: as philosophers, 240, 253, 330
Physics, 118; defined, 8; and Dewey, 246; versus human and social sciences, 272; mechanism as prelude to, 254; and metaphysics, 266; as model of knowledge, 239, 263; and nature, 250; paradigm shift in, 240, 245, 250, 251, 265, 291, 332; and warfare, 38; and World War I, 252, 253
Pickering, Andrew, 291-92
Plato, 23
Plekhanov, George, 95, 105-10 passim, 203
Political science, 279, 280
Popper, Karl, 8, 111, 115, 147, 240, 259-60, 281, 329
Population ecology. See Social ecology
Positivism, 240, 241, 248, 257, 275, 277-78, 283, 326, 340, 341
Positivist Dispute in German Sociology, 193
Poverty of Philosophy, 42-45, 97-98, 101
Preface to the Contribution to the Critique of Political Economy, 105, 186

Principia, 94, 226
Prison Notebooks, 124
Production, 188; and artisans, 52, 58, 79-80; defined, 44-45; and efficiency, 79; and labor, 46-47, 49, 68, 79, 87; machine, 55-56, 58; and Marxist theory, 39, 42, 46, 50, 66, 69, 73, 76, 85, 116; mental versus material, 37; and scientific knowledge, 37, 38, 299; and social relations, 31, 35-36, 38, 43, 44, 47-48, 64-65, 69, 75, 78, 79; and socialism, 36, 76
Proletariat, 52, 71, 73, 77, 98, 101, 102-4, 119-20. *See also* Working class
Protestantism, 9, 152
Proudhon, Pierre-Joseph, 44
Psychoanalysis: critique of, 111, 115; versus experimental science, 114; Freudian-Lacanian, 174; and sociology, 276
Psychology, 112-13
Ptolemic science, 149, 268

Quantum mechanics, 17, 28, 240-41, 251, 252, 255, 258, 272, 291, 330, 348
Quark, 291-92
Quest for Certainty, The, 246
Quine, W. V. O., 240, 242-45, 247-48

Rancière, Jacques, 124-25
Reductionism, 301, 307; anti-, 309; in biology, 304, 306; in social sciences, 301
Rée, Jonathan, 184
Reich, Wilhelm, 127
Reification, 47, 103-4
Relativity theory, 240-41, 251, 252, 256, 348
Religion: versus scientific rationality, 8, 9, 11, 17, 60; waning of, 10
Research, scientific. *See* Scientific research
Revolution: and Marxism, 76
Revolutionary process, 35
Rich, Adrienne, 27
Rickert, Heinrich, 140
Ricoeur, Paul, 114
Rifkin, Jeremy, 15
Robotics: in U.S. auto industry, 84
Roemer, John, 185

Russell, Bertrand, 144, 244

Sakharov, Andrei, 205
Sartre, Jean-Paul, 124, 183, 184
Schmidt, Alfred, 40
Science, empirical, 37, 239, 290
Science, history of: context for, 254; demarcation points in, 257, 258, 259, 320; externalist versus internalist, 269, 307, 320; overview, 268; and paradigm shifts, 250, 251, 260, 264-65; versus philosophy, 244, 245, 248-49, 321
Science, modern: versus art, 19, 20, 63; attack on, 14, 27; versus capital, 20, 41; defined, 8; ethics of, 16, 313; fields within, 321-22; versus Islamic science, 60; and machinery, 65; and political doctrine, 284, 286-87, 319; and production, 9; versus religion, 11, 16, 17, 27, 28, 60, 271, 295, 321, 340, 349; social study of, 30, 270, 285, 288, 290, 317; task of, 304-5. *See also* Natural science; Science and technology
Science, philosophy of, 8, 239, 240, 245, 258, 329, 330, 347-50; and Dewey, 245, 248; and Hubner, 268-71; and Kuhn, 261-65; and Marxism, 37, 41; and social theory, 287, 289, 293, 294, 295, 299, 335; versus study of science, 248-49, 347; task of, 240; and theory formation, 250, 257, 259, 333; and truths, 287, 339, 341; and Wittgenstein, 240, 260
Science and technology: 93, 107, 119, 161; and bourgeois worldview, 39, 65, 83; versus common sense, 147; critique of, 11; defined, 6; and ecosystem crises, 311; as ideology, 122; and management, 69, 80, 323-24, 343; Marxist view of, 32, 37, 38, 40, 65, 68, 70, 72, 73, 85, 121; and mutations, 270; versus nature, 150; and power, 296, 317, 319, 351; purpose of, 38; and social influences, 32, 38, 48, 81, 296; and socialism, 38-39, 72; and society, 7, 16, 19-20, 64, 322-23, 344-45; in Soviet Union, 201-36; and Stalinism, 115, 117
Science education, 10

Science of Logic, 107, 110
"Science of the Concrete, The," 125-26
*Science, Technology and Society in 17th
Century England*, 285
Scientific communuity: as a class, 297;
formation of, 262; manager in, 323-24;
philosophical differences in, 28;
politics of, 284, 289, 351; and rewards,
284; versus social context, 6, 28, 290;
study of, 284-85, 286; thought styles
within, 287-88; and U.S. military, 296
Scientific investigation. *See* Scientific
research
Scientific knowledge, 37, 38, 285-86, 287,
288, 289, 292, 293, 297, 328, 338, 339,
342, 344
Scientific laboratory: as model of social
power, 19, 80, 293, 296
Scientific Management of Society, 212
Scientific rationality, 271, 321, 328, 329; in
biology, 304, 306; versus religion, 8, 9,
11, 17, 60; in sociology, 274, 275, 276,
279; and state institutions, 8-9, 10
Scientific research, 271, 288, 292, 314,
318-19, 320, 333; and agreements with
corporations, 15, 16, 20, 293, 314, 324,
337; graduate students in, 325; in
industry, 80; organization of, 323
Scientific socialism, 105-6
Scientific-Technological Revolution, 212,
215-20
Sechenov, Ivan, 113
Second International Marxism, 30-31, 45,
100-102 *passim*, 104, 105, 124
Semiotics, 311
Simon, Herbert, 4
Skepticism, 256, 288
Smith, Adam, 150
Social ecology: and domination, 310; and
ecosystems, 315; effect on science
and technology, 18, 21, 22, 27, 38; and
human action, 310; and population,
307; and randomness of nature, 18-19
Social relations: and communities, 275,
282; and conflict, 309; in production,
31, 35-36, 38, 43, 44, 47-48, 64-65, 69,
75, 78, 79; and reductionism, 306;
study of, 281, 282; and world systems,
302

Social sciences, 136-37; fields within, 278,
297; and language, 282, 283, 294, 299;
and mathematics, 248, 273; versus
physics and biology, 272, 277, 306;
and reductionism, 301, 304, 309;
research methods in, 134-38, 149, 278,
279, 281, 302, 326. *See also* Sociology
Social system: defined, 276, 295; and
Parsons, 279; and psychoanalysis, 276;
rules within, 282; stability within, 274
Socialism, 101, 155; British and French,
76; Marx's definition of, 77; and
production, 36, 76; and science and
technology, 38-39, 72
Socialism and the Political Struggle, 105-6
Socialist Republic, The, 101
Socialization, 274
Sociology: defined, 273, 276, 277;
economic system in, 274;
methodology of, 277, 278, 290, 295,
327; objects of, 282, 290; positivist,
275, 277-78, 282, 283, 326; as science,
279, 286; and study of communities,
284-85, 286, 289; theory of, 276, 278,
301, 309, 350
Sociology of knowledge, 273, 275, 276,
279, 281, 283, 285, 286, 294, 295
Sohn-Rethel, Alfred, 127, 142-44, 189
Species: emergence of new, 305
Stalin, Joseph, 93, 114, 116, 159
Stalinism, 111, 115, 117
Star Wars, 205
Steel industry: artisan versus scientist in,
79-80
STR. *See* Scientific-Technological
Revolution
Structuralism, 126-27
Structure of Scientific Revolutions, 287
Suicide: in sociology, 275
Symposium, 23
Synthetic fibers: development of, 80-81
Synthetic statements. *See*
Analytic/synthetic statements
*Systems Theory: Philosophical and
Methodological Approaches*, 236
Szelenyi, Ivan, 206

Taylorism, 207
Technology: and culture, 344; defined,
343; and emancipation of labor, 76;

growth of, in U.S., 62; and human freedom, 7, 314; and industrial societies, 5, 9, 10, 81, 343; Marxist view of, 38, 72-73, 78; and modern science, 6, 38; nuclear, 4-5, 314; and political economy, 45; and power, 291, 293; as savior, 10-11. *See also* Computers; Science and technology

"Technology and Science as 'Ideology'," 159

Teleology, 92, 99-100, 105, 109, 247, 271, 295, 307

Textile industry, 63-64, 78-79

Theses on Feuerbach, 39, 40

Thinking and being, 110

Third International, The, 30-31, 45

Timpanaro, Sebastian, 124-25

Togliatti, Palmiero, 124

Tonnies, Ferdinand, 140

Touraine, Alain, 196

Twentieth Congress of the Communist Party of the Soviet Union, 159

Unions: and labor supply, 57

Universe, self-regulating, 253

Untermann, Ernest, 95

Ure, Alexander, 67

Utopians: and Marx, 76

Validity, theory of, 247

Vavilov, N., 212

Veblen, Thorsten, 211

Wage determination, 57

Watson, James, 15

Weber, Max, 97, 100, 128, 136-37, 140, 146, 148, 155, 160, 163, 179

Weimar culture and science, 252-53, 254, 257

Weizsacker, Karl Friedrich von, 133-34, 240, 241

Wellmer, Albrecht, 48

Weltanschauung, 95

White, Lynn, 144

Willis, Paul, 131-32

Winch, Peter, 281

Wittgenstein, Ludwig, 114, 240, 254, 259, 282, 294

Working class, 102-4, 116-17, 127-28. *See also* Proletariat

Wright, Eric Olin, 185

Stanley Aronowitz is professor of sociology in the graduate school of the City University of New York. He has also been employed as a steelworker, trade union organizer and school planner. Aronowitz recieved his Ph.D. in sociology from Union Graduate School. Among his many books are *False Promises* (1973), *The Crisis in Historical Materialism: Class, Politics, and Culture in Marxist Theory* (1981), *Working Class Hero: A New Strategy for Labor* (1983) and, with Henry Giroux, *Education Under Siege*, (1985), which was named by the American Educational Studies Association one of the most significant books in education for the year 1986. He contributes to *The Nation, Social Policy, Village Voice*, and *Social Text*.